WebGL
Insights

WebGL Insights

Edited by Patrick Cozzi

CRC Press
Taylor & Francis Group
Boca Raton London New York

CRC Press is an imprint of the
Taylor & Francis Group, an **informa** business

AN A K PETERS BOOK

CRC Press
Taylor & Francis Group
6000 Broken Sound Parkway NW, Suite 300
Boca Raton, FL 33487-2742

© 2016 by Taylor & Francis Group, LLC
CRC Press is an imprint of Taylor & Francis Group, an Informa business

Printed and bound in India by Replika Press Pvt. Ltd.

Printed on acid-free paper
Version Date: 20150504

International Standard Book Number-13: 978-1-4987-1607-9 (Pack - Book and Ebook)

Library of Congress Cataloging-in-Publication Data

WebGL insights / editor, Patrick Cozzi.
    pages cm
    Includes bibliographical references.
    ISBN 978-1-4987-1607-9 (alk. paper)
    1. Computer graphics--Computer programs. 2. WebGL (Computer program language) I. Cozzi, Patrick.

    T386.W43W43 2016
    006.6'633--dc23   2015011022

Visit the Taylor & Francis Web site at
http://www.taylorandfrancis.com

and the CRC Press Web site at
http://www.crcpress.com

For Liz, Petey, and the Captain

Contents

Section III Mobile

Section IV Engine Design

Section V Rendering

Section VI   Visualization

Section VII   Interaction

Foreword

Since the first release of the WebGL specification in February 2011, a remarkable and passionate community has developed around it. Enthusiasts from varied domains, from the demoscene to medical research scientists and everyone in between, have created beautiful visual effects, artifacts, applications, and libraries. Perhaps most remarkably, these individuals and groups have shared their work for all to see and use on the World Wide Web.

The web's culture of sharing dates back to its origins and the "view source" option in the earliest web browsers. Web page creators were encouraged to see how others achieved visual techniques, to copy them, and to help improve them. Since that time, the web has grown organically and exponentially. As web developers noticed missing functionality, new features were proposed in the HTML specification to bridge these gaps, and browsers evolved rapidly to incorporate them. A tremendous number of open-source libraries have sprung up that make it easier to create visually compelling web pages, user interfaces, and applications that scale well from a smartphone screen all the way up to oversized desktop displays.

It's been exciting to witness WebGL's adoption into the web's sharing culture. When I first began studying computer graphics, the web was in its infancy; leading-edge techniques were published in printed journals a few times a year, and source code was almost never released by the authors. Today, computer graphics researchers from around the world publish articles every day on their own websites and blogs, and include not only source code, but also live examples written using WebGL!

This book and the accompanying website represent the epitome of the web's sharing culture: a curated collection of leading-edge techniques and insights, with online source code and live demonstrations. In this book, engine authors present their strategies for achieving high-performance, good scalability, and leading-edge visuals. Educators share their experience in moving real-time computer graphics courses to the web and WebGL. Application developers and toolchain authors share their experiences both building large, new JavaScript code bases and bringing existing C++ code bases to the web. Graphics researchers show how to implement leading-edge rendering techniques in WebGL, allowing these techniques to be deployed seamlessly to hundreds of millions of devices and billions of people. Visualization researchers demonstrate how to render huge data sets with high performance and high impact. Interaction researchers provide insights into effective navigation and interaction paradigms for 3D applications. Finally, browser and GPU implementers give a look under the hood of WebGL implementations to help developers tune their code for best performance on a range of devices.

This book contains a wealth of information and will be a treasured reference for years to come. I thank Patrick Cozzi for initiating and driving this project to completion, and for the opportunity to have been involved with it.

Ken Russell
Khronos WebGL Working Group Chair
Software Engineer, Google Chrome GPU Team

Preface

WebGL Insights is a celebration of our community's accomplishments and a snapshot of the state of the art in WebGL.

There is WebGL support in every modern browser across every major platform, including Android and iOS. Given its ubiquity, plugin-free deployment, ease of development, tool support, and improved JavaScript performance, WebGL adoption from major developers and startups alike is on the rise. WebGL is used by Unity, Epic, Autodesk, Google, Esri, Twitter, Sony, *The New York Times*, and countless others. Likewise, many startups base their core technology on WebGL, such as Floored and Sketchfab, who share their experiences in this book. In addition, WebGL is playing a central role in future web technologies such as WebVR.

In academia and education, WebGL is gaining momentum due to its low barrier to entry. The introductory graphics course at SIGGRAPH moved from OpenGL to WebGL, as did the popular introductory book, *Interactive Computer Graphics: A Top-Down Approach.* The Interactive 3D Graphics Udacity course, which teaches graphics using WebGL/three.js, has had more than 60,000 students sign up. In my own teaching at the University of Pennsylvania, I've moved to WebGL and often talk with other educators who are doing the same.

The demand for skilled WebGL developers is high. As an educator, I receive more requests for WebGL developers than we can fill. As a practitioner, I find that our team of skilled WebGL developers gives us a unique advantage to develop quickly and implement solutions that are efficient and robust.

The advancement of the WebGL community, the demand for highly skilled WebGL developers, and our appetite for continuous learning have led to this book. Our community has the need to go beyond the basics of introductory WebGL books and to learn advanced techniques from developers with practical experience.

In the spirit of *OpenGL Insights,* developers from the WebGL community have contributed chapters sharing their unique expertise based on their real-world experiences. This includes hardware vendors sharing performance and robustness advice for mobile; browser developers providing deep insight into WebGL implementations and testing; WebGL-engine developers presenting design and performance techniques for several of the most popular WebGL engines; application developers explaining how WebGL is reaching beyond games into areas such as neurological data visualization researchers presenting massive model rendering and educators providing advice on migrating graphics courses to WebGL.

Throughout the chapters we see many common themes, including

- The move of desktop applications to the web either by writing new clients in JavaScript or by converting C/C++ to JavaScript using Emscripten. See Section II and Chapters 16, 17, and 21.
- Runtime or preprocessed shader pipelines to generate GLSL from higher level material descriptions and shader libraries. See Chapters 9, 11, 12, and 13.
- The use of Web Workers for compute or load pipelines to offload the main thread from CPU-intense work. See Chapters 4, 11, 14, 19, and 21.
- Algorithms that are light on the CPU and offload massively parallel work to the GPU. See Chapters 18 and 19.
- Understanding the CPU and GPU overhead of different WebGL API functions, and strategies for minimizing their cost without incurring too much CPU overhead in our application. See Chapters 1, 2, 8, 9, 10, 14, and 20.
- Incrementally streaming 3D scenes, often massive scenes, to quickly provide the user with a coarse representative followed by higher detail. See Chapters 6, 12, 20, 21.
- Organization of large JavaScript code bases and GLSL shaders into modules. See Chapters 4 and 13.
- A focus on testing both WebGL implementations and WebGL engines built on top of them. See Chapters 3, 4, and 15.
- Open source. The web has a culture of openness, so it should come as no surprise that most of the WebGL implementations, tools, engines, and applications described in this book are open source.

I hope that *WebGL Insights* inspires you, teaches you, gives you new insights into your own work, and helps bring you to the next level in your adventures.

Patrick Cozzi

Acknowledgments

First, I would like to thank Christophe Riccio (Unity) who edited *OpenGL Insights* with me. The community focus of *WebGL Insights* is the direct result of the culture Christophe created with *OpenGL Insights*. Christophe set the bar very high for *OpenGL Insights*, and I believe we have continued the tradition in *WebGL Insights*. I also thank Christophe for his support of the *WebGL Insights* book proposal and chapter reviews.

At SIGGRAPH 2014, I suggested the idea for *WebGL Insights* to Ed Angel (University of New Mexico), Eric Haines (Autodesk), Neil Trevett (NVIDIA), Ken Russell (Google), and several others. It was their support that got the project off the ground. I thank them for supporting the initial book proposal. I also thank Ken for writing the foreword.

WebGL Insights is the story of 42 contributors sharing their experiences with WebGL and related technologies. They made this book what it is and make the WebGL community the lively community that it is. I thank them for their dedication to and enthusiasm for this book.

The quality of *WebGL Insights* is due to the work of the contributors and the 25 technical reviewers. Each chapter had at least two reviews; most had three to five and a couple of chapters received seven or more.

I thank all the reviewers for volunteering their time: Won Chun (RAD Games Tools), Aleksandar Dimitrijevic (University of Niš), Eric Haines (Autodesk), Havi Hoffman (Mozilla), Nop Jiarathanakul (Autodesk), Alaina Jones (Sketchfab), Jukka Jylänki (Mozilla), Cheng-Tso Lin (University of Pennsylvania), Ed Mackey (Analytical Graphics, Inc.), Briely Marum (National ICT Australia), Jonathan McCaffrey (NVIDIA), Chris Mills (Mozilla), Aline Normoyle (University of Pennsylvania), Deron Ohlarik, Christophe Riccio (Unity), Fabrice Robinet (Plumzi, Inc.), Graham Sellers (AMD), Ishaan Singh (Electronic Arts), Traian Stanev (Autodesk), Henri Tuhola (Edumo Oy), Mauricio Vives (Autodesk), Luke Wagner (Mozilla), Corentin Wallez (Google), Alex Wood (Analytical Graphics, Inc.), and Alon Zakai (Mozilla).

Eric Haines deserves special thanks. Not only did Eric review several chapters in detail, but he also introduced me to his Autodesk colleagues, Traian Stanev and Mauricio Vives, who provided exceptional reviews.

In addition to external technical reviews, we had a great culture of contributor peer review. I especially thank Olli Etuaho (NVIDIA), Muhammad Mobeen Movania (DHA Suffa University), Tarek Sherif (BioDigital), and Jeff Russell (Marmoset) for going above and beyond.

I thank Rick Adams, Judith Simon, Kari Budyk, and Sherry Thomas from CRC Press for all their work publishing *WebGL Insights*. Rick was very supportive of the project from the start and helped bring *WebGL Insights* from idea to proposal to contract in record time.

I thank Norm Badler from the University of Pennsylvania, whose encouragement sparked my initial involvment in producing books in 2009.

Editing *WebGL Insights* on top of my developer and teaching positions made evenings, weekends, and even holidays with friends and family few and far between. For their patience, understanding, and occasional copyediting, I thank Peg Cozzi, Margie Cozzi, Anthony Cozzi, Colleen Curran Cozzi, Cecilia Cozzi, Audrey Cozzi, Liz Dailey, Petey, and the Captain.

Website

The companion *WebGL Insights* website contains source and other supplements. It is also the place to find announcements about future volumes.

    www.webglinsights.com

Please e-mail me with your comments or corrections: pjcozzi@siggraph.org

Tips

WebGL Report (http://webglreport.com/) is a great way to get WebGL implementation details for the current browser, especially when debugging browser- or device-specific problems.

For performance, avoid object allocation in the render loop. Reuse objects and arrays where possible, and avoid built-in array methods such as `map` and `filter`. Each new object creates more work for the Garbage Collector, and in some cases, GC pauses can freeze an application for multiple frames every few seconds.

Save memory and improve performance by ensuring that contexts are created with the `alpha`, `depth`, `stencil`, `antialias`, and `preserveDrawingBuffer` options set to `false`, unless otherwise needed. Note that `alpha`, `depth`, and `antialias` are enabled by default and must be explicitly disabled.

For performance, query attribute and uniform locations only at initialization.

Int precision default qualifiers aren't the same between vertex and fragment shaders. This can lead to surprising visual differences when moving computation between each.

For portability, keep space requirements of varyings and uniforms within the limits of the GLSL ES spec. Consider using `vec4` variables instead of `float` arrays, as they potentially allow for tighter packing. See A.7 in the GLSL ES spec.

Non-power-of-two textures require linear or nearest filtering, and clamp-to-border or clamp-to-edge wrapping. Mipmap filtering and repeat wrapping are not supported.

When we are using more than one draw buffer with the WEBGL_draw_buffers extension and we don't want to write to a given draw buffer, pass `gl.NONE` to the draw buffers parameter list. We must always provide all color attachments that our framebuffer has.

Always enable strict mode via the `"use strict"` directive. It slightly alters JavaScript semantics so that many silent errors turn into runtime exceptions and can even help the browser better optimize our code.

Code linters, such as JSHint (http://jshint.com/), are an invaluable tool for keeping JavaScript code clean and error free.

Continued

Create new textures, rather than changing the dimensions or format of old ones.	Chapter 1
Avoid use of `gl.TRIANGLE_FAN`, as it may be emulated in software.	Chapter 1
Flag buffer usage as `gl.STATIC_DRAW` where appropriate, to allow browsers and drivers to make use of optimizations for static data.	Chapter 1
Make sure that one of the array attributes is bound (using `gl.bindAttribLocation`) to location 0. Otherwise, high overhead should be expected when running on non-ES OpenGL platforms such as Mac OS X and desktop Linux.	Chapter 2
Pass data back and forth from Web Workers using transferable objects whenever possible.	Chapters 4 and 21
Although typed arrays have performance advantages, using JS arrays in teaching allows students to write clearer JS code with the use of array methods.	Chapter 7
Using `mediump` precision in fragment shaders provides the widest device compatibility, but risks corrupted rendering if the shaders are not properly tested.	Chapter 8
Using only `highp` precision prevents corrupted rendering at the cost of losing some efficiency and device compatibility. Prefer `highp`, especially in vertex shaders.	Chapter 8
To test device compatibility of shaders that use `mediump` or `lowp` precision, it is possible to use software emulation of lower precision. Use the—emulate-shader-precision flag in Chrome.	Chapter 8
When using an RGB framebuffer, always implement a fallback to RGBA for when RGB is not supported. Use `gl.checkFramebufferStatus`.	Chapter 8
To save a lot of API calls, use vertex array objects (VAOs) or interleave static vertex data.	Chapter 8
For performance, do not update a uniform each frame; instead update it only when it changes.	Chapters 8, 10, and 17
If shrinking the browser window results in massive speed gains, consider using a half-resolution framebuffer during mouse interaction.	Chapter 14
Load time can be improved by amortizing slow tasks across several frames.	Chapter 14
Be vigilant about using `requestAnimationFrame`—ensure that most, if not all, of your WebGL work lives inside it.	Chapter 14
The `textureProj` GLSL function, `vec4 color = textureProj(sampler, uv.xyw)`;, can be simulated with `vec4 color = texture(sampler, uv.xy/uv.w)`	Chapter 17
Avoid using common text-based 3D data formats, such as Wavefront OBJ or COLLADA, for asset delivery. Instead, use formats optimized for the web, such as glTF or SRC.	Chapter 20
Use OES_element_index_uint to draw large indexed models with a single draw call.	Chapter 21
Smooth, cinematic camera transitions can be created by a cosine-based interpolation of the camera's position and orientation. Unlike nonlinear interpolation alternatives, the cosine interpolation is computationally cheap and easy to calculate.	Chapter 23

SECTION I
WebGL
Implementations

Knowing what goes on under the hood makes us more effective WebGL developers. It helps us understand which API calls are fast, which are slow, and why. This is particularly important for engine developers, whose users rely on the engine to make efficient use of WebGL. The stack under WebGL is involved. It can include API validation, driver workarounds, shader validation and translation, compositing, and interprocess communication, and eventually calls to the native graphics API, which itself has a stack of OS and GPU vendor code. In this section, developers from Google and Mozilla provide deep insight into what happens between our WebGL calls and the native graphics API.

ANGLE is perhaps best known as an OpenGL ES 2.0 wrapper over Direct3D 9 used to implement WebGL in Chrome, Firefox, and Opera on Windows. However, it is much more. ANGLE now also has a Direct3D 11 backend and is not just used on Windows; its shader validator is used on Linux by Chrome and Firefox, and on OS X by Chrome, Firefox, and Safari. In Chapter 1, "ANGLE: A Desktop Foundation for WebGL," Nicolas Capens and Shannon Woods go under the hood of ANGLE and provide performance tips, such as why we should avoid TRIANGLE_FAN and wide lines, and debugging tips like how to step through ANGLE source code in Chrome.

In Chapter 2, "Mozilla's Implementation of WebGL," Benoit Jacob, Jeff Gilbert, and Vladimir Vukicevic dive into the details of Mozilla's WebGL implementation by explaining what happens between when a WebGL function is called in JavaScript and when Firefox calls the native graphics API. This includes datatype conversion, error checking, state tracking, texture conversion, draw call validation, shader source transformation, and compositing. Knowledge of these details helps us optimize our WebGL code; for example, by knowing how data copying, data conversion, and error checking work in texImage2D and texSubImage2D, we can call these functions in such a way as to get their fastest path.

WebGL support is consistent across an incredible combination of operating systems, browsers, GPUs, and drivers. As an engine developer, I can say with certainty that our WebGL engine has far fewer workarounds than our OpenGL engine (with that said, the state of modern OpenGL drivers is now also very good). The consistency and stability of WebGL is thanks to the testing performed by browser and hardware vendors. In Chapter 3, "Continuous Testing of Chrome's WebGL Implementation," Kenneth Russell, Zhenyao Mo, and Brandon Jones explain the continuous test environment used for Chrome. This includes the hardware and software infrastructure and many "gotchas" encountered in practice when running GPU tests on servers. The lessons learned are applicable for many graphics testing scenarios, not just WebGL implementations.

ANGLE

A Desktop Foundation for WebGL

Nicolas Capens and Shannon Woods

1.1 Introduction

WebGL is a powerful tool from a web development perspective, providing a gateway to GPU-accelerated 3D graphics through a JavaScript API. Developers can author their applications once and expect them to run across a wide variety of hardware, both mobile and desktop, with the full assistance of the GPU. It is no simple task to support such a system seamlessly, removing the need for WebGL applications to handle differences in operating system, browser, GPU, or available driver. While applications don't need to handle these differences themselves, being familiar with the ways in which WebGL calls are validated, modified, translated, and finally issued to the hardware provides developers with the tools to create efficient WebGL applications across implementations and platforms. In this chapter, we discuss ANGLE, an open-source project used by several browsers as part of this seamless multi-platform support, and cover some tools and best practices developers can exercise to ensure WebGL applications perform as expected, everywhere. We conclude with helpful tips on examining the translated shader output generated by ANGLE and building and debugging ANGLE itself as a stand-alone library or as part of the Chrome browser.

1.2 Background

Browsers handle the calls made in WebGL by interpreting them and issuing graphics commands to the underlying hardware using a native graphics API. In mobile browsers, these native commands are extremely similar to WebGL, because the vast majority of mobile graphics drivers implement OpenGL ES, on which WebGL is directly based. On the desktop, things become slightly more complicated; OpenGL ES drivers are not available for some desktop operating systems. Under Linux and OS X, the support path is clear, as desktop OpenGL is the native 3D graphics API for those platforms and is widely and robustly supported.

Windows, on the other hand, has its own challenges. While desktop OpenGL drivers exist for Windows, most games and other applications which make use of the GPU instead utilize Direct3D, Microsoft's 3D graphics API. Even if a user's machine is known to have WebGL-capable hardware, it cannot be guaranteed that the user has OpenGL drivers installed at all. By contrast, Direct3D drivers are installed with the operating system. Requiring OpenGL drivers, then, would be a barrier for a large number of Windows users, keeping them from easily being able to experience WebGL—a significant downside for an emerging web API.

For this reason, Google initiated the ANGLE project. ANGLE began as an implementation of OpenGL ES 2.0, the native 3D API on which WebGL 1.0 is directly based, built on top of Direct3D 9. This implementation is used both by Google Chrome and Mozilla Firefox on Windows as the backend for WebGL. Chrome also uses ANGLE for hardware-accelerated rendering support across the entire browser, from page rendering to hardware-accelerated video. In addition, ANGLE's shader translator, used to translate shaders from the OpenGL ES Shading Language (ESSL) into Direct3D's High Level Shading Language (HLSL), also functions as a shader validator for WebGL and is used in that capacity not just on Windows, but also on Linux by Chrome and Firefox and on OS X by Chrome, Firefox, and Safari [Koch 12].

ANGLE has continued to build upon this initial implementation. In 2013, a Direct3D 11 rendering backend was added to ANGLE, allowing WebGL implementations the ability to make use of a newer native API [ANGLE 13]. Direct3D 11 adds support for many of the texture and vertex formats included in OpenGL ES 2.0 and WebGL 1.0, which in the Direct3D 9 backend must be converted to natively supported formats. Direct3D 11 additionally provides all the necessary features for ANGLE to support OpenGL ES 3.0. ANGLE's rendering backends are selectable at runtime, meaning that a browser or other client application can choose to use the Direct3D 9 or 11 implementation depending on the particular needs of the application and the hardware on which it's running. ANGLE's backend targets a minimum feature level of 10_0, Direct3D 11 so hardware with Direct3D 10 support and above are able to make use of the new backend. However, certain features needed for ES 3.0 parity do not appear until feature level 11_0 or 11_1 and must be supported with software-side workarounds on lower feature level hardware, as discussed in Section 1.4.2.

The addition of this runtime-selectable backend opens up new possibilities for ANGLE. By treating the implementation of each renderer as an encapsulated object with its own simple interface, we can enable choosing between not just different versions of the Direct3D API, but also other native graphics APIs. This would allow a client application to target OpenGL ES, while in actuality executing on Direct3D, OpenGL, or any other rendering API for which future ANGLE renderers are written. For Chrome in particular, this is a great benefit, as Chrome currently must perform validation of OpenGL ES calls within its own graphics process before either translating them to desktop OpenGL itself or forwarding them to ANGLE, which performs its own validation. If ANGLE were to handle translation for all platforms and APIs, Chrome would be able to delegate all OpenGL ES validation to ANGLE. The ANGLE team launched an engineering effort to perform additional refactoring necessary to become entirely platform agnostic in mid 2014 [ANGLE 14a] and will be adding a backend targeting desktop OpenGL. This will enable Chrome to use ANGLE as the OpenGL ES implementation on Windows, Mac, Linux, and even mobile, with ANGLE performing any necessary translation to the platform-specific API behind the scenes. For WebGL developers, this means that ANGLE will be playing a role in enabling our applications in a growing number of situations.

Figure 1.1 illustrates where ANGLE fits in the overall architecture of a typical WebGL implementation.

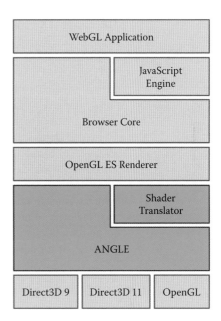

Figure 1.1

System architecture.

1.3 ANGLE Is Not an Emulator

It is important to highlight that ANGLE is not an emulator in the classical sense. Emulators typically suffer from a great loss in performance from having to account for substantial architectural differences between the guest and host platform. ANGLE instead merely provides an implementation of the OpenGL ES libraries, and happens to use another underlying graphics API to accomplish this. Very little performance is lost,* because the hardware itself is designed to efficiently support any generic rasterization graphics pipeline. A native OpenGL ES driver can be provided by the hardware vendors, but in cases where no such driver is present, ANGLE fulfills the same role.

From a bird's-eye view, all rasterization graphics APIs consist of two types of operations: state setting commands and drawing commands. Setting state includes operations such as specifying the culling mode and attribute layout, and also includes setting shaders. After setting all the states which affect how rendering calls will be processed, the real processing work is started by issuing drawing commands such as `glDrawArrays` and `glDrawElements`. ANGLE efficiently sets equivalent states so that the drawing commands are processed equally fast. Therefore it acts more like a translator than an emulator.

Direct3D, as the name implies, is in many ways a lower level graphics API than OpenGL. Thus ANGLE performs many of the tasks a native OpenGL ES driver would also have to perform. That said, translating between OpenGL ES and Direct3D is not without major challenges. We won't delve too deeply into the implementation details, as these have already been covered by articles in OpenGL Insights (Direct3D 9) [Koch 12] and

---

* Sometimes performance is gained due to using more optimal drivers, better cooperation with other APIs that use the GPU (e.g., Direct2D), or optimizations performed by ANGLE itself.

GPU Pro 6 (Direct3D 11) [Woods 15]. We will, however, discuss the major caveats from a WebGL development perspective.

If ANGLE just maps an underlying API such as Direct3D to OpenGL ES for use by yet another API, WebGL, why not translate directly between WebGL and Direct3D? This actually wasn't a viable option in the early days of WebGL, because conformance tests were nonexistent except for OpenGL ES. Conformance tests were and still are critically important to ensuring WebGL's success, or the success of any web standard for that matter. Chapter 3 details the testing effort for the Chrome browser. These days we have a robust WebGL conformance test suite [Khronos 13b], but it is still valuable to have an intermediate API layer which can be used to determine if issues are caused above or below that layer. Aside from testability, ANGLE's drop-in libraries allow it to be used for any application which wishes to target OpenGL ES, not just browsers which aim to support WebGL. The vast majority of mobile applications use OpenGL ES, and thus ANGLE allows that code to be reused when porting these applications to the desktop. Building ANGLE for this purpose, or for deep debugging of WebGL issues, is covered in Section 1.5.

1.4 ANGLE in WebGL App Development

ANGLE is a fully conformant implementation of OpenGL ES 2.0 [Kokkevis 11] and passes an ever-increasing number of WebGL tests—new ones are added almost daily—with performance highly competitive to that of one based on a native desktop OpenGL implementation. ANGLE's aim is to be invisible to the developer. Still, with the majority of desktop WebGL implementations being based in whole or in part on ANGLE, insights into some of its implementation details can result in higher performance or faster bug resolution.

1.4.1 Recommended Practices

The most important thing WebGL developers can do to ensure their applications are robust and performant is to develop for and test with pre-stable release channels of the browser of their choice and report suspected implementation bugs as they are found. When new features or optimizations are deployed in ANGLE, they will appear in the pre-stable channels of browsers which use them first, giving developers and early adopters time to encounter and report bugs and the ANGLE team time to address those bugs before most users encounter them. Chrome Dev and Firefox Developer Edition both track a regularly updated version of ANGLE.

Additionally, developers can ensure peak performance by using the types, formats, and commands in WebGL which require the least intervention from ANGLE to deliver to the underlying API. Because the particular work that ANGLE must do to translate calls varies depending on which backing renderer is being used by a given user, best practice is to be aware of, and minimize use of, computationally intensive paths for any of ANGLE's supported platforms.

- **Recommendation: Avoid Use of LINE_LOOP and TRIANGLE_FAN**
  LINE_LOOP does not exist in either of the Direct3D APIs, and TRIANGLE_FAN is not supported by Direct3D 11, so ANGLE must rewrite index buffers to support them. This is less of a problem for LINE_LOOP, as it can be easily represented by LINE_STRIP with an additional copy of the first index,

but TRIANGLE_FAN must be represented internally as TRIANGLES, with a greatly expanded index buffer, when backed by Direct3D 11. The impact of this rewriting can be nontrivial in certain situations; one of our benchmarks demonstrates that a particularly pathological case results in TRIANGLE_FAN requiring an additional 5 ms of render time per frame on one test system as compared to TRIANGLES.*

- **Recommendation: Create New Textures, Rather Than Redefining Old Ones**
  The Direct3D API family requires that the format and dimensions of textures and their mipmap chains be fully specified on creation, and it has no notion of incomplete textures. To support textures being defined one mipmap level at a time, ANGLE maintains copies of those textures in system memory, creating the GPU-accessible textures only when they are finally used by a rendering command. Once that texture is created, altering the format or dimensions of any of its constituent levels involves more overhead than for a newly created texture still being maintained in system memory. This overhead can cost precious milliseconds at draw time: In a very simple benchmark, defining the levels of a newly created texture was 3 to 6 ms faster than using identical GL calls to redefine the dimensions of a texture that had already been used in a previous draw as described previously. To avoid this potential penalty, create new textures, rather than changing the format or dimensions of an already existing one. By contrast, if only the pixel data contained in a texture need to be updated, it is best to reuse the texture—the additional overhead is only incurred when updating texture format or dimensions, because these require redefinition of the mipmap chain.

- **Recommendation: Do Not Perform Clears with Scissors or Masks Enabled**
  One of the subtle differences between the Direct3D APIs and the GL APIs is that in the former, clear calls ignore scissors and masks, while the latter applies both to clears [Koch 12]. This means that if a clear is performed with the scissors test enabled, or with a color or stencil mask in use, ANGLE must draw a quad with the requested clear values, rather than using clear. This introduces some state management overhead, as ANGLE must switch out all the cached state such as shaders, sampler and texture bindings, and vertex data related to the draw call stream. It then must set up all the appropriate state for the clear, perform the clear itself, and then reset all of the affected state back to its prior settings once the clear is complete. If multiple draw buffers are currently in use, using WEBGL_draw_buffers, then the performance implications of this emulated clear are compounded, as the draw must be performed once for each target. Clearing buffers without scissors or masks enabled avoids this overhead.

- **Recommendation: Render Wide Lines as Polygons**
  ANGLE does not support line widths greater than 1.0, commonly called "wide" lines. While this is sometimes a requested feature, wide lines are not a native primitive of any modern hardware, and therefore require emulation in either the driver or the application. There is much disagreement about how exactly they should be rendered; the handling of corners, endpoints, and

---

* See benchmarking samples at https://chromium.googlesource.com/angle/angle/+/master/samples/angle/.

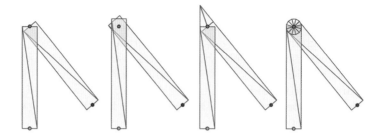

Figure 1.2

Some line joint variations.

overdraw/transparency varies widely between vendors, drivers, and APIs. See Figure 1.2 for a few common joint rendering implementations. Basically, there is no right answer; game developers mostly want speed, while CAD and mapping applications may prefer high-quality wide lines that can be slow to render. Many applications resort to using their own implementation to achieve consistent results, and this makes driver-based implementations unnecessary bloat, which comes with a minor performance cost from extra state tracking. For these reasons, wide lines are not supported by Direct3D and have also been deprecated from the desktop OpenGL 3.0 core profile [Khronos 10]. It is important to realize that any fully conformant WebGL implementation can have a maximum line width of 1.0. Therefore, there is no guarantee that when we get the desired result on one platform, it will look identical on another. ANGLE does not want to be part of the problem by giving the impression that wide lines are widely supported. One approach for rendering wide lines in the application can be found in Cesium [Bagnell 13].

- **Recommendation: Avoid Three-Channel Uint8Array/Uint16Array Data in Vertex Buffers**
Direct3D has limited support for three-channel vertex formats. Only 32-bit three-channel formats (both integer and float) are supported natively [MSDN 14a]. Other three-channel formats are expanded by ANGLE to four-channel internally when using a Direct3D backend. If the vertex buffer usage is dynamic, this conversion will be performed each time the buffer is used in a draw. To avoid the expansion, use four-channel formats with 8- or 16-bit types.

- **Recommendation: Avoid Uint8Array Data in Index Buffers**
Neither Direct3D 9 nor Direct3D 11 supports 8-bit indices, so ANGLE supports these by converting them to 16 bits on those platforms. To avoid the conversion cost, supply 16-bit index values instead.

- **Recommendation: Avoid 0xFFFF in 16-Bit Index Buffers**
In Direct3D 11, index values with all bits set to 1 are considered a special sentinel value, indicating a triangle strip-cut [MSDN 14b]. In OpenGL, this feature is known as primitive restart. Primitive restart does not appear in OpenGL ES until the 3.0 specification. In OpenGL ES 3.0, primitive restart can be toggled on and off, while by contrast, the feature is always enabled in Direct3D 11. This caused ANGLE to inadvertently break some WebGL applications that use highly

detailed geometry.* This bug reached Chrome's Stable channel before it was noticed by anyone. It caught both the ANGLE team and WebGL developers off-guard—another reminder to always test applications with Beta or Dev versions of the browser. We fixed the bug by detecting this strip-cut index and converting the whole index buffer to use 32-bit values when it's present. This cost could be avoided by using the `OES_element_index_uint` extension [Khronos 07a], but developers should have a fallback for when it is not available. Alternatively, content authoring tools could avoid using the strip-cut index value in 16-bit index buffers by creating indexed triangle strip buffers with less than 65536 vertices. Triangle lists are unaffected by primitive restart, so this offers another alternative.

- **Recommendation: Make Appropriate Use of Static Buffers and Flag Usage Correctly**
  Due to the limited set of vertex formats supported natively in Direct3D 9, ANGLE must convert much of these data before uploading them to the GPU on this platform. If the provided vertex data are not updated subsequent to their first use in a rendering command, the data need not be converted every time they are used. Static data should therefore be stored in separate, designated buffers when possible. Additionally, while ANGLE will track updates to a buffer and promote it internally to static if no updates are made, developers can avoid needless data conversions by designating STATIC_DRAW as the usage for these buffers.

- **Recommendation: Always Specify the Fragment Shader Float Precision**
  GLSL ES 1.00 and 3.00 do not specify a default precision for floating-point values in the fragment shader. This makes it a compilation error not to explicitly specify their precision. Earlier versions of ANGLE did not adhere to this rule and did not produce an error. Desktop hardware typically supports high precision, so this never posed a problem for ANGLE itself. However, this lenience caused developers to forget to set the precision, because shaders ran properly without it on their systems. This caused their code not to run at all on various other platforms which do demand the precision to be specified. Precision matters, especially on mobile platforms. We therefore decided to strictly enforce the specification. This change broke a handful of applications, but they were quickly resolved after explaining the issue to the authors. Those developing mainly on an ANGLE-powered browser will now be met with a compilation error if the precision is not set. For everyone else, we highly recommend to check if the browser enforces it or not, and if not, to test frequently with one that does. For more on shader precision, see Chapter 8.

- **Recommendation: Do Not Use Rendering Feedback Loops**
  In the OpenGL APIs, attempting to write to and sample from the same texture or renderbuffer in a rendering operation is considered a rendering feedback loop, and the results of such an operation are undefined in desktop OpenGL and OpenGL ES [Khronos 14a]. For some varieties of graphics hardware, Direct3D 9 could actually provide results for these operations which appeared to be correct.

---

* ANGLE issue 708, http://code.google.com/p/angleproject/issues/detail?id=708

In Direct3D 11, this ability disappeared—attempting to perform such rendering operations produced images with black pixels where the sampled values were used. Users, not realizing that this was an undefined behavior in OpenGL ES and WebGL, began to report this as an error in ANGLE. We decided to create consistent behavior in ANGLE and make clear that the operation is not intended to be supported, by disabling such renders in Direct3D 9 as well. Additionally, the WebGL specification has since been amended to require an error when a rendering feedback loop is in place, rather than leaving the behavior undefined [Khronos 14b].

1.4.2 Beyond WebGL: Recommendations for OpenGL ES, WebGL 2, and More

There are a number of features already supported in ANGLE which are not yet exposed in WebGL. These features also have some caveats and some performance benefits for developers. They're accessible to users of ANGLE's OpenGL ES interface, and many will become available with WebGL 2. We present our recommendations for these use cases next.

- **Recommendation: Don't Use Extensions without Having a Fallback Path**
  It is understandably very tempting to rely on extensions when a quick test indicates that they are supported across a large swath of hardware. Unfortunately Murphy's law, and the huge number of extensions and hardware variants, are not in our favor. Even YouTube has fallen victim to this.* A single-character ANGLE bug caused the OES_texture_npot extension [Khronos 07b], which enables support for textures whose dimensions are not powers of two, not to be advertised on certain hardware that did support it. Our conformance tests don't test unavailable extensions, as an implementation without extensions is still completely conformant, so this regression went entirely unnoticed for some time until YouTube broke. Expecting NPOT textures to be present without having performed an extension check, the hardware accelerated video decode path in Chrome attempted to create a pbuffer surface whose dimensions were not a power of two and encountered a failure. This was quickly remedied (the ANGLE bug in question being a single missing exclamation point in a double negation) once it was known, but some trouble could have been avoided by querying the extension string and providing a fallback path or an alert if the expected extension was not present. Issues with extensions continue to get more complicated over time with increasingly varying hardware features. We therefore recommend using them judiciously and frequently testing the fallback code path.

- **Recommendation: Use Immutable Textures When Available**
  Historically, OpenGL and WebGL textures had to be created one mip level at a time. OpenGL does this via glTexImage*, a method that allows users to create internally inconsistent textures, considered by the GL to be "incomplete." This same method is what is available to developers in WebGL, as texImage*. By contrast, Direct3D requires that users define the dimensions and format of their entire textures at texture creation time, and it enforces internal consistency.

---

* ANGLE issue 799, http://code.google.com/p/angleproject/issues/detail?id=799

Because of this difference, ANGLE must do a considerable amount of bookkeeping and maintain system memory copies of all texture data. The ability to define an entire texture at creation time did later get introduced to OpenGL and its related APIs as immutable textures, which also enforce internal consistency and disallow changes to dimensions and format. Immutable textures came to OpenGL ES 2.0 with `EXT_texture_storage` [Khronos 13a], and they are included in the core OpenGL ES 3.0 specification and the WebGL 2 Editor's Draft specification. When immutable textures are available via extension or core specification, some of ANGLE's bookkeeping can be avoided by using the `texStorage*` commands to define textures.

- **Recommendation: Use RED Textures instead of LUMINANCE**

  In WebGL and unextended OpenGL ES 2.0, the only option developers have for expressing single-channel textures is the LUMINANCE format, and LUMINANCE_ALPHA for two-channel textures. The `EXT_texture_rg` extension [Khronos 11] adds the RED and RG formats, and these formats become core functionality in OpenGL ES 3.0. The formats also appear in the WebGL 2 Editor's Draft specification. Meanwhile, Direct3D 11 has dropped all support for luminance textures, providing only red and red-green formats [MSDN 14a]. This may seem to be a trivial difference—a channel is a channel—but sampling from a luminance texture is performed differently than from textures of other formats. The single channel of a luminance texture is duplicated into the red, green, and blue channels when a sample is performed, while sampling from a RED texture populates only the red channel with data. Similarly, the second channel of a LUMINANCE_ALPHA and an RG texture will populate only the alpha and green channels in a sample, respectively. To support luminance formats against Direct3D 11, rather than alter the swizzle behavior in shaders, ANGLE instead expands the texture data to four channels. This expansion, and the associated additional memory and texture upload performance costs, can be avoided by developers keen for clock cycles by simply using RED textures in place of LUMINANCE and RG in place of LUMINANCE_ALPHA when using ANGLE with APIs that support them.

- **Recommendation: Avoid Integer Cube Map Textures**

  Cube maps with unnormalized integer formats are not supported by Direct3D 11 [MSDN 14c]. The ANGLE team hasn't encountered any uses for it, which may be the reason it was left out of D3D11, but it is a feature of OpenGL ES 3.0 and gets tested by the conformance tests. ANGLE therefore must emulate it in ANGLE's ESSL to HLSL translator. The cube texture is replaced by a six-layer 2D array texture, and the face from which to sample, and at what location, is manually computed. Rather than unnormalized integer formats, we recommend using normalized integer formats for cube maps. If integer values are expected, multiply the sampled value by the maximum integer value, and round to the nearest integer. For example, for signed 16-bit integers: `int i = int(round(32767 * f));`

- **Recommendation: Avoid Full-Texture Swizzle**

  Texture swizzling is an OpenGL ES 3.0 feature which allows a texture's components to be sampled in a different order, using the TEXTURE_SWIZZLE_R,

TEXTURE_SWIZZLE_G, TEXTURE_SWIZZLE_B, and TEXTURE_SWIZZLE_A texture parameters. This is most often used to read RGBA textures as BGRA, or vice versa, and can also be used to replicate components as with luminance textures. This feature is, however, not supported by Direct3D 11. Even though it appears a seemingly simple operation to perform during the shader translation, it is actually not feasible to determine which textures are sampled where, because samplers can be passed from function to function as parameters, and the same texture sampling function can be used to sample various different textures. ANGLE therefore swizzles the texture data itself. This consumes some memory and incurs some overhead at texture upload. These costs can be avoided by not changing the TEXTURE_SWIZZLE_R, TEXTURE_SWIZZLE_G, TEXTURE_SWIZZLE_B, and TEXTURE_SWIZZLE_A texture parameters from their defaults. If necessary, use multiple shader variants to account for different texture component orders.

- **Recommendation: Avoid Uniform Buffer Binding Offsets**
  Uniform buffer objects (UBOs), newly added in OpenGL ES 3.0, are bound objects which store uniform data for the use of GLSL shaders. UBOs offer benefits to developers, including the ability to share uniforms between programs and faster switching between sets of uniforms. OpenGL ES 3.0 also allows UBOs, much like other buffer objects, to be bound at an offset into the buffer, rather than just the buffer head. Direct3D, on the other hand, does not support referencing its analogous structure, constant buffers, until Direct3D 11.1, with the addition of the VSSetConstantBuffers1 method [MSDN 14d]. Offsets are supported with a software workaround on all hardware of lower feature levels. Developers can avoid any performance penalty associated with this workaround by binding UBOs at offset 0 only.

- **Minor Recommendation: Beware of Shadow Lookups in 2D Array Textures**
  Our final recommendation is a minor one, because the range of hardware affected is relatively small. Shadow comparison lookups are a feature introduced in OpenGL ES 3.0. These texture lookups can perform prefilter comparison of depth data contained in a texture against a provided reference value. ES 3.0 also introduces new texture types, including 2D texture arrays. Where these two features intersect, a caveat emerges. Direct3D 11 does support shadow lookups for 2D texture arrays—but not at feature level 10_0 [MSDN 14e]. For this reason, ANGLE must either exclude feature level 10_0 hardware from ES 3.0 support or implement a workaround, with potential performance penalties. If the latter approach is chosen, developers may encounter performance issues on Direct3D 10.0 hardware. If the former approach is chosen instead, then OpenGL ES 3.0 would not be available on this hardware at all.

1.5 Debugging (with) ANGLE

1.5.1 Debugging Translated Shaders

ANGLE's Direct3D backends require an intermediate translation of GLSL ES shaders into HLSL. This can complicate bug analysis because the HLSL compiler reports errors in the translated code, not the original code. This intermediate code can be inspected using the

WEBGL_debug_shaders extension [Khronos 14c]. This extension also allows looking at the translated GLSL code for a desktop OpenGL-based implementation, and even GLSL ES for mobile implementations. The browser may apply workarounds and alterations to your shader code to improve safety—for example, preventing out-of-bounds accesses.

A full example of the use of this extension can be found at http://www.ianww.com/2013/01/14/debugging-angle-errors-in-webgl-on-windows/

1.5.2 Building ANGLE Libraries

Since ANGLE binaries come as a set of drop-in dynamically linked libraries (DLLs), they are isolated from the browser. This makes it relatively straightforward to build your own version and replace the ones that came with the browser. While this isn't intended to be common practice, WebGL developers who want to push the envelope will run into implementation-dependent behavior sooner or later, and ANGLE provides a window into that. It may, for example, be useful to a developer who is experiencing unexpected behavior, or to those interested in writing tools to assist in the debugging or performance profiling of WebGL code, to build ANGLE's debug configuration so that a debugger may be attached and the code may be stepped through and examined in progress. We cover this process briefly next.

The process of downloading and building ANGLE is covered in detail in the ANGLE wiki.* This process may change from time to time, so please consult the page for the latest step-by-step guide. ANGLE may additionally be built as a part of Chromium. Information on how to build a specific version of ANGLE within the larger Chromium build is included in our wiki as well.†

1.5.3 Debugging ANGLE with Chrome

Once the ANGLE libraries are built, they can be dropped into Chrome and used in place of the libraries that shipped with the browser. On Windows, they must be copied to the folder containing the current version of Chrome's support DLLs, which is a numbered folder located alongside the Chrome executable. It's a good idea to move the original versions of the ANGLE DLLs out of the way first, because they'll be needed if something goes awry with ANGLE or to revert to the original state once debugging is complete.

Finding the correct Chrome process to attach the debugger to might seem a daunting prospect, but Chrome provides some helpful command line switches to assist with this task. The most useful of these for the purposes of ANGLE debugging is --gpu-startup-dialog. With this argument, Chrome will spawn a dialog box when the GPU process starts, containing the process ID, and will pause until the box is dismissed. This provides the developer with the opportunity to attach a debugger to the identified process. Also helpful is --use-gl=desktop, which forces Chrome to translate calls to desktop OpenGL itself, rather than translating via ANGLE. This can be useful in identifying whether a problem is specific to ANGLE or occurs somewhere else in the Chrome graphics stack. A full listing of the currently available command line switches in Chrome is maintained at http://peter.sh/experiments/chromium-command-line-switches/

---

* http://code.google.com/p/angleproject/wiki/DevSetup
† http://code.google.com/p/angleproject/wiki/BuildingANGLEChromiumForANGLEDevelopment

From time to time, changes are made to ANGLE or the Chromium codebase which break compatibility with previous versions of Chrome. If this occurs, the issue can be mitigated by working with the version of ANGLE corresponding to the Chrome release being used. Determining which ANGLE version corresponds to a Chrome release is very simple. First, choose "About Google Chrome" from the Chrome menu. This will open a tab that displays information about Chrome, including the software version. This will look something like "Version 39.0.2171.65 m." Make note of the third subset, which identifies the Chromium branch from which this version of Chrome was built. ANGLE maintains branches following the Chromium branch strategy, each named `chromium/<branch number>`, so once this branch number is known, the corresponding ANGLE branch can be checked out. For example, the ANGLE branch used by Chrome 39.0.2171.65 is checked out using the command `git checkout chromium/2171`. The version of an already built ANGLE DLL can be checked by right-clicking on it in Windows Explorer, choosing Properties, and selecting the Details tab in the dialog that appears. The SHA listed as the Product Version is the git hash representing the ANGLE repository tree from which the DLL was built. To check out that version of ANGLE, use `git checkout <SHA>`.

Once attached to the GPU process, the debugger can break at points set within the ANGLE source code, internal variable values can be examined, and code can be stepped through incrementally. Developers may notice that some of the calls that come through ANGLE in Chrome do not correspond to their own WebGL. This is because both WebGL contexts and the Chrome compositor itself make use of ANGLE, so the calls are intermingled.

It may additionally be helpful to use a graphical debugging tool to see visual results of rendering commands as they are issued, examine the contents of textures and buffers, or monitor GPU usage. A step-by-step guide to starting Chromium with one such tool, NVIDIA Nsight, can be found on the Chromium developer wiki, at http://www.chromium. org/developers/design-documents/chromium-graphics/debugging-with-nsight.

1.6 Additional Resources

There are a number of avenues for communicating with and getting help from members of the ANGLE community and team. Bugs in ANGLE can be filed at http://code.google. com/p/angleproject/issues, or at http://crbug.com if they affect Chrome. Filing bugs helps the ANGLE team to maintain a high-quality, conformant project. Our forum and mailing list are excellent resources for finding answers to questions from ANGLE team members and other ANGLE users. They can be mailed at *angleproject@googlegroups.com*, or viewed online at http://groups.google.com/group/angleproject. Developers can discuss ANGLE in real time in our IRC channel, #ANGLEproject on FreeNode.

Bibliography

[ANGLE 13] ANGLE project. "ANGLE Development Update—June 18, 2013." https:// code.google.com/p/angleproject/wiki/Update20130618, 2013.

[ANGLE 14a] ANGLE project. "M(ultiplatform)-ANGLE Effort." https://code.google. com/p/angleproject/wiki/MANGLE, 2014.

[Bagnell 13] Daniel Bagnell. "Robust Polyline Rendering with WebGL." http://cesiumjs. org/2013/04/22/Robust-Polyline-Rendering-with-WebGL/, 2013.

[Khronos 07a] The Khronos Group. "OES_element_index_uint." Contact Aaftab Munshi. https://www.khronos.org/registry/gles/extensions/OES/OES_element_index_uint.txt, 2007.

[Khronos 07b] The Khronos Group. "OES_texture_npot." Contact Bruce Merry. https://www.khronos.org/registry/gles/extensions/OES/OES_texture_npot.txt, 2007.

[Khronos 10] The Khronos Group. "The OpenGL Graphics System: A Specification (Version 3.3 (Core Profile)— March 11, 2010)." Edited by Mark Segal and Kurt Akeley. https://www.opengl.org/registry/doc/glspec33.core.20100311.pdf, 2010.

[Khronos 11] The Khronos Group. "EXT_texture_rg." Contact Benj Lipchak. https://www.khronos.org/registry/gles/extensions/EXT/EXT_texture_rg.txt, 2011.

[Khronos 13a] The Khronos Group. "EXT_texture_storage." Contacts Bruce Merry and Ian Romanick. https://www.khronos.org/registry/gles/extensions/EXT/EXT_texture_storage.txt, 2013.

[Khronos 13b] The Khronos Group. "Testing/Conformance." https://www.khronos.org/webgl/wiki/Testing/Conformance, 2013.

[Khronos 14a] The Khronos Group. "OpenGL ES Version 3.0.4." Edited by Benj Lipchak. https://www.khronos.org/registry/gles/specs/3.0/es_spec_3.0.4.pdf, 2014.

[Khronos 14b] The Khronos Group. "WebGL Specification." Edited by Dean Jackson. https://www.khronos.org/registry/webgl/specs/latest/1.0/, 2014.

[Khronos 14c] The Khronos Group. "WEBGL_debug_shaders." Contact Zhenyao Mo https://www.khronos.org/registry/webgl/extensions/WEBGL_debug_shaders/, 2014.

[Koch 12] Daniel Koch and Nicolas Capens. "The ANGLE Project: Implementing OpenGL ES 2.0 on Direct3D." *WebGL Insights*. Edited by Patrick Cozzi and Christophe Riccio. Boca Raton, FL: CRC Press, 2012.

[Kokkevis 11] Vangelis Kokkevis. "The Chromium Blog: OpenGL ES 2.0 Certification for ANGLE." http://blog.chromium.org/2011/11/opengl-es-20-certification-for-angle.html, 2011.

[MSDN 14a] Microsoft. "DXGI_FORMAT enumeration." http://msdn.microsoft.com/en-us/library/windows/desktop/bb173059, 2014.

[MSDN 14b] Microsoft. "Primitive Topologies." http://msdn.microsoft.com/en-us/library/windows/desktop/bb205124, 2014.

[MSDN 14c] Microsoft. "Load (DirectX HLSL Texture Object)." http://msdn.microsoft.com/en-us/library/windows/desktop/bb509694, 2014.

[MSDN 14d] Microsoft. "ID3D11DeviceContext1::VSSetConstantBuffers1 method." http://msdn.microsoft.com/en-us/library/windows/desktop/hh446795, 2014.

[MSDN 14e] Microsoft. "SampleCmp (DirectX HLSL Texture Object)." http://msdn.microsoft.com/en-us/library/windows/desktop/bb509696, 2014.

[Woods 15] Shannon Woods, Nicolas Capens, Jamie Madill, and Geoff Lang. "ANGLE: Bringing OpenGL to the Desktop." *GPU Pro 6: Advanced Rendering Techniques*. Edited by Wolfgang Engel. Boca Raton, FL: CRC Press, 2015.

2

Mozilla's Implementation of WebGL

Benoit Jacob, Jeff Gilbert, and Vladimir Vukicevic

2.1 Introduction

A typical implementation of WebGL in a browser has to implement this API on top of system graphics APIs such as OpenGL, OpenGL ES, or Direct3D. Thus, WebGL is effectively an additional layer on top of OpenGL or other similar APIs, potentially adding overhead. Some understanding of what kind of work takes place in a browser's WebGL implementation may help application developers write better code that suffers from less overhead.

Figure 2.1 gives a general overview of Mozilla's WebGL implementation. Over the course of this chapter, we will touch on each part.

2.2 DOM Bindings

The first step in our study of what happens in the browser *below* WebGL method calls is studying the mechanics of the call *itself.*

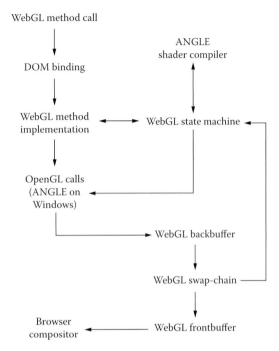

WebGL method call

DOM binding

WebGL method
implementation

OpenGL calls
(ANGLE on
Windows)

ANGLE
shader compiler

WebGL state machine

WebGL backbuffer

WebGL swap-chain

Browser
compositor

WebGL frontbuffer

Figure 2.1

Overview of Mozilla's WebGL implementation.

Take, for example, the WebGL `uniform4f` method. In a JavaScript program, a call to `uniform4f` looks like this:

```
gl.uniform4f(location, x, y, z, w);
```

Here, the `location`, `x`, `y`, `z`, `w` parameters, are JavaScript values. They could be anything: numbers, objects, arrays, strings…. The WebGL IDL* says that `location` should be a `WebGLUniformLocation` object, and `x`, `y`, `z`, `w` should be numbers. This means that the WebGL implementation must verify that the values passed for these parameters are of these types, or can be converted to these types; otherwise, the WebGL implementation must generate an exception.

In addition to that, the bit representation of these parameters is not the same on the JavaScript side and on the browser internal C++ side, so some conversion work needs to be done.

This is very repetitive work that needs to be done for every parameter of every DOM API method that is internally implemented in C++. This is automated. In Mozilla's implementation, a Python script parses all the web IDL and generates some C++ code for each method,† which we call its *DOM binding*. For each method, the generated code takes care of validating and converting the input JavaScript parameters, and then calling the corresponding C++ function in the Mozilla browser code. The execution flow looks like Figure 2.2.

From the application developer's perspective, the main takeaway should be that one should expect the cost of calling a method in a DOM API to grow roughly linearly with the number of method parameters. Methods taking more parameters should be more expensive to call. However, this overhead is low enough that it only becomes noticeable for really cheap WebGL methods.

We confirmed with a microbenchmark‡ that `uniform4f` is slower than `uniform1f` in all browsers that we tried,§ as one would expect since the DOM binding for `uniform4f` has more parameters to check and convert. The speed difference between these two

---

* IDL stands for Interface Definition Language. The WebGL IDL is the part of the WebGL specification that formally defines the WebGL interfaces and methods, describing the types of methods' parameters and return values.

† In Mozilla's source tree, which can easily be searched online (try http://dxr.mozilla.org), the WebIDL parser is `WebIDL.py`, and the corresponding C++ code generator is `Codegen.py`. We don't give full paths with directories, as that should not be needed to find these files and is subject to change. The source Web IDL for WebGL is `WebGLRenderingContext.webidl`. An additional configuration file, `Bindings.conf`, helps the code generator find the C++ implementation for a given web IDL interface. The resulting generated C++ code is less easy to find online, but the interested reader could easily build Firefox from sources and then look, in the resulting object directory, for `WebGLRenderingContextBinding.cpp`.

‡ https://github.com/WebGLInsights/WebGLInsights-1/tree/master/02-Mozillas-Implementation-of-WebGL/

§ We tried Safari, Chrome, and Firefox on Mac OS X 10.10, in late 2014.

2. Mozilla's Implementation of WebGL

methods varied significantly between different browsers, from 15%–20% in one browser to 40%–50% in another.

However, as soon as one calls more expensive WebGL methods, this effect becomes negligible. For example, comparing `uniform4f` to `uniform4fv` instead of `uniform1f`, we found that `uniform4fv` is slower than `uniform4f` in all browsers that we tried, despite taking as few parameters as `uniform1f` does. This is due to `uniform4fv` having inherently more work to do than `uniform4f`, since it can handle arbitrary uniform array uploads.

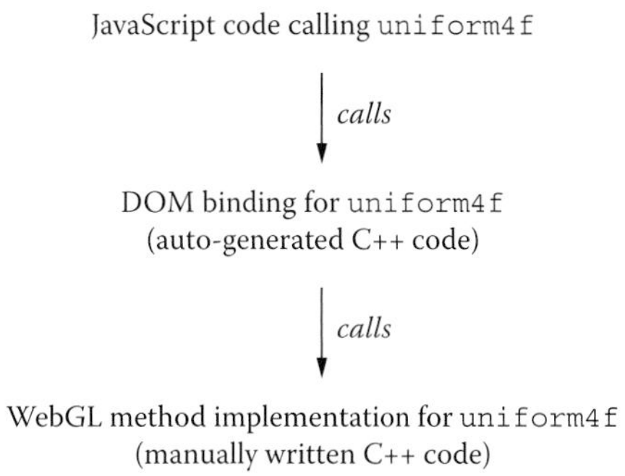

2.3 WebGL Method Implementations and State Machine

Let us now describe what kind of work takes place inside the manual C++ implementations for WebGL methods.

Take, for instance, `texSubImage2D`. It takes several parameters, one of which is the image data format, which is a `GLenum`, that is, a number. The DOM bindings work done so far has checked that the value passed for this parameter had an acceptable type, and then converted it to a native integer. But the DOM bindings don't know about anything specific to WebGL, so they have no clue as to what the allowed range is for this integer parameter.

Thus, the manually written C++ implementation for each WebGL method has the responsibility to finish checking each parameter and generating the right WebGL errors.

One might still think that there is nothing here that a WebGL implementation has to do, as all these checks should be performed already by the underlying OpenGL implementation. This isn't so, if only because WebGL is not OpenGL and there are subtle differences of semantics and capabilities that mean that WebGL validation is not the same as OpenGL validation. Additionally, even when WebGL and OpenGL semantics agree, a WebGL implementation may not want to trust the OpenGL implementation to be bug-free, especially if security is at stake.

Thus, implementations of WebGL methods have to implement most of the call validation themselves, regardless of whether some of this work is going to be duplicated in the underlying OpenGL implementation.

Like we saw before with `texSubImage2D`'s format parameter, call validation depends not only on parameter values but also on existing state. This means that the WebGL implementation needs very frequent access to current state, such as, in the case of `texSubImage2D`, the bound texture and its properties such as its format.

The WebGL implementation accesses state in two different ways. First, the WebGL implementation can query state directly from OpenGL by getter calls such as `glGetIntegerv`. Second, some of the WebGL state is manually tracked by the WebGL implementation, either because it is WebGL-specific state without a direct OpenGL equivalent, or because the OpenGL API doesn't expose it in a practical and efficient way.

For example, each WebGL texture is tracked as a `WebGLTexture` object, which contains the OpenGL id for this texture object as well as an array of structs tracking the width, height, format, etc., for each image stored on this texture object (there can be multiple

JavaScript code calling `uniform4f`

calls

DOM binding for `uniform4f`
(auto-generated C++ code)

calls

WebGL method implementation for `uniform4f`
(manually written C++ code)

Figure 2.2

Mechanics of a WebGL method call.

images for mipmaps or cube maps). Much, though not all, of that state is redundant with OpenGL state, as the underlying OpenGL texture already stores internally attributes such as width, height, format, etc. However, at least in OpenGL ES 2.0, there isn't a practical way to query the width of a texture image.

At this point we can draw two conclusions. First, a cause of inevitable speed overhead in WebGL compared to OpenGL is the extra validation work that is needed. Second, in addition to speed overhead, there is moderate memory overhead due to having to duplicate large parts of the OpenGL state—though that is mostly negligible in practice.

2.4 Texture Uploads and Conversions

So far we've talked about general considerations that apply across all of WebGL. Let's now turn to specific areas that are particularly complex and prone to causing overhead, starting with texture uploads (i.e., the `texImage2D` and `texSubImage2D` entry points).

Already in OpenGL, because of the way that these entry points are specified, they must perform a full copy of the input texture image before returning. Indeed, the contents of the texture must not be affected by any change to the input image data after `tex[Sub]Image2D` has returned.

WebGL inherits this overhead from OpenGL, and then adds some more. One cause of additional WebGL overhead that affects all `texImage2D` overloads, but does not affect `texSubImage2D`, is that, in WebGL, `texImage2D` must check if the underlying call to `glTexImage2D` generated an error. Indeed, `glTexImage2D` typically has to allocate memory, and this could fail with a `GL_OUT_OF_MEMORY` error. We care because a WebGL implementation has to track state by itself, and when updating state, it has to keep its tracked state in sync with actual OpenGL state. So if an OpenGL call that is supposed to update state fails to do so, the WebGL implementation needs to know.

Unfortunately, in OpenGL, calling `glGetError` is costly, for two reasons. First, calling `glGetError` means that a larger part of `glTexImage2D`'s work must be immediately executed instead of being run asynchronously later. Second, `glGetError` also requires the immediate execution of the entire OpenGL command stream that may have been accumulated so far, typically on mobile "deferred" GPUs.

The overhead described before is specific to `texImage2D` and does not affect `texSubImage2D`. However, the other types of overhead described below equally affect `texImage2D` and `texSubImage2D`.

The `tex[Sub]Image2D` overloads taking an HTML element, such as a `<img>` element, typically require the entire image to be converted to a different format. Indeed, the pixel format of a decoded image is a browser implementation detail that isn't exposed in any API, and can vary at any time. So the WebGL specification had to solve the problem of how to prevent such nonstandardized details to leak into the interface. The solution adopted by WebGL was to specify that, no matter what the original format of the decoded image is, it must be converted to the format described by the format parameters taken by `tex[Sub]Image2D`.

This comes at the cost of introducing substantial overhead, as now `tex[Sub]Image2D` calls taking an HTML element or a canvas ImageData must convert it to the requested format. In particular, the pressure on memory bandwidth gets worse, because now there is an additional copy with format conversion being made in the WebGL implementation even before calling OpenGL.

The pixel format is only one of several factors that could require a conversion. Other factors are controlled by `pixelStorei` parameters. They are the stride (`UNPACK_ALIGNMENT`), the alpha channel premultiplication status (`UNPACK_PREMULTIPLY_ALPHA_WEBGL`), and the vertical flipping (`UNPACK_FLIP_Y_WEBGL`).

For the `tex[Sub]Image2D` overloads taking an HTML element or canvas `ImageData`, another case needs to be mentioned: the case of opting out from colorspace conversion. This is achieved by setting another `pixelStorei` parameter, `UNPACK_COLORSPACE_CONVERSION_WEBGL`, to the value `NONE`. This is implemented by re-decoding the image from its original stream, from scratch. This, of course, is costly.

There is one particularly nasty corner case with `UNPACK_PREMULTIPLY_ALPHA_WEBGL` that may also require re-decoding an image from scratch. This is the case when the image source is an HTML `<img>` element that was already premultiplied in the browser's memory (which is very commonly done by browsers), and `UNPACK_PREMULTIPLY_ALPHA_WEBGL` has its default value of `false`. Un-premultiplying a previously premultiplied element doesn't exactly recover the original image, so the WebGL spec requires implementing this by re-decoding the image from scratch, like we described above for the case where `UNPACK_COLORSPACE_CONVERSION_WEBGL` is set to `NONE`.

For the `tex[Sub]Image2D` overloads taking an `ArrayBufferView`, things are a lot simpler: There is no possibility of format or stride or colorspace conversion, no possibility of un-premultiplication, and in the default state there is no conversion at all. Nondefault `pixelStorei` parameters can still require conversions: Setting `UNPACK_FLIP_Y_WEBGL` or `UNPACK_PREMULTIPLY_ALPHA_WEBGL` to `true` will still require a flipping or a premultiplication, respectively.

A practical takeaway from this conversation is that the default state of `UNPACK_PREMULTIPLY_ALPHA_WEBGL`, set to the value `false`, is best for `tex[Sub]Image2D` overloads taking an `ArrayBufferView` or other image sources that are known not to be premultiplied, such as canvas `ImageData`, but can be very painful for the overloads taking an HTML element, which typically are premultiplied. For those, setting `UNPACK_PREMULTIPLY_ALPHA_WEBGL` to `true` allows for cheaper texture uploads with more accurate results.

Table 2.1 summarizes the preceding discussion* of `tex[Sub]Image2D` overhead.

2.5 Null and Incomplete Textures

In some cases, differences between WebGL and OpenGL or between different flavors of OpenGL require a less straightforward approach to implementing WebGL `tex[Sub]Image2D`, which comes with specific overhead characteristics.

One such case, which we have avoided discussing in the preceding section, is when passing `null` for the input image data. In OpenGL, passing null image data means that the texture image must only be allocated, leaving its contents uninitialized. In WebGL,

---

* For the reader interested in checking the Mozilla source code for this, a good approach would be to use a code search tool such as the online one at http://dxr.mozilla.org and search for identifiers such as `UNPACK_PREMULTIPLY_ALPHA_WEBGL`.

Table 2.1 Overhead of Texture Image Specification Calls

	texImage2D with ArrayBufferView	texImage2D with element or ImageData	texSubImage2D with ArrayBufferView	texSubImage2D with element or ImageData
Inherent OpenGL overhead (has to copy data immediately)	Yes	Yes	Yes	Yes
Has to call glGetError immediately	Yes, except perhaps when replacing existing same-size image	Yes, except perhaps when replacing existing same-size image	No	No
Has to convert input image data before passing it to OpenGL	Only if nondefault pixelStorei parameters require it	Yes, except if exact same format and no pixelStorei parameter requires it	Only if nondefault pixelStorei parameters require it	Yes, except if exact same format and no pixelStorei parameter requires it
Has to re-decode image from scratch	No (not applicable)	Only if pixelStorei parameters require it	No (not applicable)	Only if pixelStorei parameters require it

uninitialized texture data would be unacceptable for security and portability reasons. So WebGL avoids exposing uninitialized memory and instead specifies that null texture images must behave as if all bytes were set to 0 (i.e., as transparent black textures).

The straightforward implementation would be to allocate a buffer, fill it with zeroes, and pass it to glTextImage2D. That works, but adds overhead, so Mozilla's implementation avoids this approach insofar as possible. The key observation is that in WebGL rendering, all such transparent black textures are indistinguishable from each other. Thus, one can use a single global transparent black texture, of size 1×1, and use it for all WebGL textures with null image data. One just has to carefully track this state and correctly revert to allocating the texture's own storage as needed, such as when doing a texSubImage2D call on it. However, in most cases then, the texSubImage2D call will replace the entire contents of the texture image, so at this point there will be no need anymore to allocate a temporary buffer of zeroes!

In summary, at least in Mozilla's current implementation, texImage2D with null is very fast, and subsequent texSubImage2D runs at normal speed provided that it covers the entire image. The slow corner case is the first partial texSubImage2D call on a null texture image.

Null texture images are only one of two corner cases that are handled with dummy black textures in Mozilla's implementation. The other case is that of incomplete textures.

In OpenGL and WebGL terminology, an incomplete texture is one that has an illegal combination of texture parameters and texture images. The precise rules vary between different flavors of OpenGL. WebGL follows the core OpenGL ES 2.0 rules. For example, in this setting, a non-power-of-two-size texture with a REPEAT wrap mode is incomplete. Mozilla's WebGL implementation may run against any flavor of OpenGL, ES, or regular, and must offer WebGL-conformant behavior everywhere. The only way to achieve that is by manually implementing all the texture-completeness logic.

Thus, Mozilla's WebGL implementation tracks each texture's completeness status, and when it finds that drawing is about to sample from an incomplete texture, it implements the correct behavior for incomplete textures, which is to sample as opaque black (RGBA 0,0,0,255), by manually binding a dummy 1×1 opaque black texture.

This, at least, has no particularly surprising overhead behavior. Under typical circumstances, drawing doesn't actually sample from null or incomplete textures, so there is just a roughly constant amount of tracking overhead[*] on each operation that can affect or depend on texture completeness, such as draw calls. In the unusual case where drawing is actually about to sample from a null or incomplete texture, some additional work is done around each draw call to bind the dummy black textures. We will get back to that when we discuss draw-call overhead in a later section.

2.6 Shader Compilation

WebGL implementations must carry their own shading language compilers for several different reasons. Mozilla's implementation uses the shader compiler from the ANGLE project[†] (Chapter 1).

There are many different flavors of the OpenGL shading language (GLSL). Non-ES OpenGL and OpenGL ES differ in this respect. Their shading language dialects are referred to as GLSL and GLSL ES, respectively. Even within GLSL ES, the specification leaves room for variation.

WebGL's shading language (WebGL GLSL) is specified by starting from GLSL ES 1.0, and tightening loose parts by mandating several restrictions that were left optional in GLSL ES. In particular, WebGL adopts restrictions on control flow, in particular disallowing `while` loops and restricting the genericity of `for` loops. WebGL also adopts the restrictions on the addressing of uniform arrays, so that in fragment shaders all array indices must be constant expressions.

In addition to validating and translating shaders, the shader compiler also prevents triggering certain classes of driver bugs and implementation-defined behavior. The most obvious way that it does so is by catching malformed shaders before they are submitted to the OpenGL implementation, but this is not the only way.

For instance, a WebGL implementation needs to guard against out-of-bounds access to uniform arrays. The previously mentioned shading language restrictions help with that, but are not enough. Addressing uniform arrays using runtime indices is still allowed in vertex shaders. In the ANGLE shader compiler, which Mozilla uses, this is addressed by injecting clamp instructions so that, at the cost of some runtime overhead, no out-of-bounds access is possible. Another possible way in which out-of-bounds access to uniform arrays could happen is in uniform array setters such as `uniform4fv`. The OpenGL specifications mandate that when the upload size exceeds the actual uniform array size, excess values must be ignored. But this is the kind of security-sensitive, memory-addressing corner case that browser implementers typically don't want to rely on drivers to implement correctly. So, while in theory not needed, Mozilla's implementation queries the uniform array sizes from the ANGLE shader compiler, and uses that to clamp uniform array uploads, adding another layer of protection against out-of-bounds uniform array uploads.

Another way in which the ANGLE shader compiler helps to avoid triggering driver bugs is long identifier shortening. Early WebGL implementations were often running

---

[*] Most of the code tracking null and incomplete textures can be seen in this file in Mozilla source code: `WebGLTexture.cpp`.

[†] ANGLE is not a Mozilla project. Its website is https://code.google.com/p/angleproject/

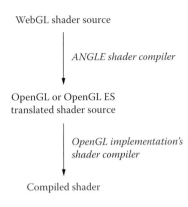

WebGL shader source

ANGLE shader compiler

OpenGL or OpenGL ES
translated shader source

*OpenGL implementation's
shader compiler*

Compiled shader

Figure 2.3

Shader compilation steps.

into driver bugs with shaders containing long identifiers; it appeared that the highest power-of-two identifier length that would be safe on all drivers in circulation was 32. Mozilla's WebGL implementation thus relies on the ANGLE shader compiler to replace long identifiers in shaders by shortened identifiers, not exceeding 32 characters in length, and obtains from ANGLE the correspondence between original and shortened identifiers.

Thus, compared to OpenGL, WebGL incurs at least one additional shader translation step,* which involves significant transformations even when the target is OpenGL ES. The cost of compiling shaders is thus significantly higher in WebGL than in a good OpenGL implementation. The shader translation diagram, at this point, looks like Figure 2.3.

This is not accounting for the special case of running on Windows, which requires additional translation described later in the chapter.

2.7 Validation and Preparation of `drawArrays` Calls

The implementation of `drawArrays` needs to iterate over all vertex attributes and, for those that are arrays and are consumed by the current program, check that they have sufficient length for the parameters passed to `drawArrays`.

The overhead of doing this on every `drawArrays` call was not negligible, so we implemented a further optimization: We keep track of when vertex attribute array fetching has already been verified and, if it has, we skip the verification step.

In addition to validation, `drawArrays` must also prepare some OpenGL state to account for differences between OpenGL and WebGL. As mentioned earlier, null and incomplete textures are swapped for 1×1 black textures. Furthermore, `drawArrays` must emulate the attribute 0 array. This is specific to the situation where WebGL has to run on top of non-ES OpenGL, as is the case on Mac OS X and on desktop Linux.

In WebGL and OpenGL ES, all vertex attributes are on the same footing. Attribute 0 is not special. Like any other attribute, it can be either an array attribute (as enabled by `enableVertexAttribArray`), or a nonarray attribute (the default state, giving the same uniform value for all vertices).

In non-ES OpenGL, attribute 0 cannot be a nonarray uniform attribute giving the same value for all vertices. Instead, when attribute 0 is not array enabled, it behaves with special semantics that are designed to allow implementing the old OpenGL 1 fixed-function vertex specification API.

Consequently, when a WebGL application tries to use attribute 0 as a nonarray, this forces the WebGL implementation to do expensive emulation work when running on non-ES OpenGL platforms. Specifically, the WebGL implementation must generate and upload an array buffer that contains N copies of the generic vertex attribute value, where N is long

---

* The reader interested in seeing the Mozilla source code for this could use a code search tool such as http://dxr.mozilla.org and search for ShCompile, which is the function offered by ANGLE to compile a shader. Mozilla's source code tree contains its own copy of ANGLE.

enough for the given draw call, and attach this buffer to attribute 0, so that it is once again an array attribute.

Takeaway performance tip: make sure that one of the array attributes is bound (using `bindAttribLocation`) to location 0. Otherwise, high overhead should be expected when running on non-ES OpenGL platforms such as Mac OS X and desktop Linux.

2.8 Validation of `drawElements` Calls

In OpenGL, `glDrawElements` has the same implementation-defined corner cases as `glDrawArrays`, plus one: `glDrawElements` chooses vertices indirectly via the bound element array, and if any of the indices read from the bound element array is out of bounds for the current vertex array attributes, the results are implementation defined.

Here, WebGL differs from OpenGL in two ways. First, WebGL implementations must prevent the adverse effects such implementation-defined behavior can have on many drivers in circulation, which includes crashing, recovering with severe lag, or dangerously exposing memory contents that shouldn't be accessible. Second, WebGL 1.0 specifically mandates an `INVALID_OPERATION` error to be generated in this case.

`drawElements` takes an `offset` and a `count` parameter and reads the corresponding contiguous subarray of the element array, starting at the given offset and containing the given count of elements. So `drawElements` must compute the maximum element in this contiguous subarray, and the rest is similar to `drawArrays` validation: One just needs to compare this value to the smallest size of all vertex attribute arrays that are to be consumed by the current shader program.

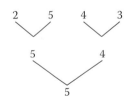

Figure 2.4

Example element array.

Thus, `drawElements` has to solve a fairly generic algorithmic problem: how to efficiently compute the maximum value in an arbitrary contiguous subarray of an array of integers. Mozilla's implementation[*] handles this efficiently enough to make this hardly matter at all, by constructing a binary tree storing precomputed maximums of parts of the array. Let's illustrate this by an example. Suppose that the element array is of length 8, with values as in Figure 2.4.

The corresponding tree is then built by grouping array entries two by two and storing the maximum of these two values in a tree leaf (thus the tree needs to have four leaves) and continuing, at each level, by storing in each node the maximum value of its two children, as in Figure 2.5.

We can always assume that the binary tree is complete, by rounding the element array size to the next power of two.

Figure 2.5

Complete binary tree storing partial maximums of elements.

Since the tree is complete, we see that it really is a binary heap, and can be stored compactly in a single array of integers, whose size is twice the size of the leaf level, as follows: Leave the first entry of this array unused, store the tree root at index 1, then store the next level at indices [2..3], then store the next level at indices [4..7], and so on. Level n is stored at indices $[2^n..2^{n+1}-1]$. Thus, our previous example is stored in an array of length 8, as in Figure 2.6.

Figure 2.6

Compact array storage of the complete binary tree.

---

[*] The relevant Mozilla source code file for this section is `WebGLElementArrayCache.cpp`.

At this point, this data structure requires exactly the same amount of space as the element array itself. How can we reduce it? Notice that each tree level is twice as small as the next level, so the bulk of the memory usage is for the last few tree levels, near the leaves. Now, notice that these last few tree levels are also the least useful, since each node in these levels corresponds only to a small number of element array entries. For example, half of the total memory overhead is caused by the leaf level, but each leaf only contains the maximum of two element array entries, so these leaves are not very useful.

Mozilla's implementation actually stores in each tree leaf the maximum of eight consecutive values in the element array, instead of two in the preceding naive approach. This reduces memory usage by a factor of four, which is usually enough to make it negligible—if not, we can simply increase the grouping.

It is easy to see that the resulting algorithm for `drawElements` validation takes $O(\log n)$ time for a `drawElements` call consuming n elements. Thus, about any usage pattern is fine. Overhead can be substantial enough to appear at the bottom of a profile, but it should never be large.

There is only one pathological use case that should be avoided as it incurs unnecessary overhead: It is when the same element array buffer is interpreted as different types by different `drawElements` calls (i.e., `UNSIGNED_BYTE` versus `UNSIGNED_SHORT` versus `UNSIGNED_INT`. Using the same element array buffer with multiple index types only requires the implementation to maintain separate trees for each type; there are three possible types so there can be up to three trees to maintain for a given element array buffer, which multiplies by three the memory usage and speed overhead. Just don't do it. There is no good reason to: Each `drawElements` call can only work with one index type anyway. Keep separate index types in separate element array buffers.

2.9 The Swap-Chain and Compositing Process

In WebGL, the browser is entirely responsible for frame scheduling and compositing. The application merely provides a callback that signals when a new frame is ready to be drawn. This callback is typically called by `requestAnimationFrame` at the browser's whim, typically scheduling frames for smooth animation, synchronizing with the graphics hardware's refresh signal.

Any WebGL canvas has a *backbuffer,* which is the default framebuffer object receiving WebGL drawing, and a *frontbuffer,* which is what the compositor shows. When the compositor is ready to update its display of the canvas, if the backbuffer has received any drawing, it is "presented" (i.e., promoted to being the frontbuffer).

However, due to asynchronicity of GPU APIs, the contents of the now-frontbuffer might not have actually finished rendering on the GPU. WebGL implementations must be sure that their presented frames are complete before letting the compositor sample from them.

Depending on browser details, and at least in Mozilla's case as of 2014, compositing involves sending frames from one thread or process, running the WebGL application producing frames, to another thread or process, running the browser compositor consuming frames.

Thus, we typically have two threads or processes, both using OpenGL, that have to synchronize their OpenGL commands with each other to prevent one side from reading the other side's rendering until it is complete.

The most trivial method for guaranteeing frame completion is to call glFinish when the canvas presents its buffer to the compositor. However, this does not provide great performance, as it forces the CPU to wait until the GPU is completely done with the command stream. Instead, we merely need to establish a dependency relation on the GPU side alone, where we don't let the GPU read from the frontbuffer until it is done rendering it.

Such GPU-side synchronization is extremely platform dependent, especially when the two sides may be in different processes. Many platforms have some variety of synchronization fence that can be inserted into the GPU command stream. On Mac, where we implement our buffers using IOSurface, we only need to call glFlush, while on other platforms glFlush does not guarantee any synchronization.

Once one has solved the problem of correctly compositing complete WebGL frames, the next problem is performance. Synchronization can harm performance by causing one side to wait idly for the other side. With only a backbuffer being rendered to and a frontbuffer being composited, synchronization would take as long as it takes for the other side to finish working.

The typical solution to this problem, which is what Mozilla's implementation does, is called triple buffering. It adds a third buffer called the "staging buffer." When a new frame has been produced by the WebGL application, the backbuffer is swapped with the staging buffer. When the compositor consumes a new frame, it swaps the frontbuffer with the staging buffer. Thus, the backbuffer and frontbuffer are never directly swapped with each other, which removes the need for expensive synchronization between the two sides. Some thread or process synchronization is still needed, but only to prevent two swaps from happening at the same time. Because these swaps are fast, this is not a problem. Figure 2.7 summarizes triple buffering.

The back/staging/frontbuffer stages in the diagram in Figure 2.7 are steps that are specific to WebGL. By contrast, a native OpenGL application would be the equivalent of the browser compositor here, so it would start right away at the last step in the figure's diagram. This means that WebGL incurs memory overhead due to having these additional frame buffers.

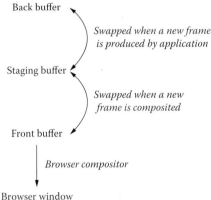

Figure 2.7

Triple-buffered swap-chain.

Triple buffering ensures that there is no significant throughput overhead on the CPU side, but depending on implementation details that are prone to changing at any time, there can be some latency overhead, and some approaches to triple buffering incur a latency of one frame's time (i.e., typically 1/60 of a second).

The first conclusion of this discussion is that if you don't need to update a frame, avoiding re-rendering it will save not only the time it takes to render it, but also a lot of internal compositing work and synchronization. So if you don't need to update a frame, don't touch it, don't even call clear.

The second conclusion is that the preserveDrawingBuffer context creation flag is best left in its default false value. Setting it to true essentially means that the browser can't just swap buffers anymore, and instead must copy buffers, which is expensive in terms of memory bandwidth.

The third conclusion is that there are additional considerations when evaluating the memory overhead of a WebGL canvas. Depth and stencil components are only needed for the backbuffer, but color and alpha components are needed for all three buffers participating in a triple-buffered swap-chain. For example, a 2 megapixel WebGL canvas with alpha, stored as 32-bit RGBA, pays for an additional two RGBA buffers of 8 MB each, resulting in 16 MB of additional memory usage compared to an equivalent OpenGL application, regardless of depth or stencil components.

Finally, a commonly asked question is whether overlaying other HTML content on top of a WebGL context causes overhead. In Mozilla's implementation, WebGL rendering always goes through the stages described earlier, ending up as one layer in the compositing process, so overlaying more HTML content isn't a problem, as the main cost is paid anyway. Consider this a fair exchange for the previously described overhead.

2.10 Platform Differences

Probably the most platform-specific part of a WebGL implementation is the swap-chain, as described in the previous section. This is by itself the most typical cause of surprising WebGL performance differences between different platforms or browsers. Aside from that, platforms fall into three broad classes as far as Mozilla's WebGL implementation is concerned: mobile OpenGL ES platforms, desktop OpenGL platforms, and Windows.

2.10.1 Mobile OpenGL ES Platforms

Mobile platforms typically implement OpenGL ES, after which WebGL is directly modeled. This makes the implementation of WebGL relatively easy. The difficulty on mobile platforms is not so much for WebGL implementations, but rather for WebGL applications. Mobile GPUs tend to have very different performance characteristics than desktop GPUs, and WebGL applications need to adapt to them while the browser's WebGL implementation merely stays out of the way.

2.10.2 Desktop OpenGL Platforms

We have discussed many differences between desktop OpenGL and OpenGL ES in earlier sections. In addition to that, desktop OpenGL also presents WebGL implementations with a dilemma, as there are very important differences between OpenGL "core profiles" and "compatibility profiles." Core profiles remove functionality considered deprecated by the OpenGL specifications, including functionality that is part of WebGL. For instance, core profiles do not support such things as the default (zero) vertex array object, which exists in WebGL. This can make WebGL harder to implement on a core profile, and thus Mozilla's WebGL implementation generally runs on top of an OpenGL implementation's compatibility profile despite its typically increased overhead.

2.10.3 Windows

Windows is a completely different situation. On Windows, the first-class, low-level graphics API is Direct3D. While some GPU vendors provide very good OpenGL implementations on Windows, in practice the proportion of Windows machines with good OpenGL drivers is not high enough for browsers to rely on exclusively. As a result, web browsers target Direct3D on Windows.

Along with the aforementioned shader compiler, the ANGLE project offers a full implementation of OpenGL ES on top of Direct3D (Chapter 1). On Windows, Google and Mozilla both implement WebGL primarily on top of ANGLE's OpenGL ES implementation.* Due to the large amount of work that ANGLE has to perform internally to implement OpenGL ES on top of Direct3D, certain operations have higher overhead on Windows than on other platforms. Most importantly, shader translation becomes more complex, as shown in Figure 2.8.

Thus, on Windows, each WebGL shader compilation involves three compilation steps—compared to two on other platforms and one in native applications. That's overhead.

2.11 Extension Interactions

With OpenGL, any extension that is supported is readily available, without requiring an "enabling" mechanism. Thus, an application can blithely use extensions without checking for their support, accidentally relying on an extension which may only be present on some subset of implementations.

Instead of such "always on" extension behavior, WebGL requires that applications explicitly enable any extensions they want to use. Before an extension is requested by the application, the WebGL implementation does not enable the extension's features, though it does advertise the availability of the extension via `getSupportedExtensions`. Only when an application explicitly requests an extension by calling `getExtension` is the extension's behavior activated.

While great for portability, this does cause issues for implementations: They must support any combination of supported extensions being enabled or not. This creates a much larger matrix of possible semantics to implement, especially in validation code. For instance, `OES_texture_float` allows creating textures with a type of `FLOAT`. However, these float textures are not allowed to be sampled with any filter other than `NEAREST` unless `OES_texture_float_linear` is enabled.

The effects of explicit extension activation extend to shader compilation as well. Since WebGL extensions (such as `EXT_frag_depth`) can enable new GLSL functionality, the extensions active at shader compilation time can change how the shader is compiled.

2.12 Closing Thoughts

While some types of overhead affect any implementation of WebGL, the resulting capabilities are relatively uncompromised. WebGL aims to strike a balance. If kept too

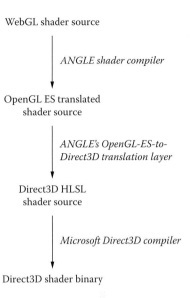

WebGL shader source

ANGLE shader compiler

OpenGL ES translated shader source

ANGLE's OpenGL-ES-to-Direct3D translation layer

Direct3D HLSL shader source

Microsoft Direct3D compiler

Direct3D shader binary

Figure 2.8

Shader compilation steps on Windows with the ANGLE renderer.

---

* Browsing one's filesystem in the Firefox application directory, one can see a few large files corresponding to the translation layers described here. `GLESv2.dll` and `EGL.dll` are the ANGLE OpenGL-ES-to-Direct3D translation layer, and `D3DCompiler_*.dll` are copies of the Microsoft Direct3D compiler.

portable, it would not be sufficiently capable. If it were to expose the full functionality and performance of the platform, it wouldn't be very portable.

WebGL's promise is to offer a standardized, portable platform for high-performance graphics on the web. Given implementations that handle driver differences sufficiently well, application developers can treat WebGL as a single platform, instead of having to worry about the minutiae of underlying implementation differences.

However, there are basic limitations to the realization of this vision. The most important one is the very different performance characteristics of existing GPUs and drivers, especially once mobile devices are factored in. Because of that, the effective portability of any high-performance graphics API is limited by uneven performance characteristics.

The WebGL specification and its implementations are designed to alleviate these performance issues as much as possible, and this chapter's primary goal was to complement that mission by helping application developers understand some of WebGL's more important performance-related specificities.

3

Continuous Testing of Chrome's WebGL Implementation

Kenneth Russell, Zhenyao Mo, and Brandon Jones

3.1 Introduction

In Chrome 28 a bad WebGL bug—crbug.com/259994, which intermittently broke Google Maps and many other WebGL applications—shipped to Stable and, embarrassingly, the issue was reported to our team primarily via Twitter. The bug slipped through automated testing for two principal reasons. First, the majority of the automated testing in the open-source Chromium project, and the Chrome browser built on top of it, occurred on virtual machines, which did not exercise the GPU-accelerated rendering path that is taken on most end users' systems. The few machines with physical hardware and real GPUs had not received enough attention from the overall team and were not reliable enough to detect intermittent failures. Second, there were not enough "try servers"—the banks of machines that developers use to certify their code changes during development and checkin time—running on physical hardware with actual GPUs. The effect was that it was impossible to detect potential breakage to Chromium's graphics stack ahead of time, even though there were dozens of developers working directly on this code.

In response, we led an effort to overhaul Chromium's GPU bots and try servers. Working closely with Chrome's infrastructure, labs, and GPU teams, as well as the Chrome team in general, we respecified the hardware, changed OS configurations, and rewrote nearly all the software and tests running on the bots in order to eliminate dozens of sources of flakiness. We built out a bank of GPU try servers: physical hardware that runs graphics tests against every incoming changelist, or CL, to the Chromium project and Blink rendering engine, in comparison to Chromium's many pre-existing VM-based test bots.

The new try servers and continuous testing, or waterfall, bots are among the most reliable on the Chromium project and typically run hundreds of consecutive builds with no errors. They test real-world web content, including Google Maps, and reliably detect breakage of not only Chromium's graphics stack, but also the browser itself. More recently, they have detected highly intermittent bugs in core browser features—`requestAnimationFrame` and video playback—affecting real web pages. They ensure that WebGL is treated as a mission-critical technology and will work as expected for the hundreds of millions of Chrome users.

This chapter describes the hardware, operating system, and software changes associated with bringing these machines to high reliability. It is our hope that this experience will help other teams replicate similar configurations for their own graphics testing. Nearly all of the code described in this article is open source and available as part of the Chromium project.

3.1.1 A Few Statistics

To help motivate the scale of the project, here are a few statistics gathered from the source code repositories of the Chromium and Blink projects.[*] Considering the 3 months between September 1, 2014, and November 30, 2014, there was an average of approximately 4,400 commits per month to Chromium—or about 144 commits per day—and approximately 1,600 commits per month to Blink, or about 51 per day. On a monthly basis there were approximately 720 unique committers to the Chromium project and 240 unique committers to the Blink project. According to OpenHub.net[†] there are approximately 13 million lines of code in Chromium.

Building Chromium from scratch takes approximately an hour. An incremental relink of the statically linked binary takes approximately 5 minutes. To speed development, Chromium's build system supports compiling the product into multiple shared libraries, which reduces incremental rebuild times to a few seconds. The shipped product is statically linked to reduce startup time.

3.1.2 Background of Chromium's Test Setup

Chromium and Chrome use the open-source Buildbot framework as the basis for the browser's continuous integration testing. Chrome's infrastructure team has written a tremendous amount of software that augments this basic setup, including integration with the Rietveld code review tool, a commit queue which automatically tests every incoming code change to the browser, systems for distributing and sharding test execution, and many others.

Developers interact with the automated test infrastructure in two primary ways: via the waterfalls and the commit queue. The waterfalls run on a continuous basis.

---

[*] See the count-chromium-commits script in the WebGL Insights github repository.
[†] https://www.openhub.net/p/chrome

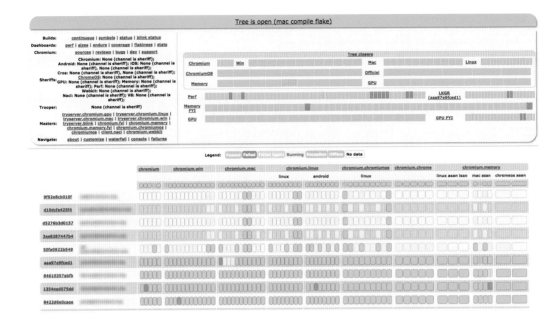

Figure 3.1

Chromium's waterfall view.

See Figure 3.1 for a visual example of Chromium's main waterfall. Whenever new code is checked in to the browser, the bots on the waterfall notice this, build the top-of-tree code, and run tests against it. It is often the case that the waterfalls group batches of commits together; there aren't enough bots to build and test the top-of-tree code at each commit. A consequence of this fact is that if a test begins to fail at a certain point, it's often necessary to scan through several or sometimes dozens of commits to find the one at fault, which is a painstaking process. It is for this reason that the best way to catch failures is with the commit queue, at code submission time.

The commit queue, in counterpoint, builds and tests each individual code change to the browser. Developers upload work in progress to the Rietveld code review tool. Once the code has been reviewed, the easiest way for the developer to submit it is to simply check the "commit" checkbox. This submits the changelist to the try servers: a large bank of machines, both physical and virtual, representing the majority of the operating systems and configurations on which Chrome runs. The try servers check out the top-of-tree code, apply the developer's patch to it, compile the product, and run automated tests. Only if all of the automated tests run successfully is the change committed to the code base. Developers may also trigger try jobs manually, without intending to also commit their changes to the code base, to reduce the amount of manual testing they must perform. Figure 3.2 shows the user interface for the code review tool on a representative CL.

The try servers run all of their jobs in parallel. It is possible that two or more code changes, touching different files in the source base, actually conflict with each other and would cause test failures when both were applied. In the worst case, two such changes might be tested in parallel, each found to pass all of the tests, and be committed. The second of

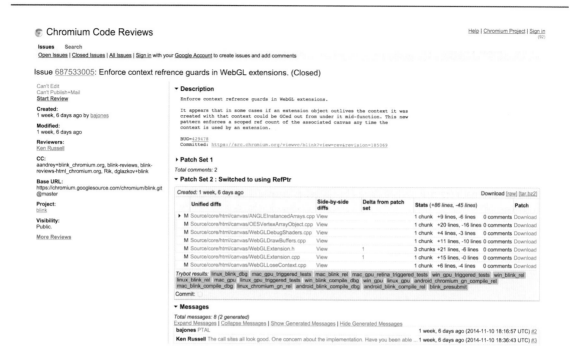

Figure 3.2

Chromium's Rietveld review tool, showing the trybot results from sending the patch through the commit queue.

these changes would cause tests to start failing on the main waterfall and "close the tree": prevent further checkins until the tests are fixed, a process requiring manual intervention. In practice, issues like this happen rarely, so the best automated way to avoid breaking tests and closing the tree is to send all incoming source changes through the commit queue.

The try servers represent the bulk of the machines in the Chromium project's automated testing infrastructure. For every configuration represented on the waterfall, such as Linux in Release (non-Debug) mode, typically 20 to 30 try servers are required in order to handle the load from developers testing their code changes. In order to save computing resources, it's critically important that the try servers be both reliable and fast. Tests which fail intermittently not only cause retries, which can double or triple the computing cycles needed to test a particular change, but also cause intermittent test failures on the main waterfall and ultimately destabilize the product. Eliminating flakiness in tests was a primary goal of the project to overhaul Chromium's GPU bots, as will be seen throughout this chapter.

3.2 Starting Points

At the start of the project, Chromium's GPU bots were a small subset of the browser's automated testing infrastructure. The configurations included Windows, Mac, and Linux, running a combination of GPUs from AMD, Intel, and NVIDIA. The GPU bots were mainly waterfall machines. There were a few GPU try servers—three each of Windows,

Mac, and Linux—that had to be triggered manually because there wasn't enough capacity to test all of the incoming changes from all of the developers on the project. This essentially meant that no Chromium developers were using the GPU try servers. Tests failed intermittently on the waterfall, seemingly for different reasons on different platforms. The intermittent test failures prevented scaling up the number of try servers for the GPU bots, because tests wouldn't run reliably on them.

The first priority, then, was to diagnose and eliminate the causes of the intermittent test failures.

3.2.1 Hardware and OS Configuration Changes

The GPU bots are necessarily physical machines with graphics cards installed, not virtual machines. Initially the bots were run headless: No monitor was plugged in, and remote desktop software (VNC on Linux, Mac OS's built-in VNC-compatible screen sharing, and Remote Desktop on Windows) was used to log in, administer the machines, and debug test failures that were seen on the machines. Attempts had been made to trick some of the machines into thinking they had a monitor plugged in, in order to activate the GPU—a dummy VGA dongle with resistors shorting some pins together.

We observed that logging on to the machines to debug test failures often caused the failures to disappear. We also found that running Chromium's GPU-accelerated rendering path over remote desktop software typically caused incorrect rendering, including, but not limited to, garbled colors and failure of the GPU code paths to initialize properly.

The first order of business was to make it possible to remotely log in to these machines without perturbing their execution. Chrome's Labs team installed an IP KVM, which plugs into the target machine's video output and USB ports, digitizes the video, and streams it to a client machine via a web browser. Because the IP KVM requires no software to be installed on the target machine, it not only avoids the interference of running remote desktop software, but also acts as a monitor, causing the GPU to activate properly. The specific model of IP KVM the Labs team used is a Raritan Dominion KX II. We evaluated other brands that claimed to speak the VNC protocol, which would avoid the need for any custom software on the client machine, but did not have success with these products. Chrome's Labs team found it necessary to provide a fake monitor's extended display identification data, or EDID, on a couple of the GPUs in order to make them use the correct screen resolution.

The switch to the IP KVM was a crucial step, eliminating strange side effects of remote desktop software and a certain class of reliability problems. All of Chromium's Linux and Windows GPU bots are now hooked up through a Raritan. On Mac OS, the built-in screen-sharing software works well. We don't hook up the Mac bots to the Raritan because doing so is not necessary for our Mac Minis, and plugging Retina display MacBook Pros into the Raritan changes the display configuration, losing the high-DPI nature of the built-in display and defeating the purpose of testing on this hardware configuration.

After switching to the IP KVM we still saw intermittent failures of our pixel tests on Linux. These tests read back the rendered image from the screen, ensuring that the browser's entire end-to-end rendering path is working correctly. Occasionally these tests would capture a black screen, a symptom of a screen saver or power saver kicking in. The problem persisted despite disabling all power-saving features from the graphical user interface's control panels. It turns out that the X server itself interfaces with the Display Power

Management System (DPMS), providing a "low-level" power-saving mode that can't be configured this way. We confirmed that this was the mechanism causing the screen to go blank by watching the test execution, waiting for the screen to blank, logging in via a new ssh terminal, and running the following command:

```
xset s reset
```

The monitor immediately awoke and the pixel tests started to pass again. After extensive web searching, we found and made the following changes to /etc/X11/xorg.conf:

```
section "ServerLayout":
    Option "BlankTime" "0"
    Option "StandbyTime" "0"
    Option "SuspendTime" "0"
    Option "OffTime" "0"

section "Monitor":
    Option "DPMS" "false"
```

After deploying this change to all of our Linux machines, they began to run the pixel tests reliably.

To summarize, the key hardware and OS configuration changes made to our GPU bots in order to improve their reliability were

- Hook up all Linux and Windows machines to an IP KVM
- Disable all power saving
- Disable DPMS on X11 systems

3.2.2 Test Harness Changes

Since their inception, Chromium's GPU bots ran the WebGL conformance tests against the browser's top of tree code on a continuous basis. Initially the tests were run in a specialized test harness called content_browsertests, which loads individual web pages in an environment similar, but not identical, to the full web browser's rendering pipeline. Unfortunately, intermittent test failures were seen in this environment that defied debugging attempts. On Windows in particular it looked like race conditions in the content_browsertests harness startup and shutdown code were causing intermittent crashes. Despite best attempts, debugging these races proved fruitless.

We decided to move away from the content_browsertests harness and start running the majority of the GPU bots' tests in the browser itself, driven by a test harness called Telemetry.* Testing the browser rather than a specialized test harness carried several benefits. First and foremost, it enabled us to test what we ship, which is a central tenet in Chromium's automated testing philosophy. Second, it ran the tests in a more realistic environment: Where content_browsertests started up and shut down a portion of the browser for each test, Telemetry simply navigated the browser to the URLs of each of the WebGL conformance tests in turn, which is how they execute when launched manually. Third, the performance of the test suite actually improved when switching to Telemetry,

---

* http://www.chromium.org/developers/telemetry

since navigating a given tab from URL to URL is much faster than closing a tab, opening a new one, and navigating to a URL, which was effectively what content_browsertests was doing.

Switching to Telemetry worked around the race conditions in the content_browsertests harness and for the first time allowed the WebGL conformance tests to run reliably on Chromium's GPU bots for hundreds of runs at a time.

The GPU bots run not only the WebGL conformance suite, but also a set of other tests which verify things like pixel accuracy of certain pages' rendering. These GPU tests must perform operations that aren't allowed in ordinary web browsers for security reasons, such as taking snapshots of the screen or querying the browser's internal state. The specialized test harnesses, including content_browsertests, have access to these facilities. In order to improve these tests' reliability they were ported from content_browsertests to Telemetry. It was necessary to expose certain privileged primitives in the Telemetry test harness, as well as port the GPU tests that had previously been written in C++ to the new framework, which is written in Python. This work took a couple of months, but allowed content_browsertests to be removed entirely from the GPU bots. See the related bugs[*] for details.

The switch from content_browsertests to Telemetry on the GPU bots significantly increased their reliability, to the point at which we could consider scaling up the infrastructure and running this set of tests against all incoming code changes to the browser— building the GPU try servers.

3.2.3 Pixel Tests

One thing we learned early on in the testing of Chrome's rendering was that bugs can appear at any step. Even capturing a texture at the end of the rendering pipeline and making sure the content looks OK would not guarantee that it is presented to the screen correctly. Therefore, pixel tests are built upon OS-level screen capturing APIs. There were a few times that a failure turned out to be a message window from the OS popping up and intersecting with Chrome's window, but that doesn't happen often and a quick inspection of the failed rendering and shutting down the offending window would fix the problem.

Reference images are stored in the cloud and indexed by OS, GPU, and other dimensions that affect rendering—for example, with antialiasing or not. We tried to generate reference images on local bots but that solution turned out to be problematic. First, incorrect reference images can go undetected for a long time and bugs may creep in. Second, a patch that changes rendering has to land and be integrated into a bot first, turning it red, before reference images on that bot can be updated. Therefore, we switched to a cloud-based solution, where inspecting and updating reference images is easier.

To compare rendering results to reference images, we use a simple pixel-by-pixel exact matching method. This is necessary because even a one pixel difference may indicate a bug in Chrome. For example, antialiasing in WebGL could be turned off incorrectly. In the future, we may look into smarter image comparison algorithms.

---

[*] http://crbug.com/278398, http://crbug.com/308675, http://crbug.com/365904

3.3 Building Out the GPU Try Servers

Once the GPU waterfall had achieved a good level of reliability, we found that code changes to the web browser caused some tests to fail nearly every day, requiring manual intervention and reverting of the associated changes. For efficiency reasons, most of Chromium's tests don't run against the full browser; the Telemetry-based GPU tests are some of the few on the waterfall that do. For this reason, a bug would occasionally slip through the commit queue and wouldn't be caught by anything else but the Telemetry tests running on the GPU bots, even though the bug was completely unrelated to graphics. One surprising failure[*] turned out to be breakage of the unbranded Chromium build of the browser; the test harnesses and full Chrome browser worked fine with this particular change. The only way to stop the flood of breaking changes was to add the GPU tests to the default set of tests run against all incoming changelists to Chromium.

3.3.1 Recipes

Early in the overhaul of the GPU bots it became clear that many configuration changes would be needed in order to achieve reliability. Buildbot's architecture is not set up to handle many incremental changes well. Normally, each change to the steps that are run on a particular bot—even something as small as changing the command line arguments passed to a particular test—requires all of the machines on that waterfall to be rebooted, which is not only a major inconvenience, but also painfully slow.

In order to reduce the number of waterfall restarts needed to change machines' configurations, Chrome's infrastructure team developed a new framework called *recipes*; see the Pointers to Code and Documentation section at the end of the chapter for links. Recipes worked within the Buildbot framework, but moved the responsibility of deciding which steps to run, and how to run them, from the computer controlling a particular waterfall to the individual bots on that waterfall. Using the new recipe framework, when a change was checked in to the script containing the steps for a particular bot, it would take effect on the next build that that bot ran, without requiring any reboots.

The productivity improvements and flexibility afforded by recipes, compared to the way Buildbot is ordinarily used, can't be overstated. It's not an exaggeration to say that, without recipes, Chromium's GPU bots would not have been overhauled successfully.

Before making any significant changes to the GPU bots, including the test harness switches described in the previous section, we converted the bots from the old-style Buildbot scripts to recipes.[†] This made it possible to make and deploy smaller and more incremental changes, and made it easier to test those changes locally. During the course of this project, a significant percentage of the work involved improvements to the recipe framework, adding support for new features and diagnosing and fixing unexpected behaviors. This work proved useful to the larger Chromium project when converting other bots to recipes.

3.3.2 Builder/Tester Split

Chromium's GPU waterfall runs the same set of tests on multiple kinds of GPUs: on the desktop, typically those from AMD, Intel, and NVIDIA. The precise GPUs were originally

---

[*] https://codereview.chromium.org/421643002/
[†] See http://crbug.com/286398 and related bugs.

selected "organically"—whatever was available in house at the time—though it was a specific design criterion to always use mid-range GPUs that were likely to be chosen by end users. Later, as the try servers were built out, it was necessary to standardize on a few GPU configurations, so that multiple machines could reliably produce the exact same pixel results. When a platform and a type of GPU are selected, usually the latest graphics driver available is installed. An exception was OS X Lion, where we had to fall back to an earlier but more stable version (10.7.5). Driver updates are only performed if a new driver with fixed bugs provides significantly more test coverage. Chromium's GPU try servers do not yet include mobile devices due to resource constraints, but some mobile devices are present on the waterfall.

At the beginning of the project, each machine on the waterfall both compiled the browser and ran the tests against it. It's clearly not efficient to compile the same code multiple times just in order to test on multiple GPUs, so one of the first changes made to the waterfall's configuration was to split the bots into separate builders and testers.[*] Once this work was complete, the physical GPU testers could be changed to relatively low-end machines, since they no longer needed to perform the heavy-duty compilation step. It was also no longer necessary for the builders to contain GPUs since they ran no tests, so they could be replaced with virtual machines.

3.3.3 Isolates

Initially, the builders zipped up the binaries they'd compiled and copied them to a server, and the testers downloaded the binaries and unzipped them into an identical Chromium workspace before running them. The data files for the tests lived in the Chromium workspace and were not included in the archives copied from builder to tester. The data files were quite large—larger than the binaries in some cases—and it would have been inefficient to copy them between machines each time.

Keeping an up-to-date copy of the Chromium workspace on the testers proved to be problematic. After the builder/tester split, the single largest percentage of the testers' cycle time was actually spent updating the Chromium workspace!

Fortunately, Chrome's infrastructure team had been developing a new mechanism for distributing tests among machines called *isolates*.[†] An isolate describes all of the files needed to run a particular test, including the binaries and any data files. Isolates differ significantly in implementation from a simple mechanism like a zip archive, however. Each file is uploaded and compressed separately, and if the contents of a particular file have been uploaded to the server recently, it is neither uploaded again by the builder nor downloaded again by the tester.

The GPU bots were converted from uploading and downloading zip archives of their binaries to using isolates[‡] in crbug.com/321878 and related bugs. This work involved significant rewrites of the tests to avoid storing any persistent data on the bots' local disk, since isolates' execution is designed to be transient. In particular, the results of the bots' pixel tests had to be placed in cloud storage.[§] While this work took some

---

[*] http://crbug.com/310926
[†] http://www.chromium.org/developers/testing/isolated-testing
[‡] http://crbug.com/321878 and related bugs.
[§] http://crbug.com/330053 and http://crbug.com/330774

time to complete, the results were tremendous. First, the bots became much easier to maintain since all of their results were stored in the cloud, and easily accessible over the web. It was no longer necessary to log in to a particular bot to examine its most recent test results. Second, the cycle time was decreased dramatically: The bots became between 1.19× faster for one of the Mac testers to 2.88× faster for one of the Windows testers. The improved hardware utilization finally allowed the GPU try servers to be built.

3.3.4 Ramping Up the Try Servers

Chrome's Labs team did a tremendous amount of work purchasing and setting up the hardware for the GPU try servers. In order to handle the load of the incoming change-lists, 20 each of Windows, Linux, Mac, and MacBook Pro Retina machines were purchased. As described earlier, the Windows and Linux try servers were hooked up to an IP KVM. The Macs were left headless. Ten beefy physical servers per platform were purchased as the builders for these testers, based on cycle time measurements on the waterfall.

Chromium's commit queue (CQ) contains an experimental framework that helps with the addition of new machines. A percentage of the load of incoming changelists can be experimentally sent to a particular try server in order to see whether it tests each change-list successfully and can keep up with that load. We helped generalize this framework to support multiple concurrent experiments at different load levels, in order to allow the Linux, Windows, and Mac GPU try servers to be brought online at different times, and we began sending CLs to the new machines.

While the try servers reliably ran jobs to completion, we quickly discovered that the bottleneck in the system was not the physical GPU testers, but the builders. The cycle time estimates we'd used to compute the number of builders and testers were off by a large margin. We originally specified that we needed 10 builders for each 20 testers, but it became clear that 30 builders for each 20 testers were necessary. It also turned out that the beefy physical builders were not fast enough to justify the additional hardware cost and that using virtual machines for the builders was more effective.

Chrome's Labs team came through once again in a pinch, reallocating the builders as virtual machines and providing 30 VMs per configuration (Windows, Linux, and Mac). In particular, the Chrome project was running low on capacity, and it was not possible to provision new Mac VMs at the time; the Labs team negotiated with other projects that were not using all of their capacity in order to fulfill our hardware requirements months ahead of schedule, for which we remain grateful.

After the reprovisioning of the builders, the load testing of the GPU try servers otherwise went smoothly, and the GPU bots were finally made Chromium tree closers in crbug. com/327170.

3.3.5 Analyzing CL Dependencies

Due to the large number of changes submitted to the Chromium project on an ongoing basis, it's particularly important to utilize the automated testing machines efficiently. If an incoming CL can't possibly affect the tests on a particular platform, the tryjobs for that CL shouldn't cause tests to be run on that platform. For example, if a CL touches Windows-only code, it definitely should not cause tests to be rerun on the Mac trybots.

Dependency analysis was added to Chromium's build system some time ago[*] and support was added to the GPU trybots more recently, once they were stabilized.[†] Experience has shown that this dependency analysis reduced load on the try servers by about 20%.[‡] The cost savings of this technique are significant, and we recommend incorporating it into any project utilizing a system similar to Chromium's trybots and commit queue.

3.4 Stamping Out Flakiness

The primary goal of building out the GPU try servers was to eliminate the sort of intermittent bugs described at the beginning of this chapter. Once the GPU bots achieved a high level of reliability, we expected that it would be easy to maintain that quality level. While the try servers have unquestionably made it easier to keep Chromium's GPU code running correctly, a few interesting and surprising intermittent failures came to light that had to be diagnosed and fixed manually. The resolution of these bugs may be useful to other projects.

3.4.1 Glib-Related Timeouts

One day in October 2013, tests suddenly started timing out on Chromium's Linux GPU bots.[§] The symptom was that a subprocess of the web browser was deadlocked inside a call to `malloc()` inside the Glib library, which sits underneath Gtk+. Glib was attempting to make a connection to the desktop bus, or Dbus, daemon, which provides information like which theme is in use by the window manager. The bug was not reproducible either on our local workstations or when logged on to the affected bots.

Members of Chrome's security team helped investigate and indicated that it looked like a typical bug with async signal-unsafe code on POSIX. It's fairly difficult to find a concise description of the problem online. The bug is related to the use of `malloc()` between calls to `fork()` and `exec()` when spawning subprocesses. What happens is specifically:

- The parent process calls `fork()`, and it happens that when it is called, another thread is in the process of calling `malloc()`.
- The internal malloc lock is then held forever in the child process, even after the parent process completes its call to `malloc()`, because the child is a copy-on-write clone of the parent.
- Before the child process calls `exec()` to overwrite its image with the new subprocess, it calls `malloc()`. The internal malloc lock is held, and this call to `malloc()` never completes. The child is then deadlocked.

The rule of thumb is that a process must never call `malloc()`, or a function which calls `malloc()` internally, between calls to `fork()` and `exec()`. Unfortunately, Glib does this when it notices that it doesn't have a connection to the Dbus daemon; it calls `opendir()` to try to find the dbus–daemon program to spawn it, but it does this after calling `fork()`. `opendir()` calls `malloc()` internally.

---

[*] http://crbug.com/109173
[†] http://crbug.com/411991
[‡] https://code.google.com/p/chromium/issues/detail?id=411991#c2
[§] http://crbug.com/309093

We assumed that a change to Chromium's initialization of Gtk+ had provoked this problem, and tried to move the Gtk+ initialization earlier, before Chromium had spawned any threads. Unfortunately, this didn't help. Studying Glib's code more deeply, it became clear that if it went down this code path at all—if Glib was initialized when the `DBUS_SESSION_BUS_ADDRESS` environment variable was not set—Glib first started threads internally, and then spawned a subprocess, in order to make a connection to the Dbus daemon. Glib inevitably raced against itself. This problem had been present all along, and only manifested when the timing of the browser's code changed subtly.

The reason this bug happened on the GPU bots and not on regular workstations, even though the bots have a real monitor and desktop session, is that the bots' environment for launching processes is slightly different from that of real users. When launching programs from a terminal window that is open on the desktop, the Dbus daemon is implicitly already started; the desktop itself is responsible for starting and connecting to it. However, the bots' scripts launch at startup, and while they properly set up an X display connection to be able to open windows, they don't make a connection to the Dbus daemon.

The solution to this problem was to modify Chromium's testing scripts to manually launch the Dbus daemon, which causes the `DBUS_SESSION_BUS_ADDRESS` environment variable to be set, before starting any tests or launching Chromium. This workaround solved the problem and was deployed across all of the Chromium project's Linux bots. Recently this problem manifested again on the Google Maps team's automated test machines, so it is a fairly common issue that should be known to anyone performing automated testing of graphical programs on Linux. Note carefully that it is necessary to manually shut down these explicitly launched dbus–daemon processes, or they will leak!

3.4.2 Google Maps Timeouts

In July 2014, the automated test of Google Maps which runs on the GPU bots suddenly started intermittently timing out on the Windows try servers.[*] The symptom was that the test's JavaScript began to execute before all of the scripts on the page were loaded, leading to exceptions when undefined symbols were referenced. The regression window was quite large and included changes in the V8 JavaScript engine, Telemetry, and DevTools, upon which Telemetry relies. Fortunately, the problem was reproducible on one of the bots using binaries compiled on our local workstations.

A manual bisect was done using "git bisect" and revealed that a change to Blink's Content Security Policy implementation had introduced the flakiness. This was quite surprising, as it seemed that this particular change did not alter the behavior of the code, but the results were indisputable. The change was reverted and the tests reached their previous levels of reliability. This is another example of a change unrelated to graphics causing broader test failures in the browser, and this bug would have inevitably caused failures on end users' machines.

3.4.3 HTMLVideoElement and `requestAnimationFrame` Timeouts

In August 2014 intermittent timeouts of several WebGL conformance tests—including context_lost_restored, premultiplyalpha_test, and webgl_compressed_texture_size_limit—spontaneously appeared and were reported.[†] The failures happened rarely,

---

[*] http://crbug.com/395914
[†] http://crbug.com/407976

and there were no obvious changes to the browser's code which could have made them start happening. They were not reproducible on our local workstations, and there were no apparent patterns in the failures. With some struggling, a crucial diagnosis was made that there was in fact a pattern among some of the timeouts: Several of the failing tests used the browser's `requestAnimationFrame` API, or rAF for short, to drive their execution. With this knowledge, it was possible to categorize the failures. We pushed to leave the failing tests in place and add logging to the browser, and two bugs were identified. First, rAF calls seemed to be intermittently ignored.[*] Second, HTMLVideoElements created on the fly from JavaScript seemed to occasionally never fire any of their event handlers.[†]

The investigation of the rAF issue dived deep into the browser's code, across the boundary between the Blink rendering engine and the embedder, into the browser's compositor and scheduler. Finally a particular piece of code was identified that changed the handling of rAF callbacks depending on whether accelerated compositing was enabled: this code was vestigial, dating back to a time before GPU-accelerated rendering of web pages was introduced into Chromium. There was a long-standing race condition in this code. Eliminating this behavioral difference fixed the bug. There is no good explanation for why this bug appeared when it did, except that the timing in the browser must have changed slightly, exposing the preexisting race condition.

Investigation of the HTMLVideoElement bug showed that the video element itself was being garbage collected before it should have been. The basic contract of the video element, as with all HTML elements, is that it may not be garbage collected until doing so would have no observable effects on the program. There was a long-standing race condition in the video element's computation of whether it had "pending activity" where, if a garbage collection was triggered between the time it had completed loading its network resources and had played its first frame, it would be prematurely collected and never trigger any of its callbacks. The affected tests would time out unless those callbacks were called.

The fixes for these two bugs addressed problems that could affect real web pages and that had been present for years, though, surprisingly, had only showed up recently. It was highly coincidental that they began happening nearly at the same time, but it has been our experience that multiple failures tend to happen at the same time, whether due to Murphy's law or because multiple developers tend to focus on the same general area of the code at the same time, occasionally causing multiple breakages. The experience of diagnosing and fixing these bugs underscored to us the importance of tracking down and fixing flaky test failures urgently, instead of allowing them to pile up.

3.4.4 Shutdown Timeouts on Windows

In October 2014, intermittent test timeouts appeared on the Windows GPU bots where it appeared that one of Chromium's subprocesses was failing to exit cleanly.[‡] This problem was not reproducible easily on developers' machines, though it was more readily reproducible on the bots. The symptom was that a child process of Chromium's main "browser"

---

[*] http://crbug.com/393331
[†] http://crbug.com/412876
[‡] http://crbug.com/424024

process failed to launch successfully, and was stuck in the suspended state before it had run any code.

Many hypotheses were put forward for why this behavior started happening. The code which managed Chromium's subprocesses had undergone some changes, but the first failure of this sort happened before the first change to that code had been committed. A couple of bugs were identified and fixed in the subprocess management code, but the stuck processes persisted.

For background, Chromium launches sandboxed processes on Windows using the following rough steps:

1. The target process is created in the suspended state.
2. The sandbox policy is applied to the target process.
3. The target process is added to a Windows job object, which will cause it to be automatically terminated when the parent process exits.
4. The target process is resumed and begins to run.

It is apparent that there is a brief race where, if the parent process is forcibly terminated between steps 2 and 3, the target process will not be automatically terminated. Logging was added to the browser that indicated that this was happening.

The Telemetry harness uses Python's subprocess management primitives in order to start up and shut down the web browser being tested. It turns out that Python's `Popen. terminate()` call uses Windows' TerminateProcess API to shut down the processes it launches; see https://docs.python.org/2/library/subprocess.html. The target process is forcibly terminated immediately, without running the process's usual exit code path. The hypothesis was that Telemetry was accidentally terminating the browser exactly in the race condition window above, leading to the leak of the suspended subprocesses.

The solution was to enumerate the target process's top-level windows from Python and send the `WM_CLOSE` message to the web browser's windows to allow them to close cooperatively.

This particular bug affected only the bots, not end users, because the browser is typically exited by either closing all windows or using the "Quit" menu option. It's once again surprising that it started occurring without any related code changes; the timing of the product simply changed in such a way that a preexisting race condition started to appear. Nonetheless, it was essential to fix the intermittent failures because they would otherwise mask real failures in the product.

3.4.5 Conclusions

On a project as large as Chromium, it's essential to catch as many bugs as possible before they reach the source tree. It is more critical for WebGL in Chromium, because all graphics commands are sent over and executed in a separate GPU process, which adds extra dimensions of complexity. Seemingly unrelated code changes could break WebGL. This turns into an even worse problem when the breakage is intermittent and happens only on certain hardware and happens rarely. Therefore, having a GPU testing farm that covers major platforms and GPUs is the only way to move forward.

It's been our experience that the GPU try servers have caught the majority of the failures that would otherwise have required manual diagnosis and reverting. We would encourage other projects to develop similar infrastructure at the beginning of the project. Delaying only increases the overall amount of work that will be required.

The intermittent failures that have slipped past the try servers have been relatively few and far between, though some of them have been difficult to diagnose and fix. We perceive a need for more stress tests that will more reliably expose these race conditions, and hope to develop and deploy such tests once more computing resources are available. Fixing these intermittent problems as they have arisen has resulted in a more reliable browser and WebGL implementation.

3.5 Testing Your WebGL App against Chromium

It is possible that your WebGL application may break when a user's Chrome browser is auto-updated. While maintaining backwards compatibility is a primary goal, occasionally a backward-incompatible change is required. For example, the invariant rules for shader varyings were enforced in a Chromium update; WebGL applications breaking these rules previously ran fine on desktop systems but not on mobile devices. Enforcing these rules made these applications portable. Also, hopefully rarely, a bug might creep into Chromium and affect your application.

Therefore, testing against Chromium builds and detecting a problem as early as possible is valuable for WebGL developers.

Setting up a testing bot like one on the Chromium's GPU waterfall requires the following steps:

1. Check out Chromium's Telemetry: https://chromium.googlesource.com/chromium/src.git/+/master/tools/telemetry/.
2. Write a test using Telemetry. You can find examples in https://chromium.google source.com/chromium/src.git/+/master/content/test/gpu/.
3. Write a Python script to check for new Chromium builds in https://common datastorage.googleapis.com/chromium-browser-continuous/index.html, download a new build, and run your test against that build.

If a bug is detected in Chromium, please file it against crbug.com.

3.5.1 Pointers to Code and Documentation

Chromium's recipes can be found in the workspace https://chromium.googlesource.com/chromium/tools/build/. The reusable modules invoked in multiple recipes are in the directory scripts/slave/recipe_modules/. The GPU bots' recipes live in scripts/slave/recipes/gpu/. Documentation for recipes overall is in scripts/slave/README.recipes.md. The GPU bots are documented in http://www.chromium.org/developers/testing/gpu-testing, and the GPU recipe specifically is documented in http://www.chromium.org/developers/testing/gpu-recipe.

The tests which run on the GPU bots are in the Chromium workspace, https://chromium.googlesource.com/chromium/src/, under content/test/gpu/gpu_tests. Some tests use page sets, which live in content/test/gpu/page_sets/, and some reference data that are contained in content/test/data/gpu/. The tests also refer to the WebGL conformance tests, a snapshot of which is contained in third_party/webgl in a full Chromium checkout. See http://www.chromium.org/developers/how-tos/get-the-code for information on checking out Chromium's code.

Acknowledgments

This project would not have been possible without the help of dozens of teammates. We thank all of the members of Chrome's Infrastructure and Labs teams for their support and guidance on this project. We especially thank Robbie Iannucci for superb help, guidance, and leadership during the long effort to fully convert the GPU bots to recipes. We also specifically thank Bryce Albritton, Sergey Berezin, Aaron Gable, Paweł Hajdan, Chase Phillips, Marc-Antoine Ruel, Peter Schmidt, Vadim Shtayura, Mike Stipicevic, Ryan Tseng, and John Weathersby for their help diagnosing and fixing dozens of issues, getting the machines purchased and set up, and keeping them running smoothly.

We thank Nat Duca, Tony Gentilcore, and David Tu from the Telemetry team for their help and willingness to accept changes to the harness to allow it to be better used for correctness testing. We thank Pavel Feldman and the rest of the DevTools team for their support in extending Telemetry's capabilities.

We thank Brian Anderson, John Bauman, Philip Jägenstedt (Opera), Jorge Lucangeli Obes, Carlos Pizano, Julien Tinnes, Ricardo Vargas, Hendrik Wagenaar, and Adrienne Walker from the extended Chromium team for their help fixing some of the thorniest bugs.

We thank John Abd-El-Malek for his help and encouragement to add the CL dependency analysis, for identifying flakiness on the trybots, and for his patience while bugs were being tracked down.

We thank the members of Chrome's GPU team for their efforts in keeping the bots running smoothly.

Finally, we thank all of the members of and contributors to Chromium for their help, support, and patience while this project was being completed.

SECTION II
Moving to WebGL

Like many WebGL developers, I'm happy to say that WebGL brought me to the web. I developed with C++ and OpenGL for years, but the lure of being able write 3D apps that run without a plugin across desktop and mobile was too great, and I quickly moved to WebGL when the specification was ratified in 2011. In this section, developers, researchers, and educators share their stories on why and how to move to WebGL. We also see similar themes throughout this book.

When our team at AGI wanted to move from C++/OpenGL to JavaScript/WebGL, we weren't sure how well a large JavaScript codebase would scale. We were coming from a C++ codebase that is now seven million lines of code. Could we manage something even 1% of that size? Thankfully, the answer was a definite yes; today, our engine, Cesium, is more than 150,000 lines of JavaScript, HTML, and CSS. In Chapter 4, "Getting Serious with JavaScript," my collaborators, Matthew Amato and Kevin Ring, go into the details. They focus on modular design, performance, and testing. For modularity, they survey the asynchronous module definition (AMD) pattern, RequireJS, CommonJS, and ECMAScript 6. Performance topics include object creation, memory allocation, and efficiently passing data to and from Web Workers. Finally, they look at testing with Jasmine and Karma, including unit tests that call WebGL and aim to produce reliable results on a variety of browsers, platforms, and devices.

Many developers, myself included, have written new engines for WebGL. However, many companies with large, established graphics and game engines may not want to rewrite their runtime engine. They want to reuse their existing content pipeline and design tools, and simply target the web as another runtime. For this, Mozilla introduced Emscripten, which translates C/C++ to a fast subset of JavaScript called asm.js that Firefox can optimize. In Chapter 5, "Emscripten and WebGL," Nick Desaulniers from Mozilla explains how to use Emscripten, including a strategy for porting OpenGL ES 2.0 to WebGL, a discussion of handling third-party code, and a tour of the developer tools in Firefox.

When moving a codebase to WebGL, there are two extremes: Write a new codebase in JavaScript or translate the existing one. Between these extremes is a middle ground: hybrid client-server rendering, where the existing codebase generates commands or images on the server. In Chapter 6, "Data Visualization with WebGL: From Python to JavaScript," Cyrille Rossant and Almar Klein explain the design of VisPy, a Python data visualization library for scatter plots, graphics, 3D surfaces, etc. It has a layered design from low-level, OpenGL-oriented interfaces to high-level, data-oriented ones, with a simple declarative programming language, GL Intermediate Representation (GLIR). GLIR allows for visualization in pure Python apps as well as JavaScript apps. A Python server generates GLIR commands to be rendered with WebGL in the browser, in a closed-loop or open-loop fashion.

WebGL adoption goes beyond practitioners, hobbyists, and researchers. Given its low barrier to entry and cross-platform support, WebGL is finding itself a prominent part of computer graphics education. Edward Angel and Dave Shreiner are at the forefront of this movement, moving both their introductory book, *Interactive Computer Graphics: A Top-Down Approach,* and the SIGGRAPH course from OpenGL to WebGL. Ed is the person who motivated me to use WebGL in my teaching at the University of Pennsylvania. In 2011, WebGL was a special topic in my course; now, it is *the* topic. In Chapter 7, "Teaching an Introductory Computer Graphics Course with WebGL," Ed and Dave explain the why and the how of moving a graphics course from desktop OpenGL to WebGL, including walking through the HTML and JavaScript for a simple WebGL app.

4

Getting Serious with JavaScript

Matthew Amato and Kevin Ring

4.1 Introduction

As we will see in Chapter 7, "Teaching an Introductory Computer Graphics Course with WebGL," the nature of JavaScript and WebGL makes it an excellent learning platform for computer graphics. Others have argued that the general accessibility and quality of the toolchain also make it great for graphics research [Cozzi 14]. In this chapter, we discuss what we feel is the most important use for JavaScript and WebGL: writing and maintaining real-world browser-based applications and libraries.

Most of our knowledge of JavaScript and WebGL comes from our experiences in helping to create and maintain Cesium,* an open-source WebGL-based engine for 3D globes and 2D maps (Figure 4.1). Before Cesium, we were traditional desktop software developers working in C++, C#, and Java. Like many others, the introduction of WebGL unexpectedly drew us into the world of web development.

Since its release in 2012, the Cesium code base has grown to over 150,000 lines of JavaScript, HTML, and GLSL; has enjoyed contributions from dozens of developers; and has been deployed to millions of end users. While maintaining any large code base presents challenges, maintaining a large code base in JavaScript is even harder.

---

* http://cesiumjs.org

Figure 4.1

Watching the sun set over the Grand Canyon in Cesium.

In this chapter, we discuss our experiences with these challenges, and our strategies for solving or mitigating them. We hope to provide a good starting point for anyone developing a large-scale application for the browser in JavaScript, or in a closely related language like CoffeeScript.

First, the lack of a built-in module system means there's no one right way to organize our code. The common approaches used in smaller applications will become extremely painful as the application grows. We discuss solutions for modularization in Section 4.2.

Second, a lot of the features and flexibility that make JavaScript approachable and easy to use also make it easy to write nonperformant code. Different browser engines optimize for different use cases, so what is fast in one browser might not be fast in another. This is especially concerning for us as WebGL developers, because interactive, real-time graphics often have the highest performance requirements of any application on the web. We give some tips and techniques for writing performant JavaScript code in Section 4.3.

Finally, dynamically typed languages like JavaScript make automated testing even more important than usual. With JavaScript's lack of a compilation step and its dynamic, runtime-resolved symbol references, even basic refactorings are unnerving without a robust suite of tests that can be run quickly to ensure that the application still works. A good approach to testing is essential to building a large-scale application and enabling it to evolve over time. We discuss strategies for testing a large JavaScript application, especially one that uses WebGL, in Section 4.4.

4.2 Modularization

Small JavaScript applications often start their lives as a single JavaScript source file, included in the HTML page with a simple `<script>` tag. The source file defines the functions and types that the application needs in the global scope. As the application grows, maybe we'll add another source file, then another, until soon enough we find ourselves with hundreds of source files and script tags.

Of course, most developers recognize the problems with this approach well before their application gets to hundreds of source files. Some of the problems are

- **Order dependency:** The `<script>` tags for the source files must be included in the HTML page in the proper order. If one file uses a symbol defined by another file before the second file has actually been loaded, the first file will see an undefined reference and throw an exception. This is especially painful in an application consisting of many HTML files, because this properly ordered list of script tags must be maintained in many places.
- **Global scope pollution:** All functions and types are added to the global scope, and each file goes to the global scope to find its dependencies. If another library uses the same names for functions and types as our application code, one or the other will fail.
- **Lack of encapsulation:** There is no obvious place to keep private details of our functions and types, such as internal helper functions.
- **Poor performance:** Loading a large number of individual JavaScript files is slow. This might be OK during development, when we're loading our code from a local web server, but the performance is terrible when the client and web server are on opposite sides of the country or world.

There are various workarounds to these problems. For example, we can avoid the poor performance and order dependency of loading many JavaScript files by creating a build step to concatenate all of our source files together prior to deployment. Of course, that build step still needs to concatenate the source files in the right order!

Early on in Cesium's life, we decided to address all of these problems by using the asynchronous module definition (AMD) pattern and RequireJS.[*]

4.2.1 Asynchronous Module Definition (AMD)

AMD is a way of structuring JavaScript modules such that they

- Explicitly state what other modules they depend upon
- Are not loaded until all of their dependencies have been loaded first
- Don't touch the global scope

A module is a small unit of functionality in our application, such as a single function or a single class. As an example, here is a slightly modified version of the Ray AMD module

---

[*] http://requirejs.org/

from Cesium. A Ray consists of an origin and direction in 3D space, and it can compute the point a given distance along the ray:

Listing 4.1 A simple asynchronous module definition.

```
define([
        './Cartesian3'
    ], function(
        Cartesian3) {
    "use strict";

  var Ray = function(origin, direction) {
    this.origin = origin;
    this.direction = direction;
  };

  Ray.getPoint = function(ray, t) {
    var offset = Cartesian3.multiplyByScalar(ray.direction, t);
    return Cartesian3.add(ray.origin, offset);
  };

  return Ray;
});
```

In the AMD pattern, our code is placed inside a function that is passed as a parameter to the define function. This "module" function gives us a place to store implementation details if desired. JavaScript's function-level scoping guarantees that anything defined inside this function will not be visible outside it unless we explicitly allow it to escape.

Our module doesn't touch the global scope at all. Instead of pulling its dependencies, such as Cartesian3, from the global scope, our Ray module expects those dependencies to be passed as parameters to the module function. The array passed as the first parameter to define specifies which modules—just Cartesian3 in this case—this module needs to have passed as parameters to its module function. Similarly, our module's export—the Ray constructor function—is never assigned anywhere in the global scope. Instead it is just returned to the caller.

But who is the caller? It's the AMD module loader.

The define function registers a module with a list of dependencies. Each of the dependencies is a module itself, usually contained in a single JavaScript source file with the same name as the module. Sometime later, when all the dependencies have been loaded, the module function is invoked. The module function returns the module back to the loader, and the loader can then load any other modules that depend on it. With knowledge of all modules and their dependencies, the loader can ensure that they are loaded in the correct order and that only the modules necessary for a given task are ever loaded.

This is where the "asynchronous" in asynchronous module definition comes from: Modules are not created immediately upon execution of the JavaScript files that contain them. Instead, they are created asynchronously as their dependencies are loaded.

With AMD, writing a web page that uses the modules is easy and doesn't require a build step. Typically, the HTML just references the main script using the RequireJS *data-main* attribute:

```
<script data-main = "scripts/main" src = "scripts/require.js"></script>
```

scripts/main.js is itself an AMD module that explicitly specifies its dependencies:

Listing 4.2 An entry point AMD module with three dependencies.

```
require(['a', 'b', 'c'], function(a, b, c) {
    a(b(), c());
});
```

When RequireJS sees the *data-main* attribute, it attempts to load the specified module. Loading that module requires all of its dependencies, *a*, *b*, and *c*, to be loaded first. Attempting to load those modules will, in turn, cause their dependencies to be loaded first. This process continues recursively. Once *a*, *b*, and *c* and all of their dependencies are loaded, *main*'s module function is called and the app is up and running.

With AMD, we get quick iteration, because no build is necessary; just reload the page! There's no need to manage an ordered list of <script> tags in each HTML page; we simply specify the entry point and RequireJS takes care of the rest. We also get ease of debugging during development, because the browser sees individual source files exactly as we've written them. This was especially important before browsers had good support for source maps.

What about deployment?

Loading all of those individual modules as separate JavaScript files can take a little while, especially over a high-latency network connection. Fortunately, the r.js optimizer[*] makes it easy to build and minify all of our modules, creating a single JavaScript source file with all the code our application requires, and no more. If our application uses libraries that are built with AMD, it's even possible to build the application and these libraries together, ensuring that only the parts of the libraries that we actually use are included in our application.

Assuming our application has a single script as its *data-main*, as shown before, we can build a combined and minified version of the application by running the following from the *scripts* directory:

```
r.js-o name=main out=../build/main.js
```

Then, we simply change the *data-main* attribute to point to the built version:

```
<script data-main = "build/main" src = "scripts/require.js"></script>
```

RequireJS has a dizzying array of options, allowing us to control how module names are resolved, to specify paths to third-party libraries, to use different minifiers, and much more. RequireJS also has a large assortment of loader plugins.[†] One that is especially useful

---

[*] https://github.com/jrburke/r.js
[†] https://github.com/jrburke/requirejs/wiki/Plugins

in WebGL applications is the *text* plugin, which makes it easy to load a GLSL file into a JavaScript string in order to pass it to the WebGL API. All the details can be found on the RequireJS website.

4.2.2 AMD Alternatives

The Cesium team has had great success with AMD, and has found RequireJS to be a robust and flexible tool. We don't hesitate to recommend its use in any serious application. However, there are some valid criticisms of AMD, most of which boil down to a distaste for its fundamental design goal: to create a module format that can be loaded in web browsers without a build step and without any preprocessing.

To that end, AMD adopts a syntax for defining dependencies that is considered by many to be ugly and cumbersome. In particular, it requires us to maintain two parallel lists at the top of each module definition and to keep them perfectly in sync: the list of required modules and the list of parameters to the module creation function. If we accidentally let these lists get out of sync, perhaps by deleting a dependency from one list but forgetting to do so from the other, our parameter named `Cartesian3` might actually be our `Matrix4` module, which would certainly lead to unexpected behavior when we try to use it.

If we accept a build step, perhaps because our code needs to be built for other reasons *anyway*, there are better options than AMD for defining modules that are easy to read and write. After all, today's web browsers support source maps, so debugging transformed code, or even combined and minified code, can look and feel just like debugging the code we actually wrote by hand. By working incrementally, a build process can often be fast enough that it will be finished before we can switch back to our browser window and hit refresh.

If we can use a simpler module pattern and make our development environment more like our production environment in the process, without sacrificing debuggability or iteration time, that's a big win. With that in mind, let's briefly survey some of the more promising alternatives to AMD.

4.2.3 CommonJS

The most popular direct alternative to AMD is the CommonJS* module format. CommonJS modules are not explicitly wrapped inside a function. The module-private scope is implied within each source file rather than expressed explicitly as a function. They also express their dependencies using a syntax that's a bit nicer and more difficult to get wrong:

Listing 4.3 Importing modules using CommonJS syntax.

```
var Cartesian3 = require('./Cartesian3');
var defaultValue = require('./defaultValue');
```

CommonJS is the module format used on the server for Node.js† modules. In Node.js, each of those calls to *require* loads a file off the local disk, so it is reasonable that it not

---

* http://wiki.commonjs.org/wiki/Modules/1.1
† http://nodejs.org/

return until the file is loaded and the module created. In the high-latency world of the browser, however, synchronous *require* calls would be much too slow.

So, to use CommonJS modules in the browser, we must convert these modules into a browser-friendly form. One approach is to transform them to AMD modules before loading them in the browser. The *r.js* tool we used earlier to create minified builds can also be used to transform CommonJS modules.

Another tool that is gaining traction, especially among the Node.js crowd, is Browserify.[*] Browserify takes Node.js-style CommonJS modules and combines them all together into a single browser-friendly JavaScript source file that can be loaded with a simple `<script>` tag. With Browserify, it's even possible to consume AMD modules. For example, when we built Australia's National Map[†] on top of Cesium, we pulled Cesium's AMD modules into our Browserify build by using the *deamdify* plugin.

One nice aspect of Browserify is that it works with the Node.js ecosystem, even enabling us to use npm for our in-browser package management. It's quite refreshing to download, install, and bundle a third-party library—and make it trivial for other developers to do the same—with little more than an `npm install`.

A very interesting approach for serious application development is to construct the application as a large number of separately developed and versioned npm packages. Each package should be useful in its own right and hosted in a separate git repo. npm elegantly manages the dependencies between these packages. While it can be challenging to see how to factor an application into these stand-alone packages, the reward is a library of packages that can be reused across applications. The stackgl[‡] project is a great example of this approach (Chapter 13).

4.2.4 TypeScript

Another approach to building a serious modularized application is to use an entirely different language that compiles to JavaScript. Two of the better-known languages in this category are CoffeeScript and Dart. Our favorite such language, though, is TypeScript,[§] in large part because of its compatibility with JavaScript. All JavaScript code is automatically valid TypeScript code, and the output of the TypeScript compiler is idiomatic JavaScript much like what we'd write by hand. TypeScript has nice syntax for defining and exporting modules, and it can be configured to produce its generated JavaScript modules using either AMD or CommonJS format. As of TypeScript 1.x, the syntax for importing modules is as follows:

Listing 4.4 Importing modules in TypeScript.

```
import Cartesian3 = require('./Cartesian3');
import defaultValue = require('./defaultValue');
```

---

[*] http://browserify.org/
[†] https://github.com/NICTA/nationalmap
[‡] http://stack.gl/
[§] http://www.typescriptlang.org/

This syntax is likely to change in TypeScript 2.0, because TypeScript is intended to closely track the upcoming ECMAScript 6 standard, discussed later.

In addition to good module support, TypeScript also supports optional type annotations, which are then enforced by the compiler. In Cesium, all public APIs and most private ones have their types explicitly documented, because doing so makes the code easier to read and the API easier to understand. We believe that having a compiler that enforces type compatibility is very beneficial to improving the documentation as well as to eliminating a certain class of bugs.

4.2.5 ECMAScript 6

The upcoming version of JavaScript, called ECMAScript 6 or ES6, will have built-in support for modules and should be an official standard by the time you read this.[*] ES6 modules avoid AMD's "synchronized lists" problem, but because they are part of the language, they can load asynchronously within web browsers. ES6 dependencies are specified as follows:

Listing 4.5 Importing modules using ES6 syntax.

```
import Cartesian3 from 'Cartesian3';
import defaultValue from 'defaultValue';
```

Even if your application targets older browsers, you can start using ES6 today by using a tool that compiles ES6 to the current version of JavaScript, ES5. A great list of such tools is maintained by Addy Osmani.[†]

4.2.6 Other Options

There are many other approaches for modularizing JavaScript code. The Google Closure compiler[‡] has support for modularization and is a popular option for building large JavaScript applications. We've even heard of folks using *#include* and the C preprocessor as a simple build process. When evaluating a toolchain for your serious application, be sure to consider how it will interface with the larger JavaScript ecosystem. Even the most beautiful module system is not so appealing if it makes it hard to leverage third-party libraries, documentation generation tools, test frameworks, test runners, etc.

4.3 Performance

Writing about JavaScript performance is tricky because it is a constantly moving target. Browser implementations continue to improve on a regular basis and what is slow now may not be slow for long. Still, even with the continually evolving nature of self-updating browsers, there is a set of common dos and don'ts that are generally applicable and unlikely to change.

---

[*] The schedule has slipped before, however. See http://ecma-international.org/memento/TC39-M.htm for the latest status.

[†] https://github.com/addyosmani/es6-tools

[‡] https://developers.google.com/closure/compiler/

It's impossible for us to write about JavaScript performance without also admitting that it varies from browser to browser. Every JavaScript engine has its strengths and weaknesses and performance characteristics can vary wildly depending on the feature being used. Websites, such as jsPerf.com, have sprung up to help compare performance across a wide variety of browsers, usually as microbenchmarks of a particular language or library feature. While jsPerf can be useful, we recommend a more direct approach. All modern browsers have amazing profiling tools built directly into their equally amazing debug environments. Whether we are using Chrome, Firefox, Internet Explorer, or Safari, we find the easiest way to determine the locations of the slow spots is to simply run the code with the profiler enabled. Unfortunately, not all performance issues show up in the profiler. Some language features or other architectural choices can have a hidden cost that is distributed throughout the entire codebase. The best way to combat these types of issues is following the best practices we lay out here, as well as staying up to date on the ever-changing state of JavaScript engines.

4.3.1 Defining and Constructing Objects

Some of the most fundamental optimizations a JavaScript engine can make depend on having type information available. Unfortunately, because JavaScript is a dynamically typed language, this information is not easily obtained. Most engines use a technique known as type inferencing to deduce types from our code at runtime [Hackett 12]. The more our JavaScript code acts as it would in a statically typed language, the easier it is for the engine to optimize it. For example, consider this `Cartesian3` constructor function:

Listing 4.6 A simple `Cartesian3` constructor.

```
var Cartesian3 = function(x, y, z) {
    this.x = x;
    this.y = y;
    this.z = z;
};
```

Later, we may need to add a w property to a particular `Cartesian3` instance. It's tempting to use the extensibility of JavaScript objects to add the property with a simple `instance.w = 1`. This is not recommended, however, because it can negatively impact performance [Clifford 12].

In a statically typed language, we couldn't dynamically add the w property to a single instance of the `Cartesian3` class at all. Instead, we'd have to define a new class and instance entirely, and copy the values of the x, y, and z properties to the new instance. While modern JavaScript runtime engines have many tricks, they fundamentally still generate instructions in the same machine code that statically typed languages compile to.

When we define a constructor function like the preceding `Cartesian3`, many JavaScript VMs create an internal representation of its type to allow for fast property access and method calls, and to make its in-memory representation as efficient as possible. Ideally, a `Cartesian3` would be represented in memory as simply three floating-point numbers, though there's likely to be a bit more overhead than that even in the best JavaScript engines. The more optimal the in-memory layout of the `Cartesian3` instance is, the more costly it will be to add that w property. It's likely that the JavaScript engine will have to reallocate our instance and copy the existing properties, just as we would manually in a statically typed language. Ultimately, the engine may decide to use a much less efficient representation for this now-dynamic instance.

In addition, when the engine generates machine code for a function that takes a `Cartesian3` instance as a parameter, it may create an optimized implementation tied to that type. For example, many engines employ inline caching, storing the results of method and property lookups in the generated code. A structural change to an instance, such as adding the w property, invalidates the inline caches. The engine will have to generate new code or, perhaps more likely, fall back on an unoptimized code path using on-stack replacement [Pizlo 14].

A better solution is to define a `Cartesian4` constructor function, including a w property from the start, and use instances of that type as needed. This gives the JavaScript engine the most information about our types and the best chance of being able to generate fast code. If something would be difficult or slow to do in a statically typed language, it will almost certainly be slow in JavaScript, too.

There are multiple ways to define and construct objects in JavaScript, but constructor functions like we used before are the fastest way. In our benchmarks, `Object.create` is three to five times slower than simply calling new, but it used to be much worse [Jones 11]. Object literal notation is almost as fast as new, but only if we cache functions in a parent scope to avoid recreating them every time. Keep in mind that these are microbenchmarks, so they may not always match real application performance. Rather than fret over object creation, a better solution is to avoid allocation altogether, which we discuss in Section 4.3.2.

Much like our method for object construction, our method for defining properties on an object can have a significant effect on overall performance. While some JavaScript engines will inline simple getters and setters, function call overhead is still significant enough in some browsers that it needs to be taken into account. This means that if we want to expose a property on an object, it's faster to make it a public field rather than abstract it behind get and set methods. Also, while `Object.defineProperty` lets us create modern properties found in languages such as C#, on some browsers they tend to have the same overhead as a function call.

The general rule we follow in Cesium is simple. If a property would normally be a simple getter or setter function in other languages, then we expose it directly. If a property requires additional work to be done on get or set, only then do we use `Object.defineProperty`. By exposing properties everywhere, this makes the API consistent and minimizes runtime overhead.

4.3.2 Garbage Collection Overhead

One of the primary issues in many high-performance JavaScript applications is garbage collection, and the nature of 3D applications compounds this problem further. For example, imagine we need to multiply two vectors for every object in the scene. Assuming we have 10,000 objects and we are targeting 60 frames per second, we end up creating 120,000 vector result objects per second. In languages such as C++ or C#, we wouldn't even consider this to be a problem because our vectors are likely allocated on the stack. In JavaScript, it can be a major bottleneck.

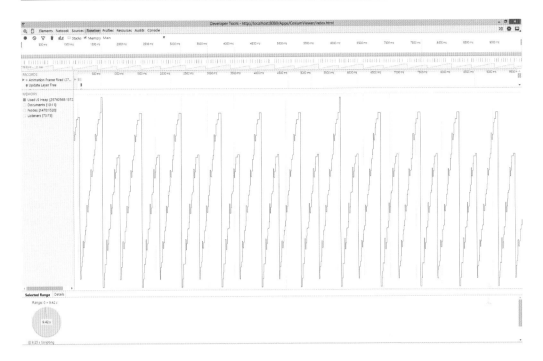

Figure 4.2

The sawtooth pattern, shown in Chrome Developer Tools, is a telltale sign of garbage collection issues.

Math operations, of the kind described previously, are unavoidable in WebGL applications. Early on in Cesium's development it was not uncommon to profile a particular use case only to discover that 50% of the time was being spent in garbage collection. A visual representation of this issue appears as a deep sawtooth pattern in browser profiling tools, as shown in Figure 4.2. The peaks are when the garbage collector kicks in, freeing memory but stealing valuable processing time from our own code. This kind of unwanted memory churn is usually created by algorithms that compute intermediate values that are quickly thrown away. For a more concrete example, see Listing 4.7, which is a simplified version of Cesium's `Cartesian3` linear interpolation implementation.

Listing 4.7 A memory-inefficient linear interpolation function.

```
Cartesian3.add = function(left, right) {
  var x = left.x + right.x;
  var y = left.y + right.y;
  var z = left.z + right.z;
  return new Cartesian3(x, y, z);
};

Cartesian3.multiplyByScalar = function(value, scalar) {
  var x = value.x * scalar;
  var y = value.y * scalar;
```

```
    var z = value.z * scalar;
    return new Cartesian3(x, y, z);
};

Cartesian3.lerp = function(start, end, t) {
    var tmp = Cartesian3.multiplyByScalar(end, t);
    var tmp2 = Cartesian3.multiplyByScalar(start, 1.0 - t);
    return Cartesian3.add(tmp, tmp2);
};
```

Every call to `lerp` allocates three objects: two intermediate `Cartesian3` instances and the result instance. While a microbenchmark of 100,000 calls takes about 9.0 milliseconds in Firefox, it doesn't expose a problem with garbage collection because the memory is not cleaned up until after our benchmark has already completed.

We can remove the extra memory allocation by using two simple techniques. First, we require users to pass in an already allocated result parameter to avoid having to create a new instance every time. Second, we use module-scoped scratch parameters in calls to add within `lerp`.

Listing 4.8 Memory-efficient linear interpolation using result parameters and scratch variables.

```
Cartesian3.add = function(left, right, result) {
    result.x = left.x + right.x;
    result.y = left.y + right.y;
    result.z = left.z + right.z;
    return result;
};

Cartesian3.multiplyByScalar = function(value, scalar) {
    result.x = value.x * scalar;
    result.y = value.y * scalar;
    result.z = value.z * scalar;
    return result;
};

var tmp = new Cartesian3(0, 0, 0);
var tmp2 = new Cartesian3(0, 0, 0);

Cartesian3.lerp = function(start, end, t, result) {
    Cartesian3.multiplyByScalar(end, t, tmp);
    Cartesian3.multiplyByScalar(start, 1.0 - t, tmp2);
    return Cartesian3.add(tmp, tmp2, result);
};
```

The modified implementation initializes two scratch variables during load but otherwise allocates no additional memory. While 100,000 calls to this version of lerp only took about 6 milliseconds in Firefox, most likely due to less object creation, it's not faster in all browsers. The important gain is only seen when profiling our application as a whole and finding that our garbage collection time has dropped significantly in the profiler, resulting in higher frame rates.

It's not uncommon for Cesium to call over 100,000 functions like this per frame. Using result parameters and scratch variables saves us several milliseconds in our per-frame budget. While we hate that we have to clutter up our code and API like this, result parameters are an absolute necessity for anyone looking to write a performant, nontrivial WebGL application.

4.3.3 The Hidden Cost of Web Workers and How to Avoid It

In Cesium we allow user-defined geometric volumes such as ellipsoids, polygons, boxes, and cylinders to be computed synchronously on the main thread or asynchronously in a background thread via Web Workers. It was much to our surprise that our initial Web Worker implementation was several orders of magnitude slower than the single-threaded version. It turns out that Web Workers have a hidden cost when posting large amounts of data between threads.

To illustrate the problem, let's assume we were working only with polygons. Polygon triangulation is a CPU-intensive task and offloading it to a worker thread prevents us from locking up our application while it's processing. It's also not uncommon for a group of polygon definitions to contain over 500,000 vertices, such as the country borders in Figure 4.3.

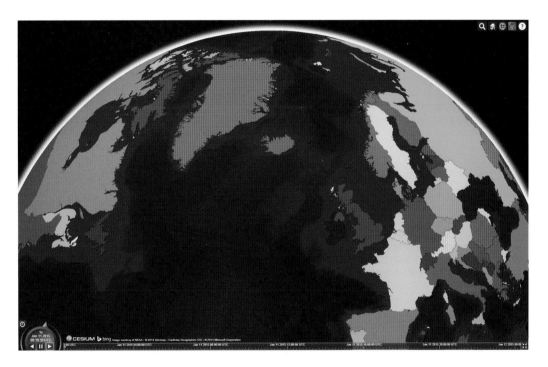

Figure 4.3

Highly detailed polygons exposed a performance issue in our Web Worker implementation.

Because JavaScript is historically a single-threaded language, working with Web Workers is very different from working with traditional APIs. A worker has no access to the DOM and executes in a different global context than the main thread.

Messages and data are always passed to workers by copy, because workers and the main thread do not share memory or any other mutable state. The HTML5 specification defines an algorithm, structured clone,[*] that is used to copy a message for a worker thread. In structured clone, the copied objects lose all prototype and function information, so it must be reconstructed by the receiving thread if needed. Structured clone is roughly equivalent to serializing an object to JSON and then deserializing it on the receiving end.

There is one way to avoid copying when passing data to a Web Worker. A Transferable[†] object, as the name implies, can optionally be transferred, instead of copied, to a worker thread. A transferred object becomes property of the receiving worker and is no longer accessible by the sender. Thus, we avoid both sharing and copying the data. While we unfortunately cannot mark our own objects as transferrable, there are two Transferable objects defined in the specification: `ArrayBuffer` and `MessagePort`.

For our use case, it turns out that the cloning operation is prohibitively slow in all browsers. This makes some sense since structured clone is a generic algorithm meant for copying almost anything. But exactly how slow is it? Suppose we had a Web Worker that simply posted the data it received back to the main thread. For an `ArrayBuffer`, it would transfer the buffer. All other data would be copied as normal. The time it would take to run this worker and receive the data back is almost entirely the overhead of the structured clone operation multiplied by two:

Listing 4.9 A simple Web Worker and timing function to measure the performance cost of structured clone.

```
//contents of worker.js
var onmessage = function(e) {
  postMessage(e.data, e.data.buffer ? [e.data.buffer] : undefined);
};

//code to spawn worker.js
function timeWorker(data) {
    var worker = new Worker("worker.js");
    var start = performance.now();

    worker.addEventListener("message", function(e) {
      console.log(performance.now() - start);
      worker.terminate();
    }, false);
    worker.postMessage(data);
}
```

In our tests, executing the preceding code with an array of 500,000 `Cartesian3` instances took an average of 3.8 to 6.2 seconds (not milliseconds!) to complete, depending

---

[*] http://www.w3.org/TR/html5/infrastructure.html#safe-passing-of-structured-data
[†] http://www.w3.org/TR/html5/infrastructure.html#transferable-objects

4. Getting Serious with JavaScript

on the browser being used. Even worse is that because posting to and receiving from a worker is synchronous, half of this time is spent with the page locked up and unable to respond to user input. In hindsight, this shouldn't have surprised us, but it was still demoralizing when we first encountered it. In many cases, the overhead of sending data to a Web Worker was much worse than just doing the work synchronously and locking up the main thread. We felt that there had to be a better way.

As we mentioned previously, `ArrayBuffers` are one of the objects that can be transferred to a worker thread without copying. What if we manually packed our data into a typed array and transferred it to the worker? Could manually packing somehow be faster than what the browser is already doing in native code? The worker would itself have to unpack the data, and the packing and unpacking code would be specific to the arguments being sent to the worker, but we felt it was worth a try. Here's the modified code from before, along with two helper functions to pack and unpack arrays of `Cartesian3` instances:

Listing **4.10** Method of packing data into a typed array for near instantaneous transfer to worker threads.

```
function packCartesian3Array(data) {
    var j = 0;
    var packedData = new Float64Array(data.length * 3);
    for (var i = 0, len = data.length; i < len; i++) {
        var item = data[i];
        packedData[j++] = item.x;
        packedData[j++] = item.y;
        packedData[j++] = item.z;
    }
    return packedData;
}

function unpackCartesian3Array(packedData) {
    var j = 0;
    var data = new Array(packedData.length/3);
    for (var i = 0; i < packedData.length; i++) {
        var x = packedData[j++];
        var y = packedData[j++];
        var z = packedData[j++];
        data[i] = new Cartesian3(x, y, z);
    }
    return data;
}

function timeWorker(data) {
    var packedWorker = new Worker("worker.js");
    var start = performance.now();
    var packedData = packCartesian3Array(data);

    packedWorker.addEventListener("message", function(e) {
        var receivedData = unpackCartesian3Array(e.data);
        console.log(performance.now() - start);
        packedWorker.terminate();
    }, false);
    packedWorker.postMessage(packedData, [packedData.buffer]);
}
```

The end result was pleasantly surprising. The manually packed version is tremendously faster than relying on default cloning, taking an average of only 60 to 600 milliseconds to complete. While it has to be manually maintained, this technique allows us to pack all objects and their properties, including strings, into a single typed array for efficient transfer.

4.3.4 Making Optimal Use of Multiple Cores

A common technique in multithreaded programming is to use as many threads as cores available to achieve maximum parallelism. If we use too many threads, performance will suffer due to excessive context switching and, with too few threads, spare cores go to waste. Unfortunately, there is no official standard for accessing the number of cores on a system via JavaScript, which we feel is a major oversight that greatly reduces the usefulness of Web Workers in general. Thankfully, a new nonstandard property has recently been added to some browsers which exposes the number of logical processors available on the client system, `navigator.hardwareConcurrency`.* Even though it is only supported in Chrome, Opera, and Safari, it is too useful not to mention here. While Firefox† and IE‡ have both decided not to implement the property at this time, a shim§ is available.

4.4 Automated Testing of WebGL Applications

We believe that automated tests are critical for any serious application. A good suite of automated tests gives us confidence that our code is working at a deep level, because we're able to test edge cases and uncommon paths. It also greatly improves our confidence when refactoring, which is critical for an application with a development plan that spans years.

While this is true in an application written in any language, there are some additional concerns in JavaScript. Web browsers are frustratingly tolerant of broken JavaScript code. We can write a JavaScript function containing code that is not even syntactically valid and the browser will not complain until that function is actually executed. Similarly, a simple typo in a code path—one that handles errors, perhaps—will likely go undetected unless that code path is executed. Automated tests with excellent code coverage are our best available tool for making sure that all of our code is executed and has the correct behavior.

There is an astounding number of JavaScript test frameworks and runners available today, each with a chart comparing itself to a subset of the others that clearly shows that it is the best of the lot. For Cesium, we settled on using the Jasmine test framework and the Karma test runner.

4.4.1 Jasmine

In some ways, Jasmine is "old school." It doesn't use a module system; instead, we include it with a script tag and it adds its functions to the global scope. It also doesn't have a ton of features. What it does have, however, is clean and elegant syntax for writing tests, and its

---

* https://wiki.whatwg.org/wiki/Navigator_HW_Concurrency
† https://groups.google.com/forum/#!topic/mozilla.dev.platform/QnhfUVw9jCI
‡ https://status.modern.ie/hardwareconcurrency
§ https://github.com/oftn/core-estimator

simplicity makes it possible to integrate into a wide variety of applications. For example, we've successfully used Jasmine in both AMD-based applications and CommonJS/Browserify-based applications.

Jasmine is a behavior-driven development (BDD) framework, which essentially means that we write our tests in a style that feels a bit like describing in English what our code is supposed to do:

Listing 4.11 A Jasmine unit test for `Cartesian3` normalization.

```javascript
describe('Cartesian3', function() {
  it('normalizes to a vector with magnitude 1', function() {
    var original = new Cartesian3(1.0, 2.0, 3.0);
    var normalized = Cartesian3.normalize(original);
    var magnitude = Cartesian3.magnitude(normalized);
    expect(magnitude).toEqual(1.0);
  });
});
```

Running Jasmine tests, referred to as specs in Jasmine, in a browser requires us to set up a `SpecRunner.html` file, using the one included in the Jasmine distribution as a template (Figure 4.4). The specifics of the SpecRunner vary depending on how our application is structured. In all cases, we include the Jasmine scripts using standard `<script>` tags. Actually running the specs, however, is dependent on our project architecture.

Figure 4.4

Cesium's customized SpecRunner.html with four failing tests.

If we forgo a module system, this is straightforward, though painful for a large application: Just include the script tags for all of our source files and spec files, in the right order, in `SpecRunner.html`.

If we have just a single combined source file with all of our specs and their dependencies, as we get with CommonJS modules built using Browserify, it's easy: We simply add the single script tag for the built JavaScript file to `SpecRunner.html`.

The asynchronous module definition (AMD) case is more complicated than the other two, simply because AMD is asynchronous. By default, Jasmine runs all of the specs it knows about in the `window.onload` event. When we're using AMD, the spec modules are not yet loaded by the time `window.onload` fires. We need to load all of our spec modules, and only launch Jasmine once they are all loaded. First, we add a `<script>` tag for RequireJS and give it a `data-main` attribute that is the entry point module for our specs.

```
//SpecRunner.html
<script data-main = "specs/spec-main" src = "../requirejs-2.1.9/
  require.js"></script>
```

Our `spec-main` module requires-in all the spec modules and then executes the Jasmine environment:

Listing 4.12 An entry point module for executing Jasmine specs.

```
//spec-main.js
define([
  './Cartesian3Spec',
  './Matrix4Spec',
  './RaySpec'
], function() {
  var env = jasmine.getEnv();
  env.execute();
});
```

We don't need to actually have a parameter for each module in our `spec-main` function because we don't need to use it; we only need to ensure that the modules are loaded.

Of course, it's unfortunate that we need to list every spec module in this way. The complete list of spec files has to be specified *somehow*, though, because the web browser can't inspect the local filesystem to determine the list of specs. In Cesium we use a simple build step to automatically generate the complete list of spec modules so that we don't have to maintain this list manually.

With our `SpecRunner.html` set up properly, we only need to serve it up through any web server and visit it from any browser to run our tests in that browser.

4.4.2 Karma

Running tests in Jasmine is a manual process. We open a web browser, navigate to our `SpecRunner.html` file, wait for the tests to run, and look for any failures. Karma lets us automate this.

With Karma, a single command can launch all the browsers on the system, run the tests in each of them, and report the results on the command line. This is critical for working with continuous integration (CI) because it gives us a way to turn test failures generated via a web browser into build process failures. Karma can also watch the tests for changes and automatically rerun them in all browsers, which is handy when practicing test-driven development or when otherwise focused on building out the test suite.

Compared to Jasmine, setting up Karma is pretty easy, even for use with AMD.[*] Install it in your Node.js environment using npm, and then interactively build a config file for your application by running the following:

```
karma init
```

Karma has out-of-the-box support for Jasmine and several other test frameworks, and more can be added via plugins. Once Karma is configured, we can run the tests in all configured browsers by running

```
karma start
```

4.4.3 Testing JavaScript Code That Uses WebGL

Most of what we've discussed so far is applicable to testing just about any JavaScript application. What unique challenges does WebGL present?

We can test most of our graphics code without ever actually rendering anything. For example, we can validate our triangulation, subdivision, batching, and level-of-detail selection algorithms with standard unit tests that invoke these algorithms and assert that they produce the data structures and numbers that we expect. Inevitably, however, some portion of our rendering code—however small—is intimately tied to the WebGL API.

Purists might argue that our unit tests should never make calls into the WebGL API directly, instead calling into a testable abstraction layer of mocks and stubs. In this perfect world, unit testing our WebGL application would be no different from unit testing any other application. We'd write tests that drive our code and then assert that the correct pattern of WebGL functions was invoked, without ever actually invoking those functions.

While we do appreciate that there is a place for this sort of testing, we also believe that every serious WebGL application will eventually have to step outside it. For one thing, mocking and stubbing the entire WebGL API—or at least a large enough subset of it to test a sophisticated piece of application logic—would be a significant undertaking.

The bigger problem is that WebGL is a complicated API. If we only ever tested against a mocked version of it, we wouldn't have great confidence that our code would work against the real one. Or, perhaps we should say real *ones*, because in some sense, each browser and GPU combination may have unique capabilities and bugs. We might choose to call these tests using the real API *integration tests* rather than *unit tests*, but the fact remains that they're an important part of our testing picture.

With that in mind, we've purposely avoided discussing cloud-based JavaScript testing solutions such as Sauce Labs[†] in previous sections. This is because none of these solutions, as of this writing, have reliable support for WebGL. It's unfortunate, because we'd love to be able to run our tests across a wide variety of operating systems and web browsers without

---

[*] http://karma-runner.github.io/0.12/plus/requirejs.html
[†] https://saucelabs.com/

maintaining any test infrastructure ourselves. But it's also understandable, because these solutions necessarily use virtualization, and GPU hardware acceleration in a virtualized environment is still in its infancy. Thus, our current approach is to use Karma to run tests on physical machines that we maintain, all driven by our CI process.

When we're writing a WebGL application, we may have hundreds or thousands of lines of code that all conspire to put a certain pattern of pixels on the screen. How can we write automated tests to confirm that the pattern of pixels is correct?

There is no easy answer to this question. On previous projects, we wrote tests to draw a scene, take a screenshot, and compare it to a "known good" screenshot. This was extremely error prone. Differences between GPUs and even driver versions inevitably caused our test images to be different from the master images. Anytime a test failed, we immediately wondered what was wrong with the driver, operating system, or test, rather than asking ourselves what might be wrong with our code. We used "fuzzy" image comparison to make our tests assert that an image "mostly" matched the master image, but even then it was a constantly frustrating balancing act between reporting failure on a new GPU where the code was actually working perfectly well, and reporting success even though something actually went wrong. We would not recommend this approach.

Others have reported better success with comparing screenshots by maintaining and manually verifying a set of "known good" images for every combination of platform, GPU, and driver [Pranckevičius 2011]. While it's easy to see how this would be effective, it also strikes us as an extraordinarily costly approach to testing.

Instead, most Cesium tests that do actual rendering render a single pixel, and then assert that the pixel is correct. For example, here's a simplified test that asserts a polygon is drawn:

Listing 4.13 Single pixel sanity checking in Cesium unit tests.

```
it('renders', function() {
  var gl = createContext();
  setupCamera(gl);
  drawPolygon(gl);

  var pixels = new Uint8Array(4);
  gl.readPixels(0, 0, 1, 1, gl.RGBA, gl.UNSIGNED_BYTE, pixels);
  expect(pixels).not.toEqual([0, 0, 0, 0]);

  destroyContext(context);
});
```

This test only asserts that the pixel is not black, which is typical of the single-pixel rendering tests in Cesium. Sometimes we may check for something a bit more specific, like nonzero in the red component or "full white." We generally don't check for a precise color, though, because differences between browsers and GPUs can make even that test unreliable.

While the preceding example test creates a unique WebGL context for the test, we try to avoid this in the Cesium tests. One reason is that context creation and setup take time, and we want our tests to run as quickly as possible. A more serious problem, though, is that web browsers don't expect applications to create and destroy thousands of contexts. We've seen bugs in multiple browsers where context creation would start failing midway through our tests. On the other hand, using a single context for all tests risks a test corrupting the context's state and causing later tests to fail. In Cesium, we've found a good balance by creating a context for each test suite. A test suite is a single source file that tests a closely related group of functionality, such as a single class, so it's usually easy to reason about the context state changes occurring in the suite.

A single-pixel rendering test like this is far from exhaustive, of course. Because it really just asserts that the polygon put *something* on the screen, there are plenty of things that could go wrong and still allow this test to pass. The opposite is not true. We should never see this test fail when the polygon, WebGL stack, and driver are working correctly.

4.4.4 Testing Shaders

Cesium has a library of reusable GLSL functions for use in vertex and fragment shaders. Some of these are fairly sophisticated, such as computing the intersections of a ray with an ellipsoid or transforming geodetic latitude to the Web Mercator coordinates commonly used in web mapping. We find it very beneficial to unit test these shader functions in much the same way we would unit test similar functions written in JavaScript. We can't run Jasmine on the GPU, however, so how do we test them?

Our technique is straightforward. We write a fragment shader that invokes the function we wish to test, checks whatever condition we're testing, and outputs white to `gl_FragColor` if the condition is true. For example, the test shader for the `czm_transpose` function, which transposes a 2×2 matrix, looks like this:

Listing 4.14 Testing reusable functions in GLSL.

```
void main() {
      mat2 m = mat2(1.0, 2.0, 3.0, 4.0);
      mat2 mt = mat2(1.0, 3.0, 2.0, 4.0);
      gl_FragColor = vec4(czm_transpose(m) == mt);
}
```

When `czm_transpose` computes the correct transpose, this shader sets `gl_FragColor` to white. If the transpose is incorrect, `gl_FragColor` is transparent black.

We then invoke this test shader from a Jasmine spec. Our spec draws a single point with a trivial vertex shader and the fragment shader above. It then reads the pixel that the shader wrote, using `gl.readPixels`, and asserts that it is white.

We've found this to be an easy, lightweight, and effective way to test shader functions. Unfortunately, it is not possible to test the entire vertex or fragment shaders in this way, nor is it straightforward to assert more than one condition per test. If either of those features is required, consider using a more full-featured GLSL testing solution such as GLSL Unit.*

For Cesium, however, we've found this to be unnecessary. By testing the building blocks of our shaders—the individual functions—and keeping the `main()` functions as simple as possible, we are able to have high confidence in our shaders without a complicated GLSL testing process.

---

* https://code.google.com/p/glsl-unit/

4.4.5 Testing Is Hard

In Cesium, we have a few types of tests:

- Tests of underlying algorithms that verify the data structures and numbers that the algorithms produce. These tests don't do any rendering.
- Rendering smoke tests, as described in Section 4.4.3. These usually render a single pixel and verify that it is not wildly wrong. We also sometimes render a full scene and verify only that no exceptions were thrown during the process.
- Shader function tests, as described in Section 4.4.4. These test the reusable functions that compose our shaders by invoking them in a test fragment shader and asserting that it produces white.

We find these types of tests to be relatively easy to write, robust, and well worth the time investment it takes to write them.

Unfortunately, we've found no way around having an actual human run the application once in a while, on a variety of systems and in a variety of browsers, to confirm that the rendered output is what we expect.

Acknowledgments

We'd like to thank everyone who reviewed this chapter and gave us valuable feedback: Jacob Benoit, Patrick Cozzi, Eric Haines, Briely Marum, Tarek Sherif, Ishaan Singh, and Traian Stanev. We'd also like to thank our families for their patience and understanding as we wrote this chapter, and our employers, Analytical Graphics, Inc. (AGI) and National ICT Australia (NICTA), for their flexibility. Finally, we'd like to thank Scott Hunter, who taught us virtually all of what we've written here, except for the parts that are wrong. NICTA is funded by the Australian Government through the Department of Communications and the Australian Research Council through the ICT Centre of Excellence Program.

Bibliography

[Cozzi 14] Patrick Cozzi. "Why Use WebGL for Graphics Research?" http://www.realtime-rendering.com/blog/why-use-webgl-for-graphics-research/, 2014.

[Hackett 12] Brian Hackett and Shu-yu Guo. "Fast and Precise Hybrid Type Inference for JavaScript." http://rfrn.org/~shu/drafts/ti.pdf, 2012.

[Jones 11] Brandon Jones. "The somewhat depressing state of Object.create performance." http://blog.tojicode.com/2011/08/somewhat-depressing-state-of.html, 2011.

[Pizlo 14] Filip Pizlo. "Introducing the WebKit FTL JIT." https://www.webkit.org/blog/3362/introducing-the-webkit-ftl-jit/, 2014.

[Pranckevičius 2011] Aras Pranckevičius. "Testing Graphics Code, 4 Years Later." http://aras-p.info/blog/2011/06/17/testing-graphics-code-4-years-later/, 2011.

[Clifford 12] Daniel Clifford. "Breaking the JavaScript Speed Limit with V8." Google I/O https://www.youtube.com/watch?v=UJPdhx5zTaw 2012.

5

Emscripten and WebGL

Nick Desaulniers

5.1 Emscripten

> I don't really know much about computer language theory. I just mess around with it in my spare time (weekends mostly). Here is one thing I've been thinking about: I want the speed of native code on the web—because I want to run things like game engines there—but I don't want Java, or NaCl, or some plugin. I want to use standard, platform-agnostic web technologies [Zakai 10].

That quote from Alon Zakai, Emscripten's creator, is quite telling of the motivations behind Emscripten. Alon simply wanted to run his favorite game engine in the web. Being a Mozilla employee, Alon recognized the prevalence and reach of JavaScript; almost everyone already had a JavaScript runtime installed on their devices. Being able to safely execute remote code is one of the web's greatest strengths; content that runs in JavaScript is not "silo-ed" to a particular browser or operating system.

Emscripten is a compiler for C and C++ code that plugs into the LLVM toolchain. The beauty of LLVM is that it's split into three logical parts: a lexer and parser (the frontend), which generates an intermediary representation (IR); an IR optimizer (middle) that performs all of the code optimizing transforms (loop invariant code motion, inlining, etc); and the code generator (backend). The frontend is called Clang, from which the command line utility is named. Emscripten reuses the frontend for parsing C and C++, and all of the optimization passes, but replaces the native code generation with a module that generates a subset of JavaScript (JS) known as asm.js.

Emscripten links against `musl` as its C runtime and `libc++abi` for its C++ runtime. Functions like `sbrk` are implemented in JS, which allows implementations of `malloc` and `free` to be compiled over from C [Zakai 13]. Not only can C/C++ be compiled over directly, but other languages whose runtimes are implemented in C/C++ can have their runtimes ported over. Thanks to Emscripten languages, Python, Ruby, and Lua can all run in the browser.

In this chapter, I hope to explain what Emscripten and asm.js are, and how we can use them today to get our OpenGL ES 2.0 based renderers up and running in all web browsers with WebGL 1.0 support.

5.2 asm.js

Originally, Emscripten emitted idiomatic JavaScript. While the code was mostly correct, it had wild variation in performance and startup time as it created objects that would trigger frequent garbage collection pauses. In a traditional executable there's a segment for static memory, containing things such as static variables and strings. There's then a separate segment for the stack, which grows downwards toward the heap (the free store of memory used for dynamic memory allocations). Emscripten emulates this approach and gets predictable speed by allocating a giant TypedArray up front, and mapping on top of a similar memory representation that a process would have. Not only is this faster than working with JavaScript objects, but also using this kind of memory representation helps a JS engine avoid having to allocate new objects during execution of the compiled code.

In late 2012, Luke Wagner, a core engineer on Firefox's JavaScript virtual machine, realized that JavaScript Just In Time compilers (JITs) could do a better job optimizing the type of JavaScript generated by tools such as Emscripten by recognizing implicit static types in the code. Even though JavaScript doesn't have actual static types, the static types in the source language (C/C++) still show through in the generated JavaScript in the form of operators (such as | and +) that force the resulting dynamic values to have a single dynamic type. Recognizing the patterned use of these operators throughout the code, the JIT can compile the code with full static type knowledge without any of the runtime profiling or chances of mis-speculation that accompany normal JS optimization. While this static-type-embedding subset of the language would be a restriction that throws out some of the best parts of the language, it would allow for JavaScript to become an excellent compiler target. And being a subset of a widely deployed language, every modern browser would be able to run this subset, known as asm.js, whether or not it specifically optimized for it.

Being a subset of the JavaScript language, pre- and postincrement and -decrement are not valid asm.js, and neither are the + = and related operators. One must use the notation x = x + 1. Does this mean ++x will not run? Absolutely not—++x certainly is valid JS; the use of pre- and postincrement and -decrement operators just will prevent the current code block from validating as pure asm.js, and so will not take the hot path to the optimizing compiler in Spidermonkey, which is Firefox's JS virtual machine.

In the code for this chapter[*] is an example of handwritten asm.js that normalizes a vector of four Float32s. We can see from that example that asm.js is cumbersome and unwieldy to write by hand. We end up referring to the asm.js spec often, both the sections on operators [asm.js §8] and value types [asm.js §2.1], to get our type conversions correct. It's about as much fun to write as assembly, hence the name. Handwriting asm.js code is difficult; instead asm.js serves us better as a compiler target. Small functions generally do not provide major improvements when converted to asm.js. Instead, when large programs or large utility functions are converted, asm.js really shines.

5.3 Hello World

Let's whet the appetite with a quick "hello world." Download the Emscripten SDK from the site, though I've also heard that "real hackers" compile their compilers from source. Grab Node.js as well, since Emscripten will use Node.js in its toolchain. Node.js provides an alternative JavaScript runtime to the one provided by a browser. It allows for command line utilities to be written in JavaScript. Cloning Emscripten's source code would be useful, too. Throughout the chapter, we refer to files in Emscripten's source as emscripten/path/to/file. Emscripten's documentation site[†] will be a useful reference.

Once we're all set and have emcc in our $PATH, write a simple hello world example in C or C++, compile with emcc hello_world.c -o hello_world.js, run with node hello_world.js, and we'll see hello world! printed to stdout. We can invoke the compiler as emcc for both C and C++ code bases. We can also use -o hello_world.html and then open the resulting file in our browser.

5.4 Working with Third-Party Code

Just like we need to have the source of a third-party library cross compiled to a new platform, third-party libraries also need to be cross compiled with Emscripten. If we use the `-o file.bc` option, Emscripten will simply emit LLVM bytecode, with its own libc/libcxx implementations. We can then statically link against them as we would native code. Emscripten does not currently support dynamically linking or calls to dlopen or equivalent. If we don't have access to the source code of a library, we should work with the vendor to provide a version that has been built with Emscripten. Emscripten already has built in glue code for most of EGL, GLUT, GLEW, GLFW, and SDL.

Emscripten is great for compiling functions to JS and allowing us to call into them through foreign function interface (FFI[‡]), though it's more common to use Emscripten to have C/C++ code call into functions defined in JS "glue code." In general, if we use Emscripten, we may end up needing JS glue code.

---

[*] https://github.com/WebGLInsights/WebGLInsights-1
[†] https://kripken.github.io/emscripten-site/index.html
[‡] https://kripken.github.io/emscripten-site/docs/porting/connecting_cpp_and_javascript/Interacting-with-code.html

Before implementing bindings to all of the HTML5 APIs, know that Emscripten has already done so. See emscripten/system/include/emscripten/html5.h and emscripten/ system/include/emscripten/emscripten.h.

There are also utilities called "embind" and "WebIDL binder" for creating C++ classes from types of JavaScript objects that can make it easier to do FFI with, though their use is more advanced than what I'll be talking about here.

5.5 OpenGL ES Support

WebGL 1.0 is very similar to OpenGL ES 2.0 (Figure 5.1), with a JavaScript-based API rather than a C-based API. In general, one should familiarize oneself with Section 6 of the WebGL 1.0 spec, which iterates 24 "Differences between WebGL and OpenGL ES 2.0" [WebGL 1.0 §6]. Most notable is that client-side arrays are not supported; one must use vertex buffer objects (VBOs). Emscripten can emulate this functionality with the compiler flag -s FULL_ES2 = 1, though Emscripten may not be able to emulate the most efficient use of VBOs. Support for some parts of the old fixed-function pipeline code can work at a severe performance penalty with the compiler flag -s LEGACY_GL_ EMULATION = 1.

5.6 Porting OpenGL ES 2.0 to WebGL with Emscripten and asm.js

The work flow for getting our OpenGL ES 2.0 code base running in WebGL using Emscripten looks like this:

1. Get it building
2. Get it rendering
3. Get it animating

The process is unique with every code base, but this is the typical work flow. I've converted Dan Ginsburg's sample code from the OpenGL ES 2.0 Programming Guide [Munshi et al. 08] from an SVN repo to a git repo. The original code is a SVN repo,[*] but if

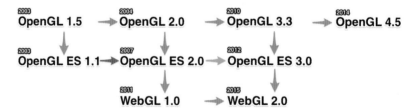

Figure 5.1

OpenGL family tree.

---

5. Emscripten and WebGL

you prefer GIT, I've uploaded the code to GitHub.* In the repo, we'll also find commits in the writing_follow_along branch that we can use to see what changes were made in this section. Please do clone the code and follow along.

Your humble author is going to be using OSX 10.8.5, but we'll be working with the Linux code base because, as we'll see, Emscripten has great support for various GNU build utilities. Go ahead and and open gles2-book/LinuxX11. If we simply run make as is, we should expect the project to fail, since OSX is missing quite a few of the required headers and libraries.

5.6.1 Get It Building

Emscripten ships with a tool called emmake. It's a Python script that invokes make with a few environment variables like $CC set to emcc (the Emscripten compiler). Let's try reinvoking the build steps using emmake, emmake make. We can see that the default make target is trying to link against EGL, which we know is definitely not available in OSX. Let's take a look at the Makefile. We'll just work with the first sample, "Hello Triangle," so we can make another make target and just edit the compiler command for Hello Triangle.

```
basic:./Chapter_2/Hello_Triangle/CH02_HelloTriangle
```

That way, for all of our changes, we need to remember to run emmake make basic. Second, let's replace all hard coded references to gcc with the environment variable $(CC). This will allow emmake to set CC = emcc.

```
./Chapter_2/Hello_Triangle/CH02_HelloTriangle: ${COMMONSRC} ${COMMONHDR}
${CH02SRC}
        $(CC) ${COMMONSRC} ${CH02SRC} -o $@ ${INCDIR} ${LIBS}
```

We should see a bunch of warnings about GLESv2, EGL, m, and X11 libraries not being found. That's OK because Emscripten has linked in implementations of the functions of these libraries. We can remove the line defining LIBS from the Makefile.

Finally, to silence all warnings, we can fix up the pointer sign conversion warning existing in the code by changing the definitions of vShaderStr and fShaderStr in Chapter_2/Hello_Triangle/Hello_Triangle.c from being of type GLbyte[] to const char* on lines 83 and 90. OK, so now we're building, but what do we get?

```
$ file Chapter_2/Hello_Triangle/CH02_HelloTriangle
Chapter_2/Hello_Triangle/CH02_HelloTriangle: data
```

It looks like CH02_HelloTriangle is binary data. The data look oddly like LLVM bytecode and, sure enough, if we run LLVM disassembler on the binary file and open the resulting. ll file, we have LLVM IR.

```
$ llvm-dis Chapter_2/Hello_Triangle/CH02_HelloTriangle
$ cat Chapter_2/Hello_Triangle/CH02_HelloTriangle.ll
; ModuleID = 'Chapter_2/Hello_Triangle/CH02_HelloTriangle'
target datalayout = "e-p:32:32-i64:64-v128:32:128-n32-S128"
target triple = "asmjs-unknown-Emscripten"
...
```

---

* https://github.com/nickdesaulniers/opengles2-book

Now we have LLVM bytecode; what can we do with that? Well, Emscripten generates different output based on the file extension of the compiler argument passed to -o. If -o is invoked without a file extension of .js or .html as is the case here, then the output is simply LLVM bytecode.

This bytecode is what we need to work with other libraries. Emscripten doesn't have support for dynamic linking ahead of time or at runtime, so for now we'll have to stick with static linking. That's essentially what we're doing here: compiling the code to an IR that can be linked statically.

In this case, we don't want an equivalent to an archive, we want a full program that we can run. Since Emscripten uses the file extension of the output option to generate different code, we need to tell it whether we want to generate code that can run in node.js (just JavaScript) or in a browser (JavaScript and HTML). In our Makefile, for the ./Chapter_2/Hello_Triangle/CH02_HelloTriangle target, let's tack on .html to $@, which is Make syntax for the target.

```
$(CC) ${COMMONSRC} ${CH02SRC} -o $@.html ${INCDIR} ${LIBS}
```

After running emmake make clean && emmake make basic, we can see that we've introduced two new warnings.

```
warning: unresolved symbol: XNextEvent
warning: unresolved symbol: XLookupString
```

Where did these come from and why didn't we see them earlier? Before telling Emscripten that we wanted html, it was creating LLVM bytecode. It wasn't running the linking phase, just the frontend of Clang and the optimization passes. Though we didn't specify any optimization levels, Emscripten actually uses compiler passes added to LLVM from NaCl to break down IR into more easily digested chunks. It's not until we tell Emscripten that we want runnable code via -o <output>.js or -o <output>.html that it actually runs the linking phase. In fact, if we look at the LLVM IR in text format (Chapter_2/Hello_Triangle/CH02_HelloTriangle.ll), we can see that a bunch of type definitions that will be resolved at link time are specified as "type opaque."

Right now, the linker is complaining that all of the symbols listed throughout had definitions except two functions, which look to me like they're related to X11, and that we pulled out from the linking options earlier. But we also removed explicitly linking against GLES2 and EGL, so why aren't we getting warnings from those? Well, Emscripten will implicitly search its system/include directory for headers for GLES 2 and EGL that are bundled with Emscripten, and definitions in JavaScript of those methods from Emscripten's src dir; see emscripten/src/library_gl.js and emscripten/src/library_egl.js in Emscripten's source code or install location. Emscripten ships with a bunch of implicit definitions to functions implemented in JavaScript. Some of these are handwritten, but it's nice to use compiled versions where it makes sense.

We don't get warnings for GLES or EGL functions because Emscripten has them implemented for us. We get a warning for the X window stuff because we did not define them, and Emscripten does not have them built in. As we will see, we won't be using the X window bits so tackling those linker warnings right now is a moot point. In general if we get compiler warnings like this, then we should make sure none of our important functions' definitions are missing.

Back to our project, we now have a .html file and a .js file. We could try to run the js file in Node.js, but we'll run into issues trying to resize the canvas since HTMLCanvasElement is not defined by the Node.js runtime. Let's see what happens if we try to run the code in the browser. We now have the code building; let's get it rendering.

5.6.2 Get It Rendering

Depending on our browser, we shall see either a slow script warning or a compositor lock up, caused by an infinite loop and resulting in no rendering. Let's try drawing just one frame, to make sure the issue is in rendering and not in animation. Let's wrap the final statement in main in #ifdefs for now to render only one frame when compiled with Emscripten:

Listing 5.1 Drawing one frame.

```
#ifdef __EMSCRIPTEN__
   Draw(&esContext);
#else
   esMainLoop (&esContext);
#endif
```

Remember to rerun emmake make basic between page loads. Now we don't get the long running script error, but still nothing renders, so we know the issue is not just with animation, but also that something is broken in our render function. Checking in our console developer tool, we can see WebGL-implementation-specific errors printed to console.error:

```
Error: WebGL: vertexAttribPointer: must have valid GL_ARRAY_BUFFER binding
   CH02_HelloTriangle.js:1937
Error: WebGL: drawArrays: no VBO bound to enabled vertex attrib index 0!
   CH02_HelloTriangle.js:1974
```

Let's try rebuilding with debug symbols. In our Makefile, let's add a simple -g4 flag to our rule for our example:

```
$(CC) ${COMMONSRC} ${CH02SRC} -o $@.html ${INCDIR} ${LIBS} -g4
```

Without any optimization flags passed to the compiler, Emscripten will not attempt to minify and compress the emitted code for anything less than -O2. We can use -g1 to preserve whitespace, -g2 to preserve function names, -g3 to preserve variable names, and -g4 to generate source maps. Source maps are an additional file, and a special comment in the emitted JavaScript, for telling the developer tools how to map lines of the emitted code back to lines of the source. If we use -g4, we should now see an additional .html.map file, and a comment at the bottom of our JavaScript file

```
//# sourceMappingURL = CH02_HelloTriangle.html.map
```

After reloading, you should be able to view the C source code, possibly in addition to the emitted JavaScript, depending on your browser and its support for source maps. See emcc—help for more compiler options.

Go ahead and set a breakpoint right before the call to `Draw(&esContext)`, reload the page, step into Draw, then step over until we find the offending line that prints the warning in the console tab. We should find that the offending line is

```
glVertexAttribPointer (0, 3, GL_FLOAT, GL_FALSE, 0, vVertices);
```

What is `vVertices`?

If we're used to working with vertex buffer objects (VBOs), we're probably expecting the sixth argument to `glVertexAttribPointer` to be an offset. But we can see a few lines up that `vVertices` is declared as being of type `GLfloat[]`. This is referred to as client-side data or rendering from client-side memory. If a zero buffer is bound with `glBindBuffer`, the sixth argument to `glVertexAttribPointer` is the address of vertex data in main memory; otherwise, it is an offset into the currently bound VBO in video memory [OpenGL ES 2.0 §2.9.1].

WebGL does not allow for client-side data. We must use VBOs in WebGL, so this code should be converted to use VBOs rather than client-side data. It's possible to recompile with the flag `-s FULL_ES2`, see something render, and then call it a day. Let's not do that. While Emscripten can emulate client-side data, it might be slow. The fix might look like the following:

Listing 5.2 Using VBOs instead of client side data.

```
#ifdef __EMSCRIPTEN__
  GLuint vertexPosObject;
  glGenBuffers(1, &vertexPosObject);
  glBindBuffer(GL_ARRAY_BUFFER, vertexPosObject);
  glBufferData(GL_ARRAY_BUFFER, 9 * 4, vVertices, GL_STATIC_DRAW);
  glVertexAttribPointer(0, 3, GL_FLOAT, GL_FALSE, 0, 0);
#else
  glVertexAttribPointer (0, 3, GL_FLOAT, GL_FALSE, 0, vVertices);
#endif
```

This is not preferable, since we're recreating a new buffer every frame, but we'll see in a bit some useful tools for helping us catch redundant calls against the GL context. Rebuild and we should see our triangle!

5.6.3 Get It Animating

Now that we've fixed issues related to rendering, let's tackle animation. Let's revert the change at the end of main that we did to render only a frame with `#ifdef`s. We should now be running `esMainLoop(&esContext)`, regardless of the compiler. Load this version up in a browser, and we'll get the slow script warning again. Go ahead and stop the script. This time, we'll notice that we at least have something rendering, though we still have to contend with an infinite loop somewhere.

First, let's talk a little bit about event loops. In our native C or C++ code, we can run code in an infinite loop and the OS will take care of making sure the system stays responsive by context switching in and out of the process. JavaScript works a little differently; instead it has an event loop. Because JavaScript is single threaded (WebWorkers can be used for multithreading), APIs for fetching data over the network (like XMLHttpRequest) are typically implemented asynchronously; they take a function as an argument to be invoked later. Such an argument is referred to as a callback function. This is similar to passing function pointers or function objects in C or C++. The callback is placed on the event queue, and won't be invoked until a few ticks of the event loop have gone by, and won't be executed until an event such as a network response. This allows for the rest of the current tasks in the event queue to run. Because JavaScript is for the most part single threaded, and runs in the main thread of the browser, long running functions or infinite loops will lock up the page. Functions such as `setTimeout` and `setInterval` push a callback onto the event queue, and indeed these were traditionally used in libraries like jQuery a long time ago to perform non-main-thread-blocking animation. There are issues with using those two functions for animation, so with the HTML5 standard we got the `requestAnimationFrame` function.

`requestAnimationFrame` is what we want to use for animation, rather than a while (true) loop. If we jump into the file Common/esUtil.c, and then down to the definition of esMainLoop on Line 290, we can indeed find the source of our infinite loop:

Listing 5.3 Main animation loop.

```
while(userInterrupt(esContext) = = GL_FALSE)
{
   . . .
   esContext->updateFunc(esContext, deltatime);
   . . .
   esContext->drawFunc(esContext);
   . . .
}
```

In order to transform our code to play nice with JavaScript's event loop, we perform the following steps:

1. Put the body of the while loop into its own function.
2. Create a struct containing variables if the while loop references variables outside its local scope.
3. Use `emscripten_set_main_loop_arg` and the address of the populated argument struct.

`emscripten_set_main_loop_arg` has the signature: `extern    void emscripten_set_main_loop_arg(void (*func)(void*), void* arg, int fps,  int simulate_infinite_loop)` and will invoke `requestAnimationFrame`. The front and back buffers will be switched by WebGL at the end of our `requestAnimationFrame` iteration. There's also a sister function `emscripten_set_main_loop` if we don't have any arguments to pass. This transformation should feel extremely familiar to anyone who has ever transformed an iterative operation to be threaded. First, let's move the body of the while loop into its own function:

Listing 5.4 Separate render function.

```
static void render (void *data)
{
   ...
   esContext->updateFunc(esContext, deltatime);
   ...
   esContext->drawFunc(esContext);
   ...
}
void ESUTIL_API esMainLoop (ESContext *esContext)
{
#ifdef __EMSCRIPTEN__
    //TODO: step 3
    //Use emscripten_set_main_loop_arg and the address of the
    //populated argument struct.
#else
   while(userInterrupt(esContext) = = GL_FALSE)
   {
      ...
      esContext->updateFunc(esContext, deltatime);
      ...
      esContext->drawFunc(esContext);
      ...
   }
#endif
}
```

Second, we'll create a struct for the variables that are initialized before the while loop:

Listing 5.5 Create a struct to house animation loop data.

```
struct loop_vars_t
{
   struct timeval t1;
   struct timezone tz;
   float totaltime;
   unsigned int frames;
   ESContext* esContext;
};
static void render (void *data)
{
   struct loop_vars_t* args = (struct loop_vars_t*) data;
   struct timeval t2;
   float deltatime;
   ...
   args->esContext->updateFunc(args->esContext, deltatime);
   ...
   args->esContext->drawFunc(args->esContext);
   ...
}
```

Third and finally, we will use `emscripten_set_main_loop_arg` and the address of our populated argument struct:

Listing 5.6 Use `emscripten_set_main_loop_arg`.

```
void ESUTIL_API esMainLoop (ESContext *esContext)
{
#ifdef __EMSCRIPTEN__
        struct loop_vars_t args = {0};
        args.totaltime = 0.0f;
        args.frames = 0;
        args.esContext = esContext;
        gettimeofday(&args.t1, &args.tz);
        emscripten_set_main_loop_arg(render, &args, 0, 1);
#else
...
```

At this point, building should give us an error that `emscripten_set_main_loop_arg` has not yet been defined. Make sure to `#include "emscripten.h"` at the top of Common/esUtil.c. Emscripten will automatically know how to include this header; there's no need for additional -I arguments during the build.

Now we should see our code properly animating as in Figure 5.2! Granted, we're not updating any vertices between frames, but we should now see the console output every 2 seconds with an average frames per second.

5.7 Texture Loading

Another aspect of working with Emscripten that is different from native development is the asynchronous loading of static assets. If we try compiling the Chapter 13 example from the OpenGL ES 2.0 Programming Guide, we should get an error: Error loading (smoke.tga) image. We have to change the Makefile to specify an .html file extensions to the -o compiler flag and use VBOs instead of client side data as before. If we add the compiler option—`preload-file./Chapter_13/ParticleSystem/smoke.tga`—we'll get an additional CH13_ParticleSystem.data file in Chapter_13/ParticleSystem/. This file is a static memory initializer that is produced in optimized builds and builds that use Emscripten's virtual filesystem. We'll still get the previous error though. From emcc—help under—reload-file: The path is relative to the current directory at compile time. So we can either edit Chapter_13/ParticleSystem/ParticleSystem.c line 165 to `"./Chapter_13/ParticleSystem/smoke.tga"` or move smoke.tga to the current working directory. Because we previously fixed our renderer, we should now see the example render and animate correctly.

5.8 Developer Tools

We can use developer tools in Firefox to help improve our code. Some of my favorites are the shader editor, the sampling profiler, the timeline viewer, and the canvas debugger.

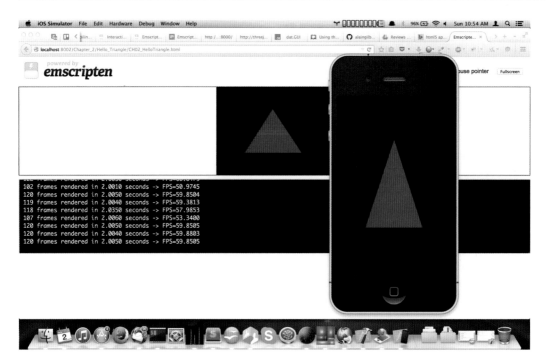

Figure 5.2

WebGL renderer side by side with OpenGL ES 2.0 renderer.

The performance tab features a sampling profiler and FPS over time graph (Figure 5.3). We will find inverting the call stack or call tree helpful in pinpointing what functions the program spent the most time in. The longer we allow the profiler to run, the more accurate its samples will be. Don't be alarmed if the profiler seems to be missing some functions; those functions typically run so fast that the sampler misses them. We'll see extra functions that we may not have written ourselves if we have a few add-ons installed; it's best to run the profiler from a fresh profile which is free of add-ons.*

We can also manually add instrumentation to our code with calls to console.time and console.timeEnd. Browsers that embed the Blink rendering engine, like Chrome, Chromium, and Opera, have chrome:tracing, which is a fantastic structural profiler to use once we've added the instrumentation to our code (see Figure 5.4). Structural profilers give us insight into exactly how much time was spent in various function calls relative to one another, but burden us with adding instrumentation. The sampling profiler will miss short running functions, but doesn't require instrumentation. It's good to start with a sampling profiler, and then switch to a structural profile if needed. Firefox's Timeline dev tool will also show console.time blocks' duration relative to other things going on within the page.

WebGL inspector† is another nice tool implemented both as a library and browser add on or extension.

---

* https://support.mozilla.org/en-US/kb/profile-manager-create-and-remove-firefox-profiles
† http://www.realtimerendering.com/blog/webgl-debugging-and-profiling-tools/

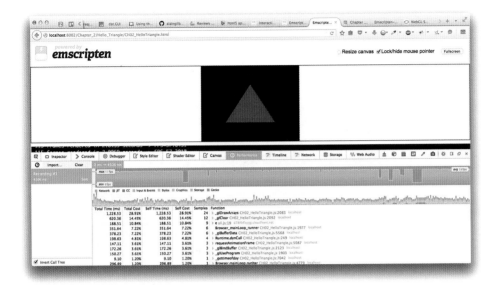

Figure 5.3

Firefox sampling profiler developer tool.

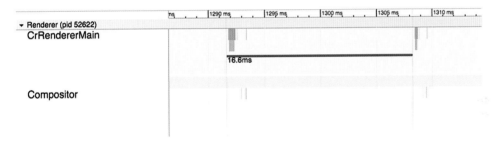

Figure 5.4

Blink's structural profiler developer tool.

From Figure 5.5, we can see that the canvas tab is a wonderful debugger for animation loops. It records, then highlights clear and draw calls in green, and usePro-gram calls in pink. If we record an iteration of our animation loop, we can see that we're repeating things that only need to be done once. We don't need to be resetting the viewport size, setting the only shader program as the active one, creating a new buffer, rebuffering our data, or re-enabling our vertex attribute pointers. Before we move those functions from Draw into Init, let's remove the browser's 60 fps limitation on request AnimationFrame to see a more dramatic before and after. In Firefox, if we open about:config in another tab, we can change the preferences layout.frame_rate from −1 to 0 and layers.offmainthreadcomposition.frame-rate from −1 to 1000, then restart the browser. It's important that we restore these to defaults when done; otherwise, every site using requestAnimationFrame we'll visit will end up doing much more work than our monitor's refresh rate can keep up with.

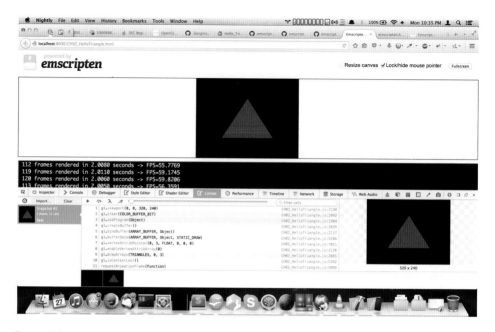

Figure 5.5

Firefox's canvas debugger developer tool.

If we pull out the -g4 compiler flag in favor of -O3, and shrink the Firefox window down to be the size of the canvas, we can update as high as 570 fps, with some large variations. This is an excellent technique that can be used for A/B testing.

It's important to use optimization levels -O2 or above to emit valid asm.js code from Emscripten. Using -O2 with asm.js means that the available heap size will be fixed. Applications which require variable and/or large amounts of memory will often fail in such cases and cannot be reliably compiled with -O2. In these cases, the runtime will print a warning on how to compile the code so that this can be worked around, for an additional runtime cost.

Some other things like new Error().stack in JavaScript can help us print a stack trace. Emscripten provides an EM_ASM macro that allows us to embed JavaScript directly in our C/C++.

5.9 Emscripten in Production

The Emscripten team worked with the folks at Epic in 2013 to bring Epic's Unreal Engine 3 (UE3) to the web, as seen in their *Citadel* and *Unreal Tournament* demos, and Unreal Engine 4 in 2014. Developers at Trendy Entertainment shipped *Dungeon Defenders Eternity* in 2014 using UE3. Unity also made a free feature in Unity 5 to export to the web, as seen with a demo of Mad Finger Games' *Dead Trigger 2*. In late 2014, Mozilla partnered with Humble Bundle to release the Humble Mozilla Bundle; releasing nine popular indie games like *FTL: Faster Than Light* and *Voxatron*, which together raised over half a million dollars in revenue. Screenshots from the games running in browser are available in Figure 5.6.

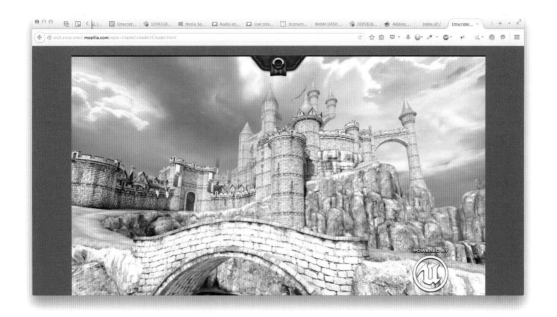

Figure 5.6

Clockwise from top left: Epic Unreal Engine 3 *Citadel* Demo, Trendy Entertainment *Dungeon Defenders Eternity* based on UE3, Mad Finger Games *Dead Trigger 2* based on Unity 5, Subset Games *FTL: Faster Than Light* as part of the Humble Mozilla Bundle. *(Continued)*

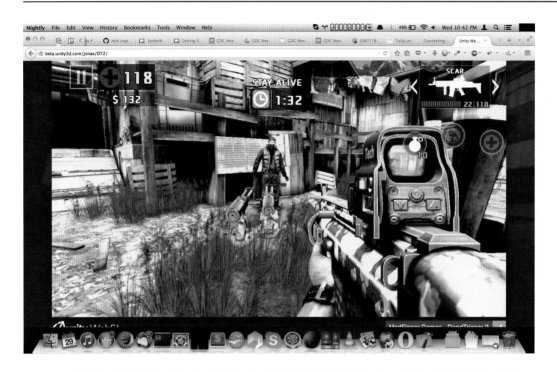

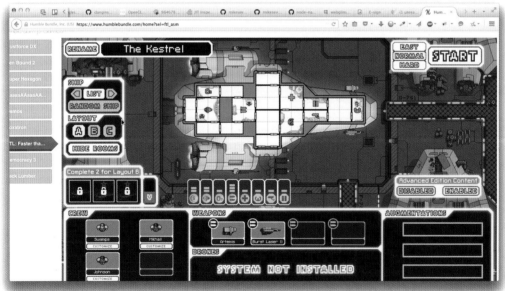

Figure 5.6 (Continued)

Clockwise from top left: Epic Unreal Engine 3 *Citadel* Demo, Trendy Entertainment *Dungeon Defenders Eternity* based on UE3, Mad Finger Games *Dead Trigger 2* based on Unity 5, Subset Games *FTL: Faster Than Light* as part of the Humble Mozilla Bundle.

5.10 More Help

For more help, it's best to read over the official documentation.[*] The irc channel #emscripten at irc.mozilla.org is also very active and all of the people involved in Emscripten's development and a bunch of developers with porting experience hang out there. It's worthwhile to read emscripten/src/preamble.js for documentation on FFI functionality and emscripten/src/settings.js for compiler options. All of the code samples and figures are available online at the book's Github code repository.[†]

Bibliography

[asm.js] Dave Herman, Luke Wagner, and Alon Zakai. "asm.js." http://asmjs.org/spec/latest/, 2014.

[Jylänki 13] Jukka Jylänki. "Emscripten Memory Profiler." https://dl.dropboxusercontent.com/u/40949268/emcc/memoryprofiler/Geometry_d.html, 2013.

[Lamminen et al. 14] Turo Lamminen, Tuomas Närväinen, and Robert Nyman. "Porting to Emscripten." https://hacks.mozilla.org/2014/11/porting-to-emscripten/, 2014.

[Munshi et al. 08] Aaftab Munshi, Dan Ginsburg, and Dave Shreiner. *OpenGL ES 2.0 Programming Guide*. Addison-Wesley Professional, 2008.

[OpenGL ES 2.0] Mark Segal and Kurt Akeley. "The OpenGL® Graphics System: A Specification." https://www.opengl.org/documentation/specs/version2.0/glspec20.pdf, 2004.

[Typed Array] Dave Herman and Kenneth Russell. "Typed Array Specification." https://www.khronos.org/registry/typedarray/specs/latest/, 2013.

[Wagner 14] Luke Wagner. "asm.js AOT compilation and startup performance." https://blog.mozilla.org/luke/2014/01/14/asm-js-aot-compilation-and-startup-performance/, 2014.

[Wagner and Zakai 14] Luke Wagner and Alon Zakai. "Getting started with asm.js and Emscripten." GDC 2014.[WebGL 1.0] Dean Jackson. "WebGL Specification." https://www.khronos.org/registry/webgl/specs/latest/1.0/, 2014.

[Zakai 10] Alon Zakai. "Experiments with 'Static' JavaScript: As Fast as Native Code?" http://mozakai.blogspot.com/2010/07/experiments-with-static-javascript-as.html, 2010.

[Zakai 13] Alon Zakai. "Emscripten: An LLVM-to-JavaScript Compiler." https://github.com/kripken/emscripten/blob/master/docs/paper.pdf?raw = true, 2013.

---

[*] https://kripken.github.io/emscripten-site/
[†] https://github.com/WebGLInsights/WebGLInsights-1

6

Data Visualization with WebGL
From Python to JavaScript

Cyrille Rossant and Almar Klein

6.1 Introduction

The deluge of data arising in science, engineering, finance, and many other disciplines calls for modern, innovative analysis methods. Whereas more and more processes can be automated, human supervision is nevertheless often required at most stages of analysis pipelines. The primary way humans apprehend data for explorative analysis is visualization. Effective big data visualization methods have to be *interactive*, *fast*, and *scalable* [Rossant 13].

Modern data sets may be large and high dimensional; thus no static, two-dimensional image can possibly convey all relevant information. A common technique is to create *interactive* visualizations, where the user explores the various dimensions and subsets of the data. For such data exploration to be most effective, the rendering frame rate needs to be optimal even with large data sets. Finally, big data visualization methods need to

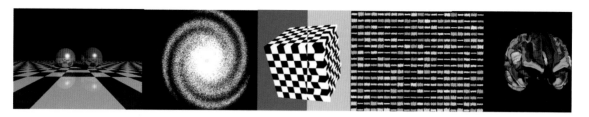

Figure 6.1

Screenshots from the VisPy gallery. From left to right: ray tracing example implemented in a fragment shader; a fake galaxy rendered with point sprites; a textured cube with custom postprocessing effects in the fragment shader; hundreds of digital signals updated in real time (the grid layout is generated entirely in the vertex shader for efficiency); a 3D mesh of a cortical surface.

support distributed and remote technologies in order to scale to huge data sets stored in cloud architectures.

We have been developing an OpenGL-based library in Python named VisPy to visualize large data sets interactively.* Using the graphics card via custom shaders allows us to display smoothly up to hundreds of millions of points. Figure 6.1 illustrates a few types of visualization supported by VisPy.

VisPy is open source, BSD-licensed, and written in Python, one of the leading open platforms for data analytics [Oliphant 07]. Although Python is widely acclaimed for its expressivity and accessibility, it lags behind the web platform when it comes to document sharing and dissemination. WebGL is an appealing technology to us because it could provide a way to embed VisPy visualizations within web documents [WebGL]. Since VisPy and WebGL are both based on OpenGL ES 2.0, the main technical barrier is the programming language. Integrating Python and JavaScript indeed remains an open problem [Kelly 13].

In this chapter, we present a set of techniques that we have been developing in order to bring VisPy to the browser thanks to WebGL. Before detailing these techniques, we give an overview of the VisPy project. VisPy's fundamental idea is to build data-oriented abstraction layers on top of OpenGL ES 2.0. VisPy brings the power of OpenGL to users who have complex visualization needs, but no knowledge of OpenGL. Although VisPy is primarily developed in Python, it is also of interest to WebGL developers who are working on data visualization projects in JavaScript.

VisPy focuses on speed and scalability. In particular, its architecture enables out-of-core visualization applications thanks to specific level-of-detail (LOD) techniques. None of these techniques are currently implemented in VisPy. Rather, VisPy provides an infrastructure that makes these use cases possible. For example, one could dynamically stream various LOD models of a huge data set from a high-performance server (with Python and VisPy installed) to a lower end WebGL-enabled device (desktop computer, smartphone, and so on). VisPy proposes a client-server architecture and a communication protocol that is specifically adapted to such use cases. The purpose of this chapter is to detail this infrastructure.

---

* http://www.vispy.org

6.2 Overview of VisPy

The power and flexibility of modern OpenGL and GLSL make them useful not only for video games and 3D modeling software, but also for 2D/3D data visualization. Examples of visualizations that can be rendered with OpenGL include scatter plots (point sprites), digital signals (polylines), images (textured quads), graphs, 3D surfaces, meshes (rasterization or volume rendering), and many others. The expressivity of GLSL makes it possible to create arbitrarily complex visualizations.

Whereas many scientists need to visualize large, complex, and high-dimensional data sets, few have the time and skills to create custom visualizations in OpenGL. Writing an OpenGL-based interactive visualization from the ground up requires knowledge of the GPU architecture, the rendering pipeline, GLSL, and the complex OpenGL API.

This high barrier to entry motivated the development of VisPy. This library offers high-level visualization routines that let scientists visualize their data interactively and effectively. Specifically, VisPy defines several abstraction layers, from low-level, OpenGL-oriented interfaces to high-level, data-oriented interfaces. Whereas the high-level interfaces allow complex visualizations to be defined with minimal code, the lower level interfaces offer finer control and customization of the visualization process. We present these interfaces in this section.

6.2.1 Dealing with Differences between Desktop GL and GL ES 2.0

VisPy targets "regular" desktop OpenGL as well as OpenGL ES 2.0 (e.g., WebGL). This is realized by limiting ourselves to the subset of OpenGL that is available in both versions. Although both versions are mostly compatible, there are a few pitfalls, some of which we account for in VisPy so that our users can create applications and write GLSL that will work on both desktop GL and WebGL. The most significant differences are[*]

- Some function names are different in both versions. Examples include the create/delete functions, and `getParameter`, which replaces `getInteger/getFloat/getBoolean`. These differences are not problematic in VisPy because we use a common representation language (GLIR) to represent GL commands (see later in this chapter).
- In desktop GL, point sprites are not enabled by default. In VisPy, we automatically enable point sprites if necessary.
- An ES 2.0 fragment shader must contain precision qualifiers. In VisPy, the medium precision level is currently used. For the future we do plan to let users specify the precision level they need.
- ES 2.0 (and WebGL) do not support 3D textures, which are used by some techniques, like volume rendering. This can be worked around by using 2D textures and implementing 3D sampling manually in the shaders. See Chapter 17.
- WebGL has a limit on the size of the index buffer, which can be problematic when visualizing large meshes. This can be worked around by using `OES_element_index_uint` or multiple buffers.

---

[*] For a more complete list see https://github.com/vispy/vispy/wiki/Tech.-EScompat

- WebGL does not allow sharing objects between contexts. This makes things simpler, but can be limiting for certain applications.
- Without a version pragma, the strictness of the GLSL compiler (e.g., allowing implicit type conversions) varies widely between various versions of desktop GL. In VisPy, the version pragma is automatically set to 120 for desktop GL to make the compiler behave similarly to ES 2.0. Further, ES 2.0 implementations are not required to support for-loops for which the number of iterations is not known at compile time.
- ES 2.0 has fewer data types for attributes and texture formats (e.g., no GL_RGB8).

Apart from allowing users to create code that runs on desktop GL as WebGL, we also want to allow users to harness the full potential of desktop GL (e.g., 3D textures and geometry shaders). These desktop-only features would be available only in the Python backends (Qt, wx, etc.) and not in the WebGL backends

6.2.2 gloo: An Object-Oriented Interface to OpenGL

The OpenGL API can be verbose. Basic operations like compiling a shader or creating a data buffer require several commands with many parameters. Nevertheless, the core concepts underlying these APIs are relatively simple and can be expressed in a more compact way.

VisPy provides an object-oriented interface to OpenGL ES, named gloo, that is implemented in both Python and JavaScript. By focusing on the central concepts and objects in OpenGL—shaders, GLSL variables, data buffers—gloo offers a natural way of creating visualizations.

For example, Listing 6.1 is a Python script displaying static random points as shown in Figure 6.2. Figure 6.3 shows the output when running the code in Listing 6.1.

Listing 6.1 A simple visualization in VisPy.

```python
from vispy import app, gloo
import numpy as np

vertex = """
attribute vec2 position;
void main() {
    gl_Position = vec4(position, 0.0, 1.0);
    gl_PointSize = 20.0;
}
"""

fragment = """
void main() {
    vec2 t = 2.0 * gl_PointCoord.xy - 1.0;
    float a = 1.0 - pow(t.x, 2.0) - pow(t.y, 2.0);
    gl_FragColor = vec4(0.1, 0.3, 0.6, a);
}
"""

canvas = app.Canvas()
```

```
program = gloo.Program(vertex, fragment)
data = 0.3 * np.random.randn(10000, 2)
program['position'] = data.astype(np.float32)

@canvas.connect
def on_resize(event):
    width, height = event.size
    gloo.set_viewport(0, 0, width, height)

@canvas.connect
def on_draw(event):
    gloo.set_state(clear_color = (0, 0, 0, 1), blend = True,
                   blend_func = ('src_alpha', 'one'))
    gloo.clear()
    program.draw('points')

canvas.show()
app.run()
```

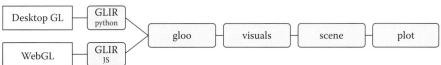

Figure 6.2

Overview of the graphics abstraction layers provided by VisPy. Levels more to the right trade flexibility for a higher level API. A declarative language called GLIR forms a common layer on top of desktop OpenGL and WebGL. The `gloo` module provides an object-oriented interface to OpenGL ES. Visuals encapsulate reusable graphical objects implemented with `gloo`. These visuals can be organized within a scene graph. Finally, a plotting API can be used to create a scene graph and visuals for common use cases.

A `Program` object is created with a vertex and fragment shader. VisPy automatically parses the GLSL code, which allows the user to easily assign data buffers to attributes and values to uniforms. The program can then be rendered with any primitive type among those provided by OpenGL (points, lines, and triangles).

In Python, data buffers are commonly created and manipulated with the NumPy library [van der Walt 11]. This widely used library provides a typed `ndarray` object similar to JavaScript's `TypedArray` [Khronos 13], except that the `ndarray` can be multidimensional.

Shaders are extensively used in VisPy. Instead of implementing rendering routines in Python or C, we leverage GLSL as much as possible. Also, this paradigm makes the conversion to WebGL easier

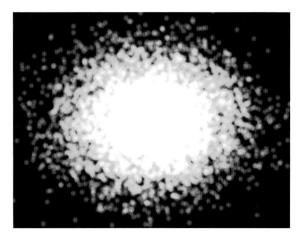

Figure 6.3

Output when running the code in Listing 6.1.

as we have less code to translate. This is because GLSL is virtually the same language between desktop OpenGL and OpenGL ES 2.0/WebGL.

As we will see later in this chapter, `gloo` plays a central role in our Python-to-WebGL translation processes.

6.2.3 Visuals

The `gloo` layer forms the foundation of VisPy's higher level graphics layers. Reusable visualization primitives are encapsulated in `Visuals`. Each visual represents a graphical object (line, image, mesh, and so on). It encapsulates several `gloo` objects and provides a Pythonic interface to control their appearance. Whereas the abstraction level of `gloo` corresponds to the OpenGL architecture, the abstraction level of a visual corresponds to a graphical object appearing in the scene. Therefore, users can create and manipulate visuals intuitively without any knowledge of OpenGL.

Examples of visuals include common geometric shapes in 2D and 3D, high-quality antialiased polylines (implemented with a GLSL port of the antigrain geometry library [Rougier 13]), 3D meshes, images, point sprites, and others. Although VisPy provides a relatively rich set of visuals, users can also create their own visuals if necessary.

Visuals can share and reuse snippets of GLSL code thanks to an internal shader composition system that organizes shaders into modular, independent components. Reusable GLSL functions can be written with placeholder variables identified by a dollar ($) prefix. These variables can be replaced at runtime by `gloo` variables (attributes, uniforms, constants), using a simple Python API. VisPy takes care of generating the final GLSL code and resolving potential name conflicts by renaming some variables if necessary.

This allows the user to modify the appearance of a visual by attaching custom GLSL coordinate transformations, color modifications, and more. For example, a pan/zoom transformation function is implemented with `$translate` and `$scale` template variables. These variables can then be bound to uniforms or attributes depending on the use cases.

6.2.4 The Scene Graph

Visuals can be used directly, or they can be organized within a scene graph. In the latter case, the visuals are represented as nodes to form a hierarchical structure. Each node has a transformation that describes the relation between the parent and the current coordinate frame. These transformations have an implementation on both the CPU and the GPU. Further, by using the scene graph, one can make use of several built-in camera/interaction models.

6.2.5 The Plotting Interface

The plotting interface is the highest level layer in VisPy. It provides ready-to-use visualization routines for common plots like scatter plots, polylines, images, and others. In essence, it provides an easy way to create visuals and set up a scene graph for common use cases.

Tip: VisPy is a data visualization library built on top of OpenGL ES. Written in Python and JavaScript, it provides several abstraction layers. End users can choose between intuitive, but slightly limited, high-level interfaces and more flexible, but more complex, low-level interfaces.

6.3 GLIR: An Intermediate Representation for OpenGL

In order to bring VisPy visualizations written in Python to the browser, we need to translate Python code defining a visualization to JavaScript code. The challenging aspect is that we require this translation to be automatic, or at least to require as little human supervision as possible. Our end users are scientists who expect to write Python code that would work indifferently in the desktop or in the browser; generally, they would be unwilling to write JavaScript code for common use cases. Also, avoiding code duplication makes code easier to maintain.

Two issues arise when it comes to translating a visualization from Python to JavaScript.

First, we have seen that visualizations may be specified in one of several abstraction levels. We need to choose a specific abstraction level in order to find a common representation that can be understood by both languages.

Second, such a representation would only define a *static* visualization. Yet, in VisPy we are focusing on *interactive* visualizations. Conceptually, interactivity means updating the OpenGL objects in the scene in response to dynamic events such as user actions (mouse, keyboard, touch) or timers. These interactivity functions are typically written in a general-purpose programming language. In VisPy, the user can implement these functions in Python using a set of modular components provided by the library. For example, the scene graph defines cameras that can be moved and rotated. Finding a way to translate these custom routines from Python to JavaScript is one of the main challenges of the process.

In this section, we focus on the issue of the representation level for translating visualizations from Python to JavaScript. In the next sections, we detail how interactivity can be translated either automatically or semiautomatically.

6.3.1 Finding an Adequate Representation Level

There is a trade-off between low-level and high-level representations when looking for an adequate translation level. Low-level representations are more expressive and require less code duplication between Python and JavaScript. However, dynamically extracting a low-level representation from a visualization defined in a high-level interface is challenging.

High-level representations are easier to translate, but require much more code duplication between Python and JavaScript. Updating and maintaining both implementations at the same time would require a significant effort.

We initially investigated the lowest possible representation level, namely the OpenGL API, but several issues arose. First, this interface is complex and verbose, and often leads to boilerplate code. These problems make the translated code less readable and harder to debug. Second, there are slight differences between the desktop OpenGL API and the WebGL API. Third, code written in this representation generally requires nonlinear control flows like loops and conditional branches (error handling, for example). Translating this code automatically from Python to JavaScript proved to be challenging.

Therefore, we decided on a new representation that corresponds to our `gloo` interface. We named this representation *GLIR*, for *GL intermediate representation*. GLIR is a simple declarative programming language that describes the dynamic creation and modification of `gloo` objects. Our architecture features two components: the GLIR generator (frontend) and the GLIR interpreter (backend). The frontend generates GLIR commands reactively in response to user or timer events. The backend interprets and executes these commands within an OpenGL context, which can be desktop OpenGL

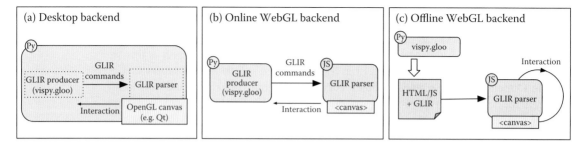

Figure 6.4

Diagram illustrating the different modes of operation of the WebGL backend: (a) with the desktop backend, the GLIR producer and interpreter operate in the same process on top of a GUI backend (Qt, for example); (b) with the online WebGL backend, the GLIR producer operates in a Python process and streams GLIR commands to the WebGL interpreter; (c) with the offline WebGL backend, a stand-alone web document containing all GLIR commands and interaction logic is generated semiautomatically.

or WebGL (Figure 6.4). Overall, the scene is initialized and dynamically updated in the OpenGL context while GLIR commands are streamed to the backend.

An important feature of GLIR is that the communication is one-way, such that the frontend will never have to wait for the backend, allowing for higher performance. Errors that occur on the backend should be reported to the frontend, but this can occur asynchronously.

Implementing this language required a minimal amount of code duplication between Python and JavaScript. Both Python and JavaScript need to parse and execute those commands. Since the GLIR command definitions are simple, we consider this representation relatively stable, and few modifications of GLIR and its implementations are expected in the future. Furthermore, our higher level interfaces are written on top of gloo, and thus are completely decoupled from the GLIR language and implementations. This means that all higher level layers of VisPy transparently work in both desktop and WebGL.

6.3.2 Example of a GLIR Representation

The Python script using gloo in Listing 6.1 would produce the GLIR representation in Listing 6.2.

Listing 6.2 GLIR representation of the scene described in Listing 6.1.

```
CREATE 1 Program
SHADERS 1 "attribute vec2 position;\n[...]" "void main() {\n[...]\n}"
CREATE 2 VertexBuffer
SIZE 2 80000
DATA 2 0 <array> (10000, 1)
ATTRIBUTE 1 position vec2 (2, 8, 0)
FUNC glViewport 0 0 800 600
FUNC glClearColor 0.0 0.0 0.0 1.0
FUNC glBlendFuncSeparate src_alpha one src_alpha one
```

　　　　　　　　　　　　　　　　　　　　　　　6. Data Visualization with WebGL

```
FUNC glEnable blend
FUNC glClear 17664
DRAW 1 points (0, 10000)
```

This representation has been automatically generated by VisPy. Before we get to the implementation details, we will explain the main principles of the GLIR representation. The full specification can be found in the VisPy wiki.[*]

Every line represents a command in the form of a tuple. The first element is the command to execute; the remaining elements are arguments. When both the frontend and backend are in the same (Python) process, the commands remain Python tuples. For the WebGL backend, the commands are serialized in JSON.

There are currently 14 different commands. Every command comes with one or several parameters. The data types for the arguments can be strings, integers, tuples of strings or integers, or array buffers.

Some commands refer to the creation or modification of gloo objects. Each gloo object is represented by an identifier, unique within the current OpenGL context. It is up to the frontend—not the backend—to generate unique identifiers for the gloo objects.

The CREATE command allows us to create a gloo object. There are currently six types of gloo objects: Program, VertexBuffer, IndexBuffer, Texture2D, RenderBuffer, and FrameBuffer. The first line of Listing 6.2 creates a Program object. In the second line, a vertex shader and a fragment shader are assigned to that program. Then, a vertex buffer is created, its size (SIZE command) is specified, and the data are uploaded into this vertex buffer (DATA command).

The data buffer may be represented in several ways. The exact representation depends on the specific frontend and backend. For example, when translating a Python visualization to JavaScript, we can use a base64 encoding of this buffer. This JavaScript string is then converted into an ArrayBuffer object and passed directly to the bufferData() WebGL command. A more efficient method consists of using the binary WebSocket protocol to send the buffer from Python to the browser.

With the ATTRIBUTE command, a program attribute is bound to the VertexBuffer objects that were just created (object #2), and the stride (8) and offset (0) are specified. Finally, a few OpenGL functions are called and we draw the program with the gl.POINTS primitive type.

Additional GLIR commands include UNIFORM (to set a program's uniform value), WRAPPING and INTERPOLATION (for textures), and ATTACH and FRAMEBUFFER (for framebuffers).

Tip: GLIR is a simple declarative programming language used to create and manipulate gloo objects dynamically. A frontend module is responsible for generating a stream of GLIR commands at initialization time, and subsequently as a function of dynamic user events (mouse movements, keystrokes, timers, and so on). A backend module receives, interprets, and executes these commands within an OpenGL or WebGL context. This two-tier architecture lets us decouple a visualization specification in Python from the rendering in the browser with WebGL.

---

[*] https://github.com/vispy/vispy/wiki/Spec.-GLIR

6.4 Online Renderer

GLIR is the foundation of our Python-to-JavaScript architecture. By writing all higher level functionality on top of `gloo` in Python, we ensure that all VisPy visualizations can be translated to WebGL.

In practice, there are several ways to bring a VisPy visualization to the browser. They differ according to whether a live Python server is available or not.

In the online renderer, the browser and a live Python server are connected in a closed loop. The browser captures user events and streams them to the Python server. In return, the server generates the rendering commands and sends them to the browser. The Python server and the browser may or may not run within the same computer.

In the offline renderer, the frontend module generates a stand-alone HTML/JavaScript document containing the entire interactive visualization. There is no need for an external Python server.

From the user's perspective, there is a trade-off between the online and offline renderer. The online renderer works in virtually all cases, and it allows one to keep some logic in Python, at the expense of a small communication overhead between the Python server and the browser. This overhead is due to the round-trip exchanges of messages over the network (user events and GLIR commands). The offline renderer allows one to create a fully stand-alone web application that does not depend on a Python server, but this process is more complex and more limited, and it may require the user to write custom JavaScript code.

6.4.1 The IPython Notebook

In order to implement an online renderer, we need the following components:

- A Python process with NumPy and VisPy installed
- A Python server
- A WebGL-compatible web browser
- A communication channel and a protocol between Python and the browser

Although directly implementing all of these components would be possible, we chose to rely on an existing architecture that is widely used in scientific computing: the *IPython notebook* [Perez 07; Shen 14]. This tool provides a web interface to a Python server. Users can type code in this interface and get the results interactively (*read-eval-print loop*, or *REPL*). Text, images, plots, and graphical widgets can also be created. Dynamic interaction between the notebook client in the browser and the underlying Python server is implemented by IPython. The architecture is based on the Tornado Python server, the ZeroMQ communication library, and the WebSocket protocol (Figure 6.5).

We decided to make use of the IPython-provided functionality to create widgets in the notebook, since this architecture implements all components we need. The custom widget that we created contains a WebGL canvas to which we can render.

6.4.2 Distributed Event Loop

With a normal desktop OpenGL backend, GLIR commands are generated and interpreted by the same Python process. With the browser backend, the GLIR commands are generated by the frontend and proxied to the browser backend in real time via the IPython communication channels.

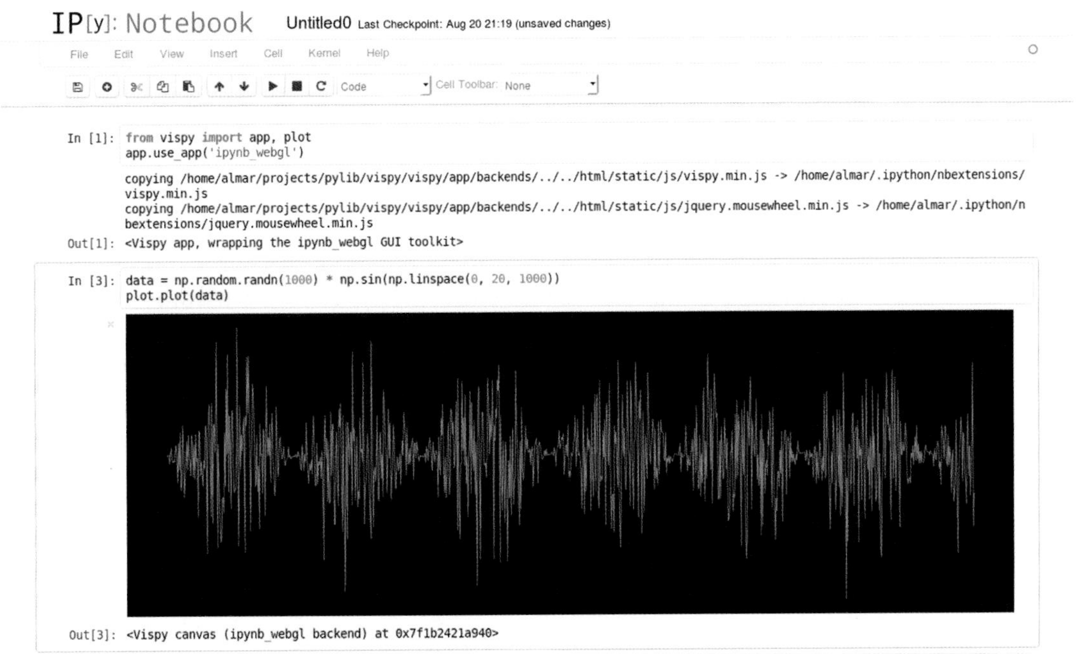

Figure 6.5

Example of VisPy in the IPython notebook.

As in all real-time rendering applications, there is an event loop that processes events and generates drawing calls. By design, this event loop is distributed between Python and the browser.

In the browser, we use `window.requestAnimationFrame()` for the WebGL event loop. In addition, two JavaScript queues are implemented. The *event queue* contains the pending user events that are yet to be sent to the Python server, while the *GLIR queue* contains pending GLIR commands that are yet to be executed by the WebGL engine (a GLIR interpreter written in JavaScript). Events are produced in JavaScript using standard event callback functions in JavaScript and jQuery.

At every WebGL frame, the following operations occur:

1. A JSON event message is produced with the pending user events from the event queue. Consecutive messages of the same type (for example, `mouse_move` events) are merged into single events for performance reasons.
2. This message is sent to Python via the IPython-provided communication channels.
3. The pending GLIR commands are executed.

At the Python side, the JSON event messages are received and the events are injected into VisPy's event system, causing the user event callback functions to be called. These functions may be implemented directly on top of `gloo`, or they may involve the machinery implemented in VisPy's higher level layers (the cameras and the scene graph, for example).

All GLIR commands that are generated by the Python process are automatically queued. Every time that a `draw` event is issued by VisPy, a JSON message with the pending GLIR commands is sent to the browser. There, these events populate the JavaScript GLIR queue; these commands will be executed during the next `requestAnimation-Frame()` iteration.

Time-based animations are implemented at the Python side using Tornado. A timer is responsible for generating `draw` events on a regular interval, which will cause programs to be drawn, GLIR commands to be sent, and so on.

6.4.3 Server-Side Offscreen Rendering

The process described before implements the online WebGL renderer in the IPython notebook. In addition, we implemented an offscreen renderer for low-end clients that don't support WebGL.

Instead of being proxied to the browser, the GLIR commands are interpreted by the Python server. What is sent to the browser is the locally rendered rasterized image instead of the GLIR commands. This process bears some resemblance to the VNC protocol.

This technique might also be used when data sets are too large to be transferred over the network or even to fit in the client's RAM. However, a custom level of detail techniques might be more effective in these cases.

Tip: With the online renderer, the browser sends user events to the Python server in real time via JSON messages. VisPy processes the events and generates a stream of GLIR commands via the `gloo` interface. These commands are sent back to the browser (again via JSON messages), and they are processed by the WebGL GLIR backend in the `requestAnimationFrame()` function. This architecture is implemented within the IPython notebook, an interactive web interface to Python that is adapted to scientific computing. Overall, this WebGL backend allows users to visualize large data sets smoothly and interactively, directly in their browser.

6.5 Offline Renderer

The online rendering methods described before work for all possible VisPy visualizations. However, they require a live Python server (to run VisPy). When IPython notebooks are viewed statically—for example, when they are exported to static, stand-alone HTML/JavaScript web documents—the visualizations do not work. It would be interesting for users to keep their interactive visualizations intact when they share notebooks with colleagues, or when they export notebooks to interactive web reports.

One approach is to run an IPython kernel in the cloud (for example, using Docker*). However, this raises questions about who will be running the server and who is paying for it. Therefore, we are currently working on an alternative approach that doesn't require a live Python server at all. Although more limited in scope than the online renderer, since all logic must be implemented on the client side, this approach has the benefit of being easier to deploy. In this section, we describe work in progress in this direction.

---

* https://www.docker.com

6.5.1 Overview

Since there are JavaScript implementations of both `gloo` and the GLIR backend, a visualization implemented directly on top of `gloo` can be manually reimplemented in JavaScript. Because this is a tedious process, we propose an assistant that can help users to convert some of their Python code to JavaScript.

Exporting the visualization occurs in two steps. First, the list of all created `gloo` objects is obtained dynamically by instantiating the scene and capturing all generated GLIR commands. Second, the event callback functions (which implement interactivity) are translated either manually or automatically. Automatic conversion cannot be done dynamically because the arguments of these functions are variable (mouse position, keystrokes, and so on). Instead, these functions are translated statically. We describe possible approaches here.

6.5.2 Python-to-JavaScript Translator

A first approach would be to use a static Python-to-JavaScript translator such as the open-source Pythonium library.[*] The translator parses the Python code with the native `ast` Python library, and generates JavaScript code on-the-fly while visiting the AST (*abstract syntax tree*).

This approach bears some resemblance to Emscripten (Chapter 5), which compiles C/C++ code to a subset of JavaScript through the LLVM compiler architecture. As far as Python is concerned, this approach is followed by several projects, including PyPy.js,[†] Pyston,[‡] and Numba.[§] We have not explored these options at this time since we are currently interested in a small subset of Python that may not require the heavy machinery involved in these projects.

In our case, the only bits of code we would need to convert are the user callback functions (`on_mouse_move()`, `on_key_press()`, etc.). Therefore, the translator is not intended to support the entire Python syntax or to cover all use cases. Rather, it would aim at simplifying the end user's task of converting a VisPy visualization to a stand-alone HTML/JavaScript document. The generated code could then be improved manually.

The features of the Python language we would need for interactivity functions include standard Python statements, conditional branches, loops, and common mathematical and array operations. Scalar operations are useful when uniform values are modified, but array operations are sometimes required with complex interaction patterns that involve modifying vertex buffers and textures. Whereas Python's `math` module and JavaScript's `Math` module provide similar functionality for scalar operations, there is currently no NumPy support in JavaScript. Therefore, we would need to implement a light version of NumPy in JavaScript as detailed later.

Alternative approaches could be based on other Python-to-JavaScript projects such as Brython[¶] or Skulpt.[**]

---

[*] https://github.com/rcarmo/pythonium
[†] https://github.com/rfk/pypyjs
[‡] https://github.com/dropbox/pyston
[§] http://numba.pydata.org/
[¶] http://www.brython.info
[**] https://github.com/skulpt/skulpt

6.5.3 JavaScript Implementation of a NumPy-like Library

VisPy is extensively based on NumPy. The `ndarray` structure can be easily manipulated in Python. First, vectorized mathematical operations can be done concisely and efficiently. For example, the sum of two vectors A and B is just A+B. Second, arrays can be uploaded to OpenGL buffers and textures efficiently without making any unnecessary copy on the GPU.

JavaScript provides a few efficient structures for WebGL, namely `ArrayBuffers` and `TypedArrays`. Whereas these objects can be easily passed to WebGL functions, no mathematical operations are provided. For this reason, we would need to write a basic JavaScript port of NumPy. Beyond vectorized mathematical operations (+, *, -, /, and so on) and functions (`exp`, `cos`, and so on), this port would also provide array creation and manipulation routines that are commonly used in NumPy. With this basic toolbox, a wide variety of interactivity routines written on top of `gloo` can be readily converted to JavaScript with little effort from the user.

6.5.4 Beyond `gloo`

Exporting all `gloo` objects by capturing the GLIR commands works indifferently whether the function is written directly on top of `gloo` or on top of VisPy's higher level interfaces. This is because using VisPy's higher layers eventually results in GLIR commands.

However, the translation process for the interactivity routines is static and does not involve any code execution in Python. For this reason, it only works when the functions are implemented directly on top of `gloo`.

When these functions use VisPy's higher level interfaces like the scene graph, an alternative approach needs to be used. One possibility would be to reimplement common cameras in JavaScript. Then, the interactivity functions could be made automatically translatable again, because they would make use of the same cameras in Python and JavaScript.

Tip: Whereas a live Python server is required with the online renderer, the offline renderer offers facilities to translate a Python visualization to an interactive, stand-alone HTML/JavaScript document that executes entirely in the client's browser. A static scene can be easily exported from Python to JavaScript by capturing all generated GLIR commands at initialization. These commands can then be executed by the JavaScript GLIR interpreter. For interactivity, a static NumPy-aware Python-to-JavaScript translator helps the user convert event callbacks implemented on top of `gloo`. This is made possible by the fact that these functions often involve basic Python constructs with mathematical operations on scalars and arrays.

6.6 Performance Considerations

We have evaluated the performance cost incurred by the architecture on a simple visualization example.[*] This example displays N normally distributed random points as white point sprites with point size `1.0`.[†]

---

[*] The code is available on https://github.com/vispy/webgl-insights/tree/master/perf

[†] All benchmarks have been performed on a local machine: a GIGABYTE laptop with an Intel(R) Core(TM) i7-4700HQ CPU @ 2.40GHz, 16GB RAM, an Intel GPU with Intel(R) Haswell Mobile Mesa 10.5.0-devel OpenGL 3.0 driver, GLFW backend, VisPy development version 0.4.dev (95a87f6 commit), and Ubuntu 14.04.1 LTS.

First, we have measured the average number of frames per second (FPS) by repeatedly updating the same scene over a 10-second period. An implementation using OpenGL API calls directly (wrapping OpenGL through Python's ctypes module) resulted in a performance of 42.53 FPS ± 2.7% FPS with N = 10,000,000 points. The same implementation using gloo and GLIR resulted in a performance of 41.98 FPS ± 3.1% with the same number of points. The performance overhead is within the error range; it corresponds to the generation of GLIR commands in Python and the execution of these commands by the Python GLIR interpreter. There is no serialization/deserialization involved in this process.

Second, we have evaluated the performance of the WebGL backend.[*] We ran the previous gloo example with N = 1,000,000 points (using N = 10,000,000 points resulted in a crash of the browser). At every trial, a draw event is generated by JavaScript. This results in this event being sent to the Python server, the script's on_draw() function to be called in the Python server, and the GLIR commands to be generated, serialized, and sent back to the browser. Then, these GLIR commands are processed by the WebGL GLIR interpreter within the next requestAnimationFrame() event. Using a single local machine, this entire process took 22 ms in average over all trials, showing that the lag overhead incurred by our architecture represents no more than one or two frames, assuming a 60 target FPS in the browser.

Although we have not made benchmarks on more complex examples, we can expect this overhead to be relatively independent on the scene complexity in many situations. Specifically, visualizations that only result in uniform updates in response to events are not expected to result in a significant lag. This is because the communication overhead essentially depends on two factors: the volume of data to transfer between Python and the browser, and the network lag.

In many scientific visualizations, the data can be transferred only once to the browser—at initialization time. Then, interaction occurs through uniform updates—for example, updating u_translate and u_scale uniform variables for pan/zoom. In this case, there is virtually no data transfer between Python and the browser during interactivity since the size of the messages is negligible.

The situation is different when significant amounts of data need to be regularly sent from Python to the browser—for example, if the data are so large that a LOD technique has to be used. In this case, there can be a significant lag when the data are transferred. The frequency of these operations can be reduced, for example, no more than one data update per second so that the visualization feels responsive most of the time. The lag could be further reduced by using a binary WebSocket protocol instead of base64 (de)serialization for transferring array buffers, which is possible in IPython ≥ 3.0.

Finally, we have only conducted benchmarks on a local machine. When different machines are used for the Python server and the WebGL client, the network lag may harm the perceived performance. In this case, it is conceivable to implement some of the interactivity in the client rather than in Python. For example, pan and zoom may be directly implemented in JavaScript so that network communication is bypassed. This could be done using some of the techniques developed for the offline WebGL backend (Section 6.5).

---

[*] Using Chrome 39.0.2171.95 (64-bit) and IPython 3.0.dev (5bcd54d commit).

6.7 Conclusion

VisPy brings the power of graphics cards to scientists who want to visualize large volumes of data efficiently. VisPy lowers the barrier of entry to OpenGL by providing several visualization-oriented abstraction layers.

Although VisPy is mainly written in Python, one of the major open source data analysis frameworks, it can also work in web browsers via the WebGL specification. The web platform is indeed highly appealing by its ease of deployment and the multiplatform capabilities. Therefore, we have implemented tools to port VisPy visualizations from Python to the browser.

This work could be extended in several ways. First, it should be possible to write a hybrid renderer where light interactivity routines are implemented in the client, while more complex routines are implemented in the browser—for example, routines that involve accessing a large, remotely located data set. This would be useful when a significant network lag hinders perceived performance—for example, while panning and zooming in a plot.

More generally, browser features like webcam support could be transparently and efficiently handled by the online renderer, without requiring round-trip transfers of large streams of data between the client and the server.

It would also be interesting to combine the generated WebGL code with existing WebGL or JavaScript libraries like three.js or dat.GUI (graphical controls in JavaScript). This would let users benefit from both Python's analytics tools and JavaScript/WebGL libraries for creating interactive visualizations in the browser.

Also, a major direction of research is to support VisPy's higher level interfaces in the WebGL offline renderer. The scene layer provides several high-level interactivity routines and cameras that could be reimplemented in JavaScript. The user could therefore write custom and complex interactive visualizations using exclusively the scene layer, and generate WebGL versions of these visualizations fully automatically. The advantage of the scene layer over the lower level `gloo` interface is that the former does not require any knowledge of OpenGL. This is a critical point, since VisPy aims at targeting scientific end users who have complex visualization needs but no OpenGL skills.

Acknowledgments

We thank the rest of the VisPy development team (Luke Campagnola, Eric Larson, and Nicolas Rougier) and all VisPy contributors for their work on the project.

Bibliography

[Kelly 13] Ryan Kelly. "pypy-js-first-steps." https://www.rfk.id.au/blog/entry/pypy-js-first-steps, 2013.

[Khronos 13] "Typed Array Specification," work in progress. https://www.khronos.org/registry/typedarray/specs/latest/, 2013.

[Oliphant 07] Travis E. Oliphant. "Python for Scientific Computing." *Computing in Science & Engineering*, 9:10–20 doi:10.1109/MCSE.2007.58, 2007.

[Perez 07] Fernando Pérez and Brian E. Granger. "IPython: A System for Interactive Scientific Computing." *Computing in Science and Engineering* 9 (3): 21–29 doi:10.1109/MCSE.2007.53. URL: http://ipython.org, 2007.

[Rossant 13] C. Rossant and K. D. Harris."Hardware-Accelerated Interactive Data Visualization for Neuroscience in Python." *Frontiers in Neuroinformatics* 7 (36) doi:10.3389/fninf.2013.00036, 2013.

[Rougier 13] Nicolas P. Rougier. "Shader-Based Antialiased Dashed Stroked Polylines." *Journal of Computer Graphics Techniques* 2.2, 2013.

[Shen 14] Helen Shen. "Interactive Notebooks: Sharing the Code." *Nature* 515:7525 doi:10.1038/515151a, 2014.

[van der Walt 11] Stéfan van der Walt, S. Chris Colbert, and Gaël Varoquaux. "The NumPy Array: A Structure for Efficient Numerical Computation." *Computing in Science & Engineering* 13:22–30 (2011), DOI:10.1109/MCSE.2011.37, 2011.

[WebGL] https://www.khronos.org/webgl/

7

Teaching an Introductory Computer Graphics Course with WebGL

Edward Angel and Dave Shreiner

7.1 Introduction

For almost 20 years, OpenGL has been the standard API used in teaching computer graphics. Regardless of the approach used in a course, OpenGL has allowed instructors to support their courses with an API that is simple and close to the hardware.

The deprecation of the fixed-function pipeline in OpenGL 3.1 forced instructors to make some difficult decisions as to which version to use. In *OpenGL Insights* [Angel 13], we argued for using a fully shader-based approach in a first course. Now, with the popularity of WebGL, instructors must decide whether or not to switch to yet another version.

In this chapter, we argue that WebGL has major advantages and few disadvantages for teaching an introductory course in computer graphics for students in computer science and engineering. These conclusions are based on one of the authors' experiences in teaching the course for over 20 years, both authors' experience teaching SIGGRAPH courses, and with the latest edition of our textbook [Angel 15]. In the following sections, we start with a comparison of a basic 3D application in WebGL with its shader-based desktop OpenGL counterpart. We then make this application interactive and add some other features, in particular texture mapping, which illustrate the advantages of WebGL over desktop OpenGL for teaching the first course.

7.2 The Standard Course

Computer graphics has been a standard course in almost all Departments of Computer Science and Engineering since the 1970s. Even though there have been enormous advances in both hardware and software, the core topics have changed little over 40 years. They include

- Geometry and modeling
- Transformations
- Viewing
- Lighting and shading
- Texture mapping and pixel processing
- Rasterization

Even as most courses have moved to a fully shader-based version of OpenGL, these topics remain the core of the standard course. We take that approach here and focus on the adjustments that have to be made in using WebGL. At the end, we discuss some alternatives that become attractive with the WebGL API.

7.3 WebGL and Desktop OpenGL

As readers of this work know, WebGL 1.0 is a JavaScript (JS) implementation of OpenGL ES 2.0. It is fully shader-based, lacks a fixed-function pipeline, and does not contain any of the functions deprecated with OpenGL 3.1. Although it may appear to be a simple conversion to take an application from desktop OpenGL to OpenGL ES and then to WebGL, there are a number of factors that make the task nontrivial but interesting.

First, because OpenGL is concerned primarily with rendering, with desktop OpenGL we have to provide either platform-dependent functions for input and connection to the window system or to use libraries such as GLUT and GLEW. With WebGL we have to interact with the browser and the web. For teaching computer graphics, this change is a good thing. As OpenGL has developed, it has become more and more difficult to support students using a variety of platforms and hardware. Although an OpenGL application is platform independent in that it can be recompiled on almost any platform, the surrounding libraries have become more of a problem. Most instructors have used GLUT or freeglut and, depending on the platform, GLEW. Finding a set of libraries that work with Windows, OS X, and various versions of Linux has become increasingly difficult. Some instructors find that they can get a set that works with 32-bit architectures but not on 64-bit architectures.

Much more problematic is that applications that worked with previous versions of OpenGL cannot work with a core profile in recent versions because the raster and bitblt functions have been deprecated. For example, menus were easy to add in older versions of OpenGL using GLUT, but GLUT menus need the raster and bitblt functions. One of the major advantages of going to WebGL is the ease with which we can add interactivity to applications. We will return to this issue in Section 7.10.

Going from C or C++ to JavaScript and HTML requires some work. Although we can make our JavaScript code appear very similar to C/C++ code, there are some "gotchas" where there are significant differences that may not be obvious. Books such as *The Good Parts* [Crocket 08] and *JavaScript: The Definitive Guide* [Flanagan 11] are excellent references for students who are already adept in high-level languages. There are also many good tutorials on the web, for example, that by McGuire [McGuire 14]. A more serious issue at many schools can be the antipathy toward JS in computer science and engineering departments. Nevertheless, as we show, the benefits are significant and students have few problem learning and using JS.

Having looked at some of the potential problem areas, before looking at the details of a simple example, let's consider briefly the benefits of going to WebGL for teaching. These benefits include

- Cross-platform broad deployment—web and mobile
- Low barrier to entry
- Fast iteration
- Variety of available tools
- Performance
- A more modern API
- Integration with other web APIs

WebGL is supported on all recent browsers, including those in most new smart phones. Because all these browsers can interpret JavaScript code, there is no explicit compilation stage; no need for the libraries such as GLUT and GLEW that had to be recompiled for each architecture; no system-specific interfaces such as xgl, wgl, and agl; and no dealing with changing versions of the operating system or any of the libraries. From an instructor's perspective, getting the course started is a breeze. From the student's perspective, she can work on the same project on multiple devices with no changes to code.

Because the code is interpreted, students can work in an environment as simple as a standard text editor, thus allowing them to modify and rerun code almost instantaneously. In addition, developer tools within the browser such as profilers and debuggers make working with WebGL code much easier than working with desktop OpenGL.

Although WebGL code is interpreted rather than compiled, the differences in performance between WebGL and desktop OpenGL are less than one might expect. JavaScript engines in the browsers have improved markedly. More important to most classes is that not only do we not tend to assign problems where performance is an issue, but also, once the data are put on the GPU, the performance is not very dependent on how the data got there.

Although, in many ways, WebGL inherits a somewhat dated programming model from the original version of OpenGL, there are a few added features that make the API more modern. For example, as we shall show later, we can use images in the standard web image formats for textures. Also, JS often allows us to write clearer, more concise application code.

But perhaps the greatest advantage of using WebGL for teaching is that students can integrate their graphics with any other code that is compatible with HTML5. For example, they can use standards such as CSS for page design and jQuery for interaction. We have found that even though our students come to the class with almost no knowledge of computer graphics, they arrive with experience in these and other web packages. Consequently, the interactive side of computer graphics, which was dropping out of the course for reasons including the recent changes in desktop OpenGL, can be returned to the course with ease.

7.4 Application Organization and Ground Rules

A typical WebGL application is a combination of JS and HTML code. In addition, most real-world applications use CSS to design the pages and a package such as jQuery for interaction. An instructor must make some key decisions as to which packages to use and how to organize applications. Although many students come to the graphics class with web experience and are familiar with CSS and jQuery, many are not. Consequently, we decided to use only JS and HTML, although students were welcome to use CSS and jQuery if they preferred. Note that HTML and JS are sufficient to develop applications with mouse input, buttons, sliders, and menus.

The next issue is how to organize the code. Minimally, every application must have a description of the web page that will display the graphics and any interactive tools, the JS code for the graphics, and two shaders. Although, all four components can be put into a single HTML file, putting everything into a single file masks the different jobs of the constituent parts and hinders development of an application. We decided that each application should have an HTML file and a separate JS file. The HTML file contains the page description (including the canvas we will draw on), the shaders,* and the location of the other files needed. The JS file contains the geometry and rendering code (i.e., the graphics part of the application).

Finally we had to decide what, if any, helper code to give students. We decided, as we did in the examples in *OpenGL Insights*, to give students a function `initShaders()` that takes the identifiers of two shaders and produces a program object as

```
var program = initShaders(vertexShaderId, fragmentShaderId);
```

Our reason for doing so is that the various functions to read, compile, and link the shaders contribute little to understanding computer graphics and can be discussed later in the course.† We later provide a package with higher level functions, which is not needed initially and will be discussed later.

7.5 Modeling a Cube

Let's consider the example of rendering a colored cube. The example is simple but incorporates most of the elements needed in any WebGL application. We need both a vertex

---

* Shaders can also be in separate files. However, if we do so, some browsers will complain about cross origin requests if the application is run locally without a web server and then fails to load the shaders.
† A second version of `initShaders` that reads the shaders from files is also made available.

shader and a fragment shader. Although the output in Figure 7.1 shows both rotation and interaction, we can avoid introducing transformations at this point by specifying vertices in clip coordinates and aligning the faces of the cube with the coordinates axes. Thus, the vertex shader can simply pass through the positions and the fragment shader needs only to set the color. We will add interaction and rotation later.

We will build a cube model in the standard way through a vertex list. Consider the cube in Figure 7.2 with its vertices numbered as shown. The basic code for forming the cube data looks something like the following:

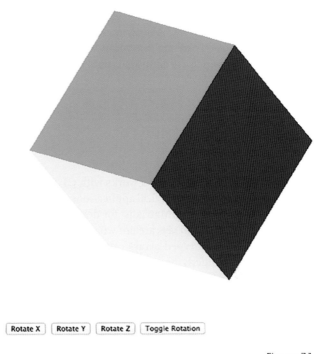

```
function colorCube()
{
    quad(1, 0, 3, 2);
    quad(2, 3, 7, 6);
    quad(3, 0, 4, 7);
    quad(6, 5, 1, 2);
    quad(4, 5, 6, 7);
    quad(5, 4, 0, 1);
}
```

Rotate X   Rotate Y   Rotate Z   Toggle Rotation

Figure 7.1

Rotating cube with button control.

The quad() function puts the vertex locations into an array that is sent to the GPU. The colorCube() function is the same whether we use WebGL or desktop OpenGL, JavaScript, or C/C++. However, when we examine the WebGL version of quad(), we see the differences from OpenGL and some of the key decisions we made in redesigning our course. The first major issue we must confront is how to handle arrays.

7.6 JavaScript Arrays

JavaScript has only three atomic types: a single numeric type (a 64-bit floating-point number), strings, and Booleans. Everything else is an object. Objects inherit from a prototype and have methods and attributes. Consider what happens if we use a JS array for our data as in the code

Figure 7.2

Representing a cube.

```
var vertices = [
  -0.5, -0.5,
  -0.5,  0.5,
   0.5,  0.5,
   0.5, -0.5
];
```

If we attempt to send these data to the GPU by

```
gl.bufferData(gl.ARRAY_BUFFER, vertices, gl.STATIC_DRAW);
```

where gl is our WebGL context, we will get an error message because WebGL expects a simple C-like array comprising floating-point numbers.

There are two main ways to get around this problem. One is to use JS typed arrays, which are equivalent to C-type arrays, as in the code:

```
var vertices = new Float32Array([
  -0.5, -0.5,
  -0.5,  0.5,
   0.5,  0.5,
   0.5, -0.5
]);
```

This method works with gl.bufferData(). Typed arrays have become standard for doing numerical applications with JavaScript as they are more efficient. There are packages such as glMatrix.js* for doing basic linear algebra with typed arrays. However, we believe typed arrays are not what we want to use for teaching computer graphics. Code developed with typed arrays often looks like C code with many loops, which tend to obscure the underlying geometric operations we try to stress in the course. Using JS arrays allows us to employ array methods to produce clean, concise code. We get around this problem of sending data to the GPU by using a function flatten() that takes as input a JS array and produces a typed array of floats as in the following:

```
gl.bufferData(gl.ARRAY_BUFFER, flatten(vertices), gl.STATIC_DRAW);
```

Consider now the implementation of the quad() function in the JS file. The vertices and vertex colors can be specified by JS arrays as[†]

```
var vertices = [
  [-0.5, -0.5,  0.5, 1.0],
  [-0.5,  0.5,  0.5, 1.0],
  [0.5,  0.5,  0.5, 1.0],
  [0.5, -0.5,  0.5, 1.0],
  [-0.5, -0.5, -0.5, 1.0],
  [-0.5,  0.5, -0.5, 1.0],
  [0.5,  0.5, -0.5, 1.0],
  [0.5, -0.5, -0.5, 1.0]
];
var vertexColors = [
  [0.0, 0.0, 0.0, 1.0], //black
  [1.0, 0.0, 0.0, 1.0], //red
  [1.0, 1.0, 0.0, 1.0], //yellow
  [0.0, 1.0, 0.0, 1.0], //green
  [0.0, 0.0, 1.0, 1.0], //blue
  [1.0, 0.0, 1.0, 1.0], //magenta
  [0.0, 1.0, 1.0, 1.0], //cyan
  [1.0, 1.0, 1.0, 1.0] //white
];
```

Now if the point and color arrays are initialized as empty JS arrays

---

* https://github.com/toji/gl-matrix
[†] Note that using our flatten() function allows us to nest the data in the two following examples—something that is possible with JS arrays but is not possible with typed arrays.

```
var points = [];
var colors = [];
```

and quad() uses two triangles for each quadrilateral, then quad() becomes

```
function quad(a, b, c, d)
{
    var indices = [a, b, c, a, c, d];

    for (var i = 0; i < indices.length; ++i) {
        points.push(vertices[indices[i]]);
        colors.push(vertexColors[a]);
    }
}
```

If we use a triangle strip, we can use the more efficient

```
var indices = [b, c, a, d];
```

Although typed arrays are generally more efficient, we prefer to teach with JS arrays. In an intro course, we generally use small examples and don't change the data very often. Consequently, once the data get to the GPU, it does not matter much whether we used typed arrays or JS arrays to form the data. For teaching computer graphics, this code is much clearer than the equivalent in desktop OpenGL or with typed arrays in WebGL.

7.7 The HTML File

The HTML file has three main parts. The first part contains the shaders. We can identify each type of script in the HTML file and we assign an identifier to each shader so that we can refer to them in the JS file. Although every implementation of OpenGL ES 2.0, and thus WebGL 1.0, must support medium precision in the fragment shader (see Chapter 8), we are still required to have a precision declaration in the fragment shader:

```
<!DOCTYPE html>
<html>
<head>

<script id = "vertex-shader" type = "x-shader/x-vertex">

attribute vec4 vPosition;
attribute vec4 vColor;
varying vec4 fColor;

void main()
{
  fColor = vColor;
  gl_Position = vPosition;
}
</script>

<script id = "fragment-shader" type = "x-shader/x-fragment">
precision mediump float;

varying vec4 fColor;
```

```
void
main()
{
  gl_FragColor = fColor;
}
</script>
```

The second part of the HTML reads in four text files. The first is a set of standard utilities available on the web* that allow us to set up the WebGL context. The second file contains a JS function that will read, compile, and link two shaders into a program object. The third file contains the flatten() function. The fourth file is our JS application file.

```
<script src = "../Common/webgl-utils.js"></script>
<script src = "../Common/initShaders.js"></script>
<script src = "../Common/flatten.js"></script>
<script src = "cube.js"></script>
</head>
```

Finally, we set up the HTML5 canvas and give it an identifier so that it can be referred to in the JS file:

```
<body>
<canvas id = "gl-canvas" width = "512" height = "512">
Oops... your browser doesn't support the HTML5 canvas element
</canvas>
</body>
</html>
```

7.8 The JS File

We now return to the JS file, which is organized much like a desktop OpenGL application. Thus, there is a lot of initialization to set up buffers, form a program object, and get data to the GPU. Once all that is done, the rendering function is very simple. We focus on the key parts that differ from desktop OpenGL. The first of these is the setting of the WebGL context.

The first step is to set up a WebGL context using the setupWebGL() function in the utility package. This function takes as input the canvas identifier we specified in the HTML file:

```
var gl;//WebGL context

window.onload = function init()
{
    var canvas = document.getElementById("gl-canvas");

    gl = WebGLUtils.setupWebGL(canvas);
    if (!gl) {alert("WebGL isn't available");}
```

The initialization function is executed once all the files have been loaded through the onload event. The WebGL context is a JS object that includes all the WebGL functions and identifiers. Note that the window object is global.

---

* https://code.google.com/p/webglsamples/source/browse/book/webgl-utils.js?r=41401f8a69b1f8d32c6863ac8c1953c8e1e8eba0

Next, we add the vertex position data as we did in the previous section. We specify a viewport and clear color using the standard OpenGL/WebGL functions. There is one important difference, however. The functions and parameters are members of the WebGL context whose name we chose (var gl) when we set up the context.

```
gl.viewport(0, 0, canvas.width, canvas.height);
gl.clearColor(0.0, 0.0, 0.0, 1.0);
```

Using the initShaders() function, we create a program object:

```
var program = initShaders(gl,"vertex-shader", "fragment-shader");
gl.useProgram(program);
```

Loading the data onto the GPU and associating JS variables with shader variables is the same as with desktop OpenGL:

```
var bufferId = gl.createBuffer();
gl.bindBuffer(gl.ARRAY_BUFFER, bufferId);
gl.bufferData(gl.ARRAY_BUFFER, flatten(vertices), gl.STATIC_DRAW);

var vPosition = gl.getAttribLocation(program, "vPosition");
gl.vertexAttribPointer(vPosition, 4, gl.FLOAT, false, 0, 0);
gl.enableVertexAttribArray(vPosition);
```

and we do the same for the vertex colors. Finally, we call our render function:

```
    render();
};
function render() {
  gl.clear(gl.COLOR_BUFFER_BIT);
  gl.drawArrays(gl.TRIANGLES, 0, numVertices);
  requestAnimationFrame(render);
}
```

Note that because this example is static, we don't need the requestAnimation-Frame(). We include it here as it will be needed in almost all other applications.

7.9 MV.js

We provide students with a matrix-vector package, MV.js, which is included into the HTML5 file:

```
<script src = "../Common/MV.js"></script>
```

MV.js defines the standard types contained in GLSL (vec2, vec3, vec4, mat2, mat3, mat4) and functions for creating and manipulating them. Thus, we can create the data for our cube program by

```
var vertices = [
  vec4(-0.5, -0.5, 0.5, 1.0),
  vec4(-0.5, 0.5, 0.5, 1.0),
  vec4( 0.5, 0.5, 0.5, 1.0),
  vec4( 0.5, -0.5, 0.5, 1.0),
```

```
    vec4(-0.5, -0.5, -0.5, 1.0),
    vec4(-0.5,  0.5, -0.5, 1.0),
    vec4( 0.5,  0.5, -0.5, 1.0),
    vec4( 0.5, -0.5, -0.5, 1.0)
];
```

without altering the `quad()` function and send the data to the GPU as before.

```
gl.bufferData(gl.ARRAY_BUFFER, flatten(vertices), gl.STATIC_DRAW);
```

Although this example is simple and may not illustrate the advantages of using MV.js, it lays the foundation for dealing with geometric types. Later, when we discuss algorithms for various graphics operations, we examine choices as to where to carry out various graphics operations. For example, we can compute per-vertex lighting in the application or in the vertex shader. We can also carry out per-fragment lighting in the fragment shader. In all three cases, by using MV.js, students can test the options with virtually identical code.

Also included in MV.js are functions for producing the matrices that were part of the fixed-function pipeline, including transformation functions (`rotate()`, `translate()`, `scale()`) and viewing functions (`ortho()`, `frustum()`, `perspective()`, `lookAt()`). These functions form the matrices, which can then be sent to the shaders, for example:

```
var modelViewMatrix = lookAt(eye, at, up);
var projectionMatrix = ortho(left, right, bottom, ytop, near, far);

gl.uniformMatrix4fv(modelViewMatrixLoc, false,
        flatten(modelViewMatrix));
gl.uniformMatrix4fv(projectionMatrixLoc, false,
        flatten(projectionMatrix));
```

or we can create matrices starting with an identity matrix as in

```
var instance = mat4();
instance = mult(instance, translate(displacement));
```

An instructor might want to delay using MV.js until the topics have been covered, but using MV.js frees students from writing a lot of repetitious core.

7.10 Input and Interaction

Thus far, we have only shown that WebGL enables us to run code that is very similar to desktop OpenGL in a browser. That in itself is a major plus for using WebGL for teaching computer graphics. At least as important is the ability to teach *interactive* computer graphics rather than just rendering—something that was becoming more and more difficult as desktop OpenGL evolved.

The underlying problem is that desktop OpenGL requires additional libraries to support interaction and to provide the glue between OpenGL and the local window system. These libraries are constantly changing, are not standardized, and became less useful once many functions were deprecated, starting with OpenGL 3.1. With WebGL, we can use a variety of packages, such as jQuery, that have become standard for developing web applications and are not coupled to WebGL.

Even without extra packages, it is very simple to bring in basic interactive devices including menus, sliders, text boxes, and buttons with just HTML and JS. For example a button for a toggle can be done with one line in the HTML file:

```
<button id = "myButton">A Toggle</button>
```

and in the JS file we change a Boolean variable `toggle` by

```
var a = document.getElementById("myButton");
a.addEventListener("click", function(event){toggle = !toggle;});
```

In fact, we can do the equivalent with just a single line in the HTML, file but we prefer to separate the page description from the action.

A slider is almost as simple. Here's one where a variable `speed` in the JS file goes from 0 to 100 and requires a minimum step size of 10 and starts with a value of 50. In the HTML file, use the line

```
<input id = "slider" type = "range"
       min = "0" max = "100" step = "10" value = "50"/>
```

and then in the JS file

```
document.getElementById("slider").onchange = function(event) {
       speed = 100 - event.target.value;};
```

7.11 Textures and Render-to-Texture

The texture mapping functions in WebGL are those from ES with one important addition—namely, that we can easily use images that are in the standard web formats (GIF, JPEG. PNG). For example, the texture mapped cube in Figure 7.3 uses the image specified using the image tag

```
<img id = "texImage" src = "SA2011_black.gif"
hidden></img>
```

in the HTML file and then configuring the texture in the JS file by

```
var image = document.getElementById
  ("texImage");
gl.texImage2D(gl.TEXTURE_2D, 0, gl.RGB,
        gl.RGB, gl.UNSIGNED_BYTE, image);
```

Some browsers support video formats allowing animated texture mapping using MPEG files. It is also possible to use a canvas element as a texture.

When we combine the ease of bringing in images with off-screen rendering, especially using render-to-texture, it opens up many possibilities for assignments and projects, even in a first class. Some examples include image processing, simulation, agent-based modeling, and GPGPU.

Figure 7.3

Texture mapped cube.

7.12 Discussion

Compared with 20 years of teaching with desktop OpenGL, the course has been a huge success as measured by student feedback and by the quality of the required term projects. For example, "traditional" term projects such as building small CAD systems were enhanced by the ability to use web packages such as jQuery rather than trying to put together a user interface with the GLUT and GLEW OpenGL libraries. What was more interesting was that students were able to explore GPGPU in areas such as image processing and algorithm design, thus enhancing their understanding of both computer science fundamentals and the power of modern GPUs. There are some questions that came up during and after the course that merit some discussion. Some of these issues were presented at SIGGRAPH 14 [Cozzi 14].

7.12.1 Should the Standard Curriculum Be Modified?

Although the core topics in an introductory course are well established, over the years, not only has more material become important but the emphasis also has changed. With programmable GPUs many topics that were considered advanced or non-realtime are now easily programmed in shaders. One of the unexpected outcomes of our class was that some students did projects that were less traditional computer graphics and more GPGPU. What these students gained was a realization of the computing power of the GPU, which was far more than for the students who did more traditional projects. Our approach at this point is to get through much of the traditional material on geometry, shading, and lighting a little faster so as to leave more time for interaction and discrete methods. A better long-term goal is to move toward two standard courses, one close to the traditional course and a second on GPU computing.

7.12.2 What Other JS Issues Haven't We Addressed?

When planning the switchover to WebGL, a major concern was using JavaScript rather than C or C++. JS is still not well regarded in academic CS and CE departments. Unfortunately, much of that bias is based on earlier versions of the language. In practice, the students had little trouble with picking up the language (although the same cannot be said for many instructors!). Note that students who chose to program in Java and used translators to convert the code to JS wound up creating large amounts of inefficient code.

There are some "gotchas" in JS that should be discussed. One is that everything other than the three basic types is an object, and objects inherit from prototypes. A related issue is that JS is a very large language and there are many ways to construct objects. These topics should be discussed, perhaps with a discussion of scope in JS. These issues are well addressed in *JavaScript: the Good Parts* [Crockford 08], which can be either a reference or a required textbook for the course.

7.12.3 What about Other Types of Shaders?

Because WebGL is an implementation of OpenGL ES 2.0, it lacks geometry, tessellation, and compute shaders. Nor will any of these shaders be included in WebGL 2.0. For a first course in computer graphics, most instructors will not need any of these shaders. At worst, they can be discussed and sometimes be replaced by clever code.

For now, implementing algorithms with fragment shaders for applications, such as image processing, using render-to-texture can demonstrate the power of GPU computing without having compute shaders.

7.12.4 Why Not three.js?

three.js (threejs.org) is a powerful scene graph API built on top of WebGL and is an alternate API for a graphics course.* Some students chose to use it for their term projects. However, there are some good reasons that it is not the best choice for a course in computer graphics for students in computer science and engineering. A scene graph API is primarily for constructing models as opposed to working with the underlying rendering concepts. Consequently, three.js and other scene graph APIs are better suited for CAD-like courses and survey courses.

7.12.5 A Final Word

How we can teach computer graphics is rapidly changing. Within the past 3 years, we have moved our course first to shader-based desktop OpenGL and then to WebGL. Although we cannot deny that these changes involved a lot of work, the course is greatly improved. The ability to run across virtually all platforms and devices with the same application is a powerful incentive for making the change, as is the ability to integrate with other web applications and to develop the application within almost any browser.

As for some of the weaknesses of the present API, we look forward to some upcoming changes that will go a long way toward overcoming any objections to using WebGL. One is the new versions of JavaScript (ES6), which will bring JavaScript closer to other programming languages. Second, there is a clear path forward for WebGL. WebGL 2.0 should be available by the time this book is published. Looking ahead slightly further, we can look at OpenGL ES 3.1, which has been released. Among its new features is compute shaders. Combined with ES6, WebGL will provide an even more powerful platform for teaching computer graphics in the coming years.

Bibliography

[Angel 13] E. Angel, "Teaching Computer Graphics with Shader-Based OpenGL in OpenGL Insights." In *OpenGL Insights,* P. Cozzi and C. Riccio (Ed.), CRC Press, Boca Raton, FL, 2013, 3–16.

[Angel 15] E. Angel and D. Shreiner, *Interactive Computer Graphics* (7th ed.), Pearson Education, New York, 2015.

[Cozzi 14] https://github.com/pjcozzi/Articles/blob/master/SIGGRAPH/2014/Teaching-Intro-and-Advanced-Graphics-with-WebGL-Small.pptx

[Crockford 08] D. Crockford, *JavaScript: the Good Parts*, O'Reilly, Sebastopol, CA, 2008.

[Flanagan 11] D. Flanagan, *JavaScript: The Definitive Guide*, O'Reilly, Sepastopol, CA, 2011.

[McGuire 14] http://casual-effects.blogspot.com/2014/01/an-introduction-to-javascript-for.html.

---

* www.udacity.com/course/cs291

SECTION III
Mobile

One of WebGL's biggest strengths is that it runs on both desktop and mobile devices, including iOS, Android, and Windows Phone. This allows us to reach the widest audience with a single codebase. In fact, in many cases, our applications will run just fine on mobile just by adding touch events or relying on the emulated mouse events. However, a deeper understanding of WebGL on mobile can help us to write more reliable and faster code. In Chapter 8, "Bug-Free and Fast Mobile WebGL," Olli Etuaho draws from his experience at NVIDIA to present tips for mobile WebGL development, including browser tools for mobile testing, profiling, and debugging; the pitfalls of shader precision and framebuffer color attachments formats; improving performance and power by reducing WebGL calls, optimizing shaders, and reducing bandwidth; and selecting a mobile-friendly WebGL engine.

8

Bug-Free and Fast Mobile WebGL

Olli Etuaho

8.1 Introduction

WebGL availability on mobile browsers has recently taken significant leaps forward. On leading mobile platforms, WebGL has also reached a performance and feature level where it is a feasible target for porting desktop apps. Often, content written on desktop will simply work across platforms. Still, special care needs to be taken to make a WebGL application run well on mobile devices. There are some nuances in the WebGL specification that may go unnoticed when testing only on desktop platforms, and limited CPU performance means that JavaScript and API usage optimizations play a much larger role than on desktops and high-end notebooks. Mobile GPU architectures are also more varied than desktop ones, so there are more performance pitfalls to deal with (Figure 8.1).

This chapter focuses on shader precision, which is not widely understood in the community and thus has been a frequent source of bugs even in professionally developed WebGL libraries and applications. I will also touch upon some other common sources of bugs, and provide an overview of optimization techniques that are particularly useful when writing applications with mobile in mind. As ARM SoCs make inroads into notebooks, such as the recent Tegra K1 Chromebooks, this material will become relevant even for applications that are only targeting a mouse-and-keyboard form factor.

Figure 8.1

An example of what is possible with WebGL on mobile platforms: a physically based rendering demo running at 30 FPS in Chrome on an NVIDIA Shield Tablet (Tegra K1 32-bit).

8.1.1 Developer Tools

Modern browsers include many developer tools that help with improving a WebGL app's mobile compatibility and performance. In Chrome 42, *JavaScript CPU Profile, Device mode,* and *Capture Canvas Frame* tools are particularly useful for mobile WebGL development. They also have equivalent alternatives in Firefox, where we can use the *Profiler* tab of dev tools, the *Responsive Design Mode*, and the *Canvas* tab or the *WebGL Inspector* extension. Some equivalent tools are also found in Internet Explorer's F12 developer tools and Safari's Web Inspector. Familiarize yourself with these tools in your preferred development environment, as I'll be referring to them throughout the rest of the chapter. I'm going to use the tools found in Chrome as an example, but often enough the basic processes can be generalized to tools provided by other browsers.

The Device mode tool mostly helps with getting viewport settings right and making sure the web application's UI can be used with touch. It does not emulate how the GPU behaves on the selected device, so it does not completely eliminate the need for testing on actual hardware, but it can help when getting started with mobile development. Firefox's Responsive Design Mode does not have the touch features of Chrome's Device mode, but can still also help.

The JavaScript CPU Profile and Capture Canvas Frame tools help with finding application bottlenecks and optimization opportunities. In Chrome 39, Capture Canvas Frame is still experimental and needs to be enabled from the flags page[*] with `#enable-devtools-experiments` and from developer tools settings. It's possible to use these inside the Inspect Devices tool[†] to perform the profiling on an Android device connected to your development workstation using USB.

---

[*] chrome://flags
[†] chrome://inspect

Firefox also has some tools that don't have exact counterparts in Chrome. In the about:config settings page in Firefox, there are two WebGL flags that are particularly useful. One is `webgl.min_capability_mode`. If that flag is enabled, all WebGL parameters such as maximum texture size and maximum amount of uniforms are set to their minimum values across platforms where WebGL is exposed. This is very good for testing if very wide compatibility with older devices is required. More on capabilities can be found from the "Capabilities Exposed by getParameter" section of this chapter.

The other interesting Firefox flag is `webgl.disable-extensions`. This is self-explanatory; it is good for testing that the application has working fallbacks in case some WebGL extension it uses is not supported.

8.2 Feature Compatibility

8.2.1 Shader Precision

Shader precision is the most common cause of issues when WebGL content developed on desktop is ported to mobile. Consider the GLSL example in Listing 8.1, which is designed to render a grid of randomly blinking circles as in Figure 8.2.

It looks like the developer is aware that some mobile devices do not support highp precision in fragment shaders [ESSL100 §4.5.2], so the first three lines intended to set mediump precision when running on mobile devices have been included. But there is already the first small error: including `#ifdef GL_ES` is unnecessary in WebGL, where the GL_ES macro is always defined. It is only useful if the shader source is directly used in both OpenGL and WebGL environments.

However, the critical error is much more subtle, and it might come as a surprise that the pseudorandom number generator is completely broken on many mobile devices. See Figure 8.3 for the results the shader produces on one mobile chip.

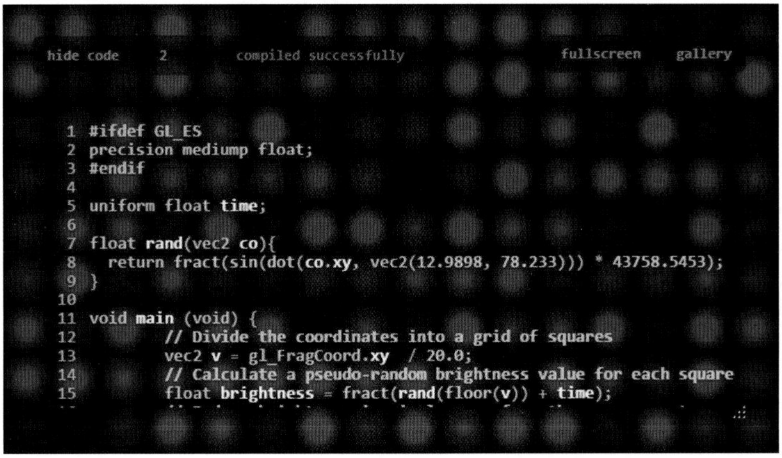

Figure 8.2

Intended effect of Listing 8.1, captured on a notebook with an NVIDIA GeForce GPU.

Figure 8.3

The example shader running on a device which includes 16-bit floating-point hardware.

The reason for the different rendering is that the lowest floating-point precision most desktop GPUs implement in hardware is 32 bits, and all floating-point calculations regardless of GLSL precision run on that same 32-bit hardware. In contrast, the mobile device represented in Figure 8.3 includes both 32-bit and 16-bit floating-point hardware, and calculations specified as lowp or mediump get performed on the more power-efficient 16-bit hardware.

The differences between mobile GPUs get a lot more varied than that, so on other devices we can see other kinds of flawed rendering. The GPUs use different bit counts in their floating-point representation, use different rounding modes, and may handle subnormal numbers differently [Olson 13]. This is all according to the specification, so the mobile devices are not doing anything wrong by performing the calculations in a more optimized way. The WebGL specification is just very loose when it comes to floating-point precision.

The calculation that goes wrong in this case is on line 8. The result from $\sin(x)$ is in the range [−1, 1], so $\sin(x)$ * 43758.5453 will be in the range [−43758.5453, 43758.5453]. The float value range in the minimum requirements for mediump precision is [−16384, 16384], and values outside this range might be clamped, which is the first possible cause of rendering issues. Even more importantly, the relative precision in the minimum requirements for mediump floats is 2^{-10}, so only the first four decimals will count. For example, the closest possible representation of 1024.5 in minimum mediump precision is 1024.0, so fract(1024.5) computed in mediump precision returns 0.0, and not 0.5. To get a better feel for how lower precision floating-point numbers behave, see this low-precision floating point simulator.*

---

* http://oletus.github.io/float16-simulator.js/

8. Bug-Free and Fast Mobile WebGL

There are two ways to fix Listing 8.1. The simplest way is to replace the first three lines with:

```
precision highp float;
```

and be done with it. As a result, the shader either works exactly as on common desktop GPUs, or does not compile at all if the platform does not support highp in fragment shaders. This might raise worries about abandoning potential users with older devices, but the majority of mobile devices that have reasonable WebGL support also fully support highp. This includes recent SOCs from Qualcomm, NVIDIA, and Apple. In particular, all OpenGL ES 3.0 supporting devices fully support highp [ESSL300 §4.5.3], but many older devices also include support for it. In data collected by WebGL Stats in the beginning of December 2014, 85% of mobile clients reported full support for highp in fragment shaders.

Listing 8.1 Example of a shader with precision issues.

```
#ifdef GL_ES
precision mediump float;
#endif

uniform float time;

float rand(vec2 co) {
   return fract(sin(dot(co.xy, vec2(12.9898, 78.233))) * 43758.5453);
}

void main (void) {
   //Divide the coordinates into a grid of squares
   vec2 v = gl_FragCoord.xy /20.0;
   //Calculate a pseudo-random brightness value for each square
   float brightness = fract(rand(floor(v)) + time);
   // Reduce brightness in pixels away from the square center
   brightness * = 0.5 - length(fract(v) - vec2(0.5, 0.5));
   gl_FragColor = vec4(brightness * 4.0, 0.0, 0.0, 1.0);
}
```

In vertex shaders, it is recommended to use highp exclusively, since it is universally supported and vertex shaders are the bottleneck less often than fragment shaders. The better performance that can be attained on a minority of mobile GPUs in particularly vertex-heavy applications usually isn't worth the risk of bugs from using mediump in vertex shaders.

Going the highp route does sacrifice some performance and power usage along with device compatibility. A better solution would be to change the pseudorandom function. In this case, the fix needs to be verified on a platform that supports performing floating-point calculations in a lower precision. One option is using suitable mobile hardware. To find out which kind of hardware you have, see the *shader precision* section of the khronos webgl Information page.*

Another convenient option is to use software emulation of lower precision floats. The ANGLE library implements such emulation by mutating GLSL shaders that are passed to WebGL. When the mutated shaders are run, they produce results that are very close to

---

* https://www.khronos.org/registry/webgl/sdk/tests/extra/webgl-info.html

what one would see on most mobile devices. The emulation carries a large performance cost and may prevent very complex shaders from being compiled successfully, but is typically fast enough to run content targeting mobile platforms at interactive rates on the desktop. Chrome 41 was the first browser to expose this emulation through a command line flag--emulate-shader-precision. You'll find more information on precision emulation on the GitHub repository of WebGL Insights.[*]

In the case of the example code, simply reducing the last coefficient in the pseudorandom function will make the shader work adequately in mediump precision. The code with the fixes in place is found in Listing 8.2.

Listing 8.2 The example shader with precision-related issues fixed.

```
precision mediump float;

uniform float time;

float rand(vec2 co) {
    return fract(sin(dot(co.xy, vec2(12.9898, 78.233))) * 137.5453);
}

void main (void) {
    //Divide the coordinates into a grid of squares
    vec2 v = gl_FragCoord.xy /20.0;
    //Calculate a pseudo-random brightness value for each square
    float brightness = fract(rand(floor(v)) + time);
    // Reduce brightness in pixels away from the square center
    brightness * = 0.5 - length(fract(v) - vec2(0.5, 0.5));
    gl_FragColor = vec4(brightness * 4.0, 0.0, 0.0, 1.0);
}
```

The OpenGL ES Shading Language spec mandates that if both a vertex shader and a fragment shader reference the same uniform, the precision of the uniform declarations must match. This can sometimes complicate making precision-related fixes, especially if the WebGL shader code is not written directly but rather generated by some tool.

If a project is hit by a compiler error because of uniform precision mismatch, it is best to fix this by making the precision of the two uniform declarations the same by any means available, rather than using two different uniforms with different precision settings to work around the issue. Using two uniforms hinders the maintainability of the code and adds CPU overhead, which is always to be avoided on mobile platforms. Keep this issue in mind when making decisions about shader development tools. Code written in ESSL is much easier to convert to other languages than vice versa, so ESSL is a good choice for a primary shading language.

Tip: Using mediump precision in fragment shaders provides the widest device compatibility, but risks corrupted rendering if the shaders are not properly tested.

Tip: Using only highp precision prevents corrupted rendering at the cost of losing some efficiency and device compatibility. Prefer highp especially in vertex shaders.

---

[*] https://github.com/WebGLInsights/WebGLInsights-1

Tip: To test device compatibility of shaders that use mediump or lowp precision, it is possible to use software emulation of lower precision.

Tip: `fract()` is an especially risky function when being run at low precision.

Tip: Remember to define sampler precision when sampling from float textures.

8.2.2 Render Target Support

Attempting to render to an unsupported color buffer format is another common source of bugs when running WebGL content on mobile platforms. Desktop platforms typically support many formats as framebuffer color attachments, but 8-bit RGBA is the only format for which render target support is strictly guaranteed in WebGL [WebGL]. In the mobile world, one may find GLES 2.0 devices that only implement this bare minimum required by the spec. With devices that support GLES 3.0 the situation is improved, and both 8-bit RGBA and RGB are guaranteed to be supported in WebGL [GLES3.0 §3.8.3].

However, one needs to tread even more carefully when using floating-point render targets, which are available in WebGL through `OES_texture_float` and `WEBGL_color_buffer_float`. The `OES_texture_float` extension, which includes the possibility of implicit render target support, is somewhat open to interpretation, and the level of support varies. Some mobile devices do not support rendering to floating-point textures at all, but others are better: some OpenGL ES 3.0 devices can expose rendering to 16-bit float RGBA, 16-bit float RGB, and 32-bit float RGBA, just like browsers on desktop. Note that support for rendering to 32-bit float RGB textures was recently removed from the WebGL specifications altogether to unify the platform, but some browsers on desktop may still include this legacy feature.

The level of render target support on different platforms is summarized in the following table for the most interesting formats. It is organized by the native APIs that WebGL implementations use as backends on different platforms. On older mobile devices made before 2013, the API supported is typically OpenGL ES 2.0. On mobile devices starting from late 2013, the API supported is typically OpenGL ES 3.0 or newer. On Apple computers, starting from some 2007 models, and recent Linux PCs, the API is OpenGL 3.3 or newer, and on Windows, browsers typically implement WebGL using a DirectX 9 or DirectX 11 backend. The Windows render target support data here are from ANGLE, which is the most common backend library for WebGL on Windows.

Format/platform	OpenGL ES 2.0	OpenGL ES 3.0/3.1	OpenGL 3.3/newer	ANGLE DirectX
8-bit RGB	Maybe	Yes	Yes	Yes
8-bit RGBA (1)	Yes	Yes	Yes	Yes
16-bit float RGB	Maybe (3)	Maybe (3)	Maybe	Yes (5)
16-bit float RGBA	Maybe (3)	Maybe (3, 4)	Yes	Yes
32-bit float RGB (2)	No	No (6)	Maybe	Maybe (5)
32-bit float RGBA	No	Maybe (4)	Yes	Yes

Notes:

(1) Guaranteed by WebGL specification, even if it's not guaranteed by GLES2.

(2) Support was removed from `WEBGL_color_buffer_float` recently.

(3) Possible to implement using `EXT_color_buffer_half_float`.

(4) Possible to implement using `EXT_color_buffer_float`.

(5) Implemented using the corresponding RGBA format under the hood.

(6) Possible to implement by using RGBA under the hood, but browsers don't do this.

To find out which texture formats your device is able to use as framebuffer color attachments, use the Khronos WebGL information page.*

Tip: When using an RGB framebuffer, always implement a fallback to RGBA for when RGB is not supported. Use `checkFramebufferStatus`.

8.2.3 Capabilities Exposed by getParameter

The WebGL function `getParameter` can be used to determine capabilities of the underlying graphics stack. It will reveal the maximum texture sizes; how many uniform, varying, and attribute slots are available; and how many texture units can be used in fragment or vertex shaders. The most important thing is that sampling textures in vertex shaders may not be supported at all—the value of `MAX_VERTEX_TEXTURE_IMAGE_UNITS` is 0 on many mobile devices.

The `webgl.min_capability_mode` flag in Firefox can be used to test whether an application is compatible with the minimum capabilities of WebGL. Note that this mode does not use the minimums mandated by the specification, as some of them are set to such low values that they are never encountered in the real world. For example, all devices where WebGL is available support texture sizes at least up to 1024×1024, even if the minimum value in the spec is only 64×64. OpenGL ES 3.0 devices must support textures at least up to 2048×2048 pixels in size.

8.3 Performance

When it comes to performance, one thing in particular sets mobile GPUs apart from their desktop counterparts: the limited main memory bandwidth that is also shared with the SoC's CPU cores [Pranckevičius 11; Merry 12]. This has led Imagination, Qualcomm, and ARM to implement a tiled rendering solution in hardware, which reduces the main memory bandwidth usage by caching a part of the framebuffer on-chip [Merry 12]. However, this approach also has drawbacks: It makes changing the framebuffer or blending state particularly expensive. The NVIDIA Tegra family of processors has a more desktop-like GPU architecture, so there are fewer surprises in store, but as a result the Tegra GPUs may be more sensitive to limited bandwidth in some cases.

In the case of WebGL, the CPU also often becomes a bottleneck on ARM-based mobile devices. Running application logic in JavaScript costs more than if the same thing was implemented in native code, and the validation of WebGL call parameters and data done by the browser also uses up CPU cycles. The situation is made worse by the fact that WebGL work can typically be spread to at most a few CPU cores, and single-core performance is seeing slower growth than parallel-processing performance. The rate of single-core performance gains started slowing down on desktop processors around 10 years ago [Sutter 09], and recently mobile CPUs have started to show this same trend.

Even before the past few years, the performance of mobile GPUs has grown faster than performance of mobile CPUs. To use performance figures Apple has used in marketing the iPhone as an example, their graphics performance has increased nearly 90% per year on average, whereas their CPU performance has grown only by 60% per year on average.

---

* https://www.khronos.org/registry/webgl/sdk/tests/extra/webgl-info.html

This means that optimizations targeting the CPU have become more important as new hardware generations have arrived.

8.3.1 Getting Started with Performance Improvements

Performance improvements should always start with profiling where the bottlenecks are. For this, we combine multiple tools and approaches. A good starting point is obtaining a JavaScript CPU profile from a target device using browser developer tools. Doing the profiling on a few different devices is also useful, particularly if it is suspected that an application could be CPU limited on some devices and GPU limited on others. A profile measured on desktop doesn't help as much: The CPU/GPU balance is usually very different, and the CPU bottlenecks can be different as well.

If idle time shown in the profile is near zero, the bottleneck is likely to be in JavaScript execution, and it is worthwhile to start optimizing application logic and API usage and seeing whether this has an effect on performance. Application logic is unique to each application, but the same API usage patterns are common among many WebGL applications, so see what to look for in the next section. This is going to be especially beneficial if the CPU profile clearly shows WebGL functions as bottlenecks. See Figure 8.4 for an example. Each of the optimizations detailed in the following sections can sometimes yield more than 10% performance improvements in a CPU-bound application.

The GitHub repository for WebGL Insights contains a CPU-bound drawing test to demonstrate the effect of some of these optimizations. The test artificially stresses the CPU to simulate CPU load from application logic and then issues WebGL draw calls with

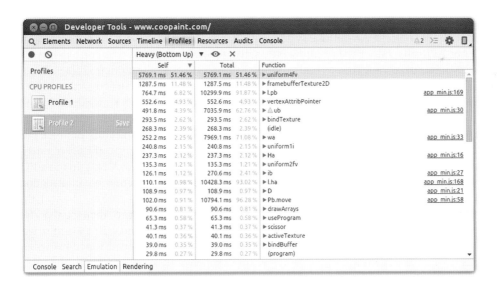

Figure 8.4

Example of a JavaScript CPU profile measured while doing heavy operations in the WebGL painting application CooPaint on a Nexus 5 (2013 phone). Some WebGL functions consume a large portion of CPU time. This application would very likely benefit from API usage optimizations such as trimming the number of `uniform4fv` calls.

options for different optimizations. The test results for each optimization given later in this chapter were measured on a Nexus 5 (2013 phone with Qualcomm Snapdragon 800) and a Shield Tablet (2014 tablet with NVIDIA Tegra K1 32-bit).

If we see plenty of idle time on the CPU executing JavaScript, our application may be GPU bound or main memory bandwidth bound. In this case, we should look for ways to optimize shaders or think of different ways to achieve the desired rendering result. Full-screen effects are usually a good place to start optimizing in this case [McCaffrey 12].

8.3.2 Optimizing API Usage

Every WebGL call has some CPU overhead associated with it. The underlying graphics driver stack needs to validate every command for errors, manage resources, and possibly synchronize data between threads [Hillaire 12]. On top of that, WebGL duplicates a lot of the validation to eliminate incompatibility between different drivers and to ensure stricter security guarantees. On some platforms, there is also overhead from translating WebGL calls to a completely different API, but on mobile platforms this is less of a concern since the underlying API is typically OpenGL ES, which WebGL is derived from. For more details, see Chapter 2.

Since every API call has overhead, performance can often be improved by reducing the number of API calls. The basics are simple enough: An application should not have unnecessary "get" calls of any kind, especially `getError`, or frequent calls requiring synchronization like `readPixels`, `flush`, or `finish`. Drawing a chunk of geometry with WebGL always has two steps: Setting the GL state required for drawing and then calling `drawElements` or `drawArrays`.

Setting the GL state can be broken down further into choosing a shader program by calling `useProgram`, setting flags and uniforms, and binding resources like textures and vertex buffer objects. An optimal system will change the GL state with the minimal number of API calls, rather than redundantly setting parts of the state that don't change for every single draw call. To start with a simple example, if an application needs blending to be disabled for all draw calls, it makes a lot more sense to call `gl.disable(gl.BLEND)` once at the beginning of drawing rather than repeatedly before every single draw call.

For binding resources, there are a few distinct ways to reduce API calls. Often the simplest way is to use vertex array objects, or VAOs, that can significantly improve performance on mobile devices. They are an API construct developed specifically to enable optimization, and are not to be confused with vertex buffer objects, or VBOs. In our CPU-bound drawing test, FPS increase from using VAOs was 10%–13%. For more information, see [Sellers 13].

A VAO encapsulates the state related to bound vertex arrays—that is, the state that's set by `bindBuffer`, `enable/disableVertexAttribArray`, and `vertexAttrib-Pointer`. By using VAOs we can replace calls to the aforementioned functions with a single `bindVertexArrayOES` call. VAOs are available in WebGL 1.0 with the `OES_vertex_array_object` extension, which is widely supported on mobile devices. As of early 2015, more than 80% of smartphone and tablet clients recorded by WebGL Stats have it.*

---

* http://webglstats.com/

Adding a fallback for devices without VAO support is also straightforward. Let's call the code that binds buffers and sets vertex attrib pointers related to a specific mesh the *binding block*. If VAOs are supported, the code should initialize the VAO of each mesh using the binding block. Then, when the mesh is drawn, the code either binds the VAO if VAOs are supported, or executes the binding block if VAOs are not supported. The only case where this becomes more complicated is when there's a different number of active vertex attribute arrays for different meshes—then the code should add `disable-VertexAttribArray` calls where appropriate. For a complete code example, see an explanation of VAOs[*] or an implementation of a fallback path in SceneJS.[†]

Lowering the number of vertex buffers helps to reduce CPU usage if VAOs are not a good fit for the code for some reason. This can be done by interleaving different types of vertex data for the same object: If we have, for example, positions, texture coordinates, and normals for each vertex, they can all be stored in the same vertex buffer in an interleaved fashion. In our CPU-bound drawing test that uses four vertex attributes, interleaving the attributes increased the FPS around 4%. The downside is that interleaving the data needs to be either handled by the content creation pipeline or done at load time; the latter may marginally slow down loading. Interleaving three attributes for a million vertices in a tight JS loop had a cost of around 200 ms on a Nexus 5 (2013 phone).

In some special cases, geometry instancing can also be used, but that is typically much harder to integrate into a general-purpose engine than using VAOs, for example. It is only feasible if drawing objects is decoupled enough from the application logic for each object. Yet another way to reduce resource binding overhead is to combine textures of different objects into large texture atlases, so that less texture binding changes are needed. This requires support from the content authoring pipeline and the engine, so implementing it is also quite involved.

Uniform values are a part of the shader program state. If we are using the same shader program to draw multiple objects, it is possible that some of the objects also share some of the same uniform values. In this case, it makes sense to avoid setting the same value to a uniform twice. Caching every single uniform value separately in JavaScript may not be a performance win, but if we can group, for example, lighting-related uniform values together, and only update them when the shader used to draw or the lighting changes, this can yield a large performance improvement. In our CPU-bound drawing test, updating one `vec3` uniform had a cost on the order of 1 microsecond on the mobile devices tested. This is enough to make extra uniform updates add up to a significant performance decrease if they are done in large quantities.

Reducing the number of `useProgram` calls is often also possible by sorting draw calls so that the ones using the same program are grouped together. This also helps to reduce redundant uniform updates further. We may even benefit from using parts of the uniform state as a part of the sort key.

Software-driven deferred rendering is still a long shot on most mobile devices, but using `WEBGL_draw_buffers` brings it a bit closer to within reach. The performance improvement from using `WEBGL_draw_buffers` varies, but it reduces CPU usage, since draw

---

[*] http://blog.tojicode.com/2012/10/oesvertexarrayobject-extension.html
[†] https://github.com/xeolabs/scenejs/blob/v4.0/src/core/display/chunks/geometryChunk.js

calls only need to be issued once, and also reduces vertex shader load and depth buffer accesses. There should be clear benefits at least in complex scenes [Tian 14].

To find out which of these optimizations we might be able to make, it's useful to get a complete picture of our application's API usage by using Chrome's Capture Canvas Frame tool. The tool will list the sequence of WebGL calls done by our application to render a single frame. It helps to see which commands to set GL state are common between multiple draw calls and how many calls are spent doing vertex array setup that might be better accomplished by using VAOs.

Tip: Sort draw calls according to shader program state, and reap the benefits by removing redundant `useProgram` calls and uniform updates.

Tip: Use vertex array objects (VAOs) and interleave static vertex data. This will save a lot of API calls.

8.3.3 Optimizing Shader Execution

When we suspect that our application is shader-bound, we can always perform a simple test to see if this really is the case: Replace all of the shaders with trivial ones that only render a single recognizable color and measure the performance. If the performance is significantly changed, the application is likely shader-bound—either by GPU computation or by texture fetches performed by the shaders.

There are a few different ways to optimize a shader-bound application. Many of the optimizations presented here are discussed in Pranckevičius's excellent SIGGRAPH talk [Pranckevičius 11]. The talk has some more details about micro-optimization, but here we will concentrate on more general guidelines.

If geometry is drawn in a random order, fragment shaders might be run unnecessarily for fragments that eventually get obscured by something else in the scene. This is commonly known as overdraw, and it can be reduced with the help of early z-test hardware built into the GPU. This requires sorting opaque geometry front-to-back, so that the fragments that are obscured by something else get processed last [McCaffrey 12]. The sorting should be done on a relatively coarse level in order to avoid spending too much time on it and moving the bottleneck to the CPU. A too fine-grained depth sort can also increase the number of shader state changes needed to render the scene, so it's often a balancing act between sorting by shader state and sorting by depth. Also see Chapter 10.

The front-to-back sorting helps to improve performance on all but Imagination's PowerVR series of GPUs, which implement deferred fragment shading in hardware [Merry 12]. Also note that if shaders modify the fragment depth value or contain discard statements, the early z-test hardware won't help. Another possible technique is to use a z-prepass, it's typically not worth the overhead [McGuire 13].

Another optimization strategy is to switch to a simpler lighting model. This may mean sacrificing some quality, but are often alternative ways to do things, such as using look-up textures instead of shader computation. However, since mobile GPUs are typically even more bandwidth-bound than they are computation-bound, this kind of change may not always be a performance win. Recent trends are also toward higher computational performance in mobile GPUs, with bandwidth being the more fundamental limit [McCaffrey 12].

Recent Tegra processors perform especially well in shader computation, so they become texture-bandwidth-bound more easily than other mobile GPUs.

Finally, if we want to squeeze out the last bits of performance, we can try lowering the precision of our shader computations. Just remember to test any shaders that use mediump or lowp carefully for correctness. The performance benefits that can be gained from this vary greatly between platforms: on NVIDIA Tegra K1, precision does not matter at all, and on the Qualcomm Adreno GPU line, precision only makes a small difference, but large differences have been reported on some other mobile GPUs.

8.3.4 Reducing Bandwidth Usage

There are a few different ways to reduce bandwidth usage in a WebGL app, and some of them are WebGL-specific. First of all, we should make sure that extra copies don't happen when the WebGL canvas is composited. Keep the `preserveDrawingBuffer` context creation attribute in its default value false. In some browsers, also having the `alpha` context creation attribute as false may enable more efficient occlusion culling; it communicates to the browser compositor that the canvas element is opaque and does not need to be blended with whatever is underneath it on the page. However, if we need a static background other than a flat color for our WebGL content, it may make sense to render the background using other HTML elements and only render the parts that change on every frame on the main WebGL canvas, keeping `alpha` as true.

The more obvious ways to reduce bandwidth are reducing texture or framebuffer resolution. Reducing texture resolution can sometimes be done without sacrificing too much visual quality, especially if the textures have mostly low-frequency content [McCaffrey 12]. Many mobile device displays have extreme pixel densities these days, so it's worthwhile to consider whether rendering at the native resolution is worth losing performance or reducing the visual quality in other ways. Also see Chapter 14.

Implementing full-screen effects in an efficient way or avoiding them altogether can also enable huge bandwidth savings [McCaffrey 12; Pranckevičius 11]. In particular, it is better to combine different postprocessing filters into a single shader or add simple post-processing effects directly into the shaders used to render geometry, when possible.

Using lots of small polygons also costs bandwidth on tiler architectures, since they need to access the vertex data separately for each tile [Merry 12]. Optimizing models to minimize the vertex and triangle count helps on these GPUs.

8.3.5 Choosing a WebGL Engine for Mobile

Using a higher level library or engine instead of writing directly against the WebGL API makes development more efficient. This also applies to mobile applications. The downside is that there's some overhead from using the library, and if the library is not a good fit for the content, its effects on performance on mobile platforms may be disastrous. There are other use cases for WebGL besides rendering 3D scenes, but here I'll give a brief overview of libraries that are specifically targeting 3D rendering from the mobile perspective.

three.js is one of the most popular WebGL libraries. However, it's not usually the best fit for developing applications for mobile. three.js offers a very flexible API, but this comes at a cost: Lots of dynamic behavior inside the library adds a whole lot of CPU overhead. At the time of writing, it had only a few of the optimizations detailed in this chapter. It often does redundant or inefficient updates to the GL state, particularly to uniform values, and due to how it's structured, this is unlikely to improve much without some very significant changes. Still, if we're planning to render only relatively simple scenes with few objects, it's possible to get good performance from three.js, and it allows us to easily use advanced effects using built-in and custom shaders. It does have depth sorting to help with avoiding overdraw. With three.js, it is recommended to use the `BufferGeometry` class instead of the Geometry class to improve performance and memory use.

Some slightly less widely used libraries may be a better fit for your application, depending on what kind of content you're planning to render. Babylon.js (Chapter 9) and Turbulenz (Chapter 10) have both been demonstrated to be able to run fairly complex 3D game content in an ARM SoC environment. Babylon.js has a robust framework for tracking GL state and making only the necessary updates, which saves a lot of CPU time. Turbulenz does some of this as well, and also uses VAOs to improve performance. Be aware of bugs though: Demos using older versions of Babylon.js do not do justice to the newer versions, where lots of issues appearing on mobile platforms have been fixed. Turbulenz still has some open issues that affect mobile specifically, so more work is required to ship a complete product with it.

SceneJS is another production-proven WebGL library, which was heavily optimized in version 4.0. It lacks some of the flexibility of its peers and may not be suited to all types of 3D scenes, but makes up for this by using VAOs and state tracking to optimize rendering. If the application content consists of large amounts of static geometry, SceneJS may be an excellent pick.

Native engines that have been Emscripten-compiled to JavaScript (Chapter 5) are sadly not yet a viable alternative when targeting mobile devices. The memory cost is simply too high on devices with less than 4 GB of memory, and the runtimes have a lot of overhead. Future hardware generations, improvements to the typed array specification, JavaScript engines, and the 3D engines themselves might still change this, but so far better results can almost always be had by using JavaScript directly.

8.4 Resources

See this chapter's materials in the WebGL Insights GitHub repo for additional resources, such as:

- A JavaScript calculator that demonstrates the behavior of low-precision floats
- Tools that emulate lowp and mediump computations on desktop
- A WebGL test that demonstrates CPU optimizations

Acknowledgments

Thanks to Shannon Woods, Florian Bösch, and Dean Jackson for providing useful data for this chapter.

Bibliography

[ESSL100] Robert J. Simpson. "The OpenGL ES Shading Language, Version 1.00.17." Khronos Group, 2009.

[ESSL300] Robert J. Simpson. "The OpenGL ES Shading Language, Version 3.00.4." Khronos Group, 2013.

[GLES3.0] Benj Lipchak. "OpenGL ES, Version 3.0.4." Khronos Group, 2014.

[Hillaire 12] Sébastien Hillaire. "Improving Performance by Reducing Calls to the Driver." In *OpenGL Insights*. Edited by Patrick Cozzi and Christophe Riccio. Boca Raton, FL: CRC Press, 2012.

[McCaffrey 12] Jon McCaffrey. "Exploring Mobile vs. Desktop OpenGL Performance." In *OpenGL Insights*. Edited by Patrick Cozzi and Christophe Riccio. Boca Raton, FL: CRC Press, 2012.

[McGuire 13] Morgan McGuire. "Z-Prepass Considered Irrelevant." Casual Effects Blog. http://casual-effects.blogspot.fi/2013/08/z-prepass-considered-irrelevant.html, 2013.

[Merry 12] Bruce Merry. "Performance Tuning for Tile-Based Architectures." In *OpenGL Insights*. Edited by Patrick Cozzi and Christophe Riccio. Boca Raton, FL: CRC Press, 2012.

[Olson 13] "Benchmarking Floating Point Precision in Mobile GPUs." http://community.arm.com/groups/arm-mali-graphics/blog/2013/05/29/benchmarking-floating-point-precision-in-mobile-gpus, 2013.

[Pranckevičius 11] Aras Pranckevičius. "Fast Mobile Shaders." http://aras-p.info/blog/2011/08/17/fast-mobile-shaders-or-i-did-a-talk-at-siggraph/, 2011.

[Resig 13] John Resig. "ASM.js: The JavaScript Compile Target." http://ejohn.org/blog/asmjs-javascript-compile-target/, 2013.

[Sellers 13] Graham Sellers. "Vertex Array Performance." OpenGL SuperBible Blog. http://www.openglsuperbible.com/2013/12/09/vertex-array-performance/, 2013.

[Sutter 09] Herb Sutter. "The Free Lunch Is Over: A Fundamental Turn toward Concurrency in Software." Originally appeared in *Dr. Dobb's Journal* 30(3), 2005. http://www.gotw.ca/publications/concurrency-ddj.htm, 2009.

[Tian 14] Sijie Tian, Yuqin Shao, and Patrick Cozzi. "WebGL Deferred Shading." Mozilla Hacks Blog. https://hacks.mozilla.org/2014/01/webgl-deferred-shading/, 2014.

[WebGL] Khronos WebGL Working Group. "WebGL Specification." https://www.khronos.org/registry/webgl/specs/latest/1.0/, 2014.

SECTION IV
Engine Design

Most graphics developers—and I believe most WebGL users—build applications on top of game and graphics engines that hide the details of the graphics API and provide convenient abstractions for loading models, creating materials, culling, level of detail, camera navigation, and so on. This allows developers to focus on their specific problem domain and not, for example, have to write shaders and model loaders by hand.

Given how many developers build on top of engines, it is important that engine developers make efficient use of the WebGL API and provide simple and flexible abstractions. In this section, the creators of some of the most popular WebGL engines and platforms describe engine design and optimizations.

Common themes throughout this section are understanding the CPU and GPU overhead of different WebGL API functions such as setting render state and uniforms, strategies for minimizing their cost without incurring too much CPU overhead, and shader pipelines to generate GLSL from higher-level material descriptions and shader libraries. Given the wide array of use cases, we see both ubershaders and shader graphs, as well as offline, online, and hybrid shader pipelines.

Babylon.js is an open-source WebGL engine from Microsoft focused on simplicity and performance. It was used, for example, to build UbiSoft's *Assassin's Creed Pirates Race*. In Chapter 9, "WebGL Engine Design in Babylon.js," David Catuhe, the engine's lead developer, explains the design philosophy, public API, and internals, including its render loop, shader generation with an ubershader for optimized shaders and fallbacks for low-end hardware, and caches for matrices, states, and programs.

I first learned about Turbulenz, an open-source WebGL game engine, at WebGL Camp Europe 2012. David Galeano presented it, including an impressive demo of rendering Quake 4 assets with culling, state sorting, and many other optimizations. In Chapter 10, "Rendering Optimizations in the Turbulenz Engine," David goes into detail on the engine's optimizations. Using the Oort Online game as a use case, David covers culling, organization of the renderer, and efficiently sorting draw calls to minimize WebGL implementation and overdraw overhead.

Given the popularity of Blender as a 3D modeling and animation tool, it should be no surprise that an open-source WebGL engine, Blend4Web, was developed to easily bring Blender content to the web. In Chapter 11, "Performance and Rendering Algorithms in

Blend4Web," Alexander Kovelenov, Evgeny Rodygin, and Ivan Lyubovnikov highlight parts of Blend4Web's implementation. Performance topics include runtime batching, culling, LOD, and performing physics in Web Workers with time synchronization. Blend4Web's shader pipeline is detailed with a focus on using a custom preprocessor to create new directives for code reuse. Finally, ocean simulation and shading is covered including LOD, waves, and several fast and plausible shading techniques for refraction, caustics, reflections, foam, and subsurface scattering.

As for publishing 3D models to the web, Sketchfab has gained a lot of traction in this space. It supports a variety of model formats and converts them to a common 3D scene format for rendering in their WebGL viewer. In Chapter 12, "Sketchfab Material Pipeline: From File Variations to Shader Generation," Cedric Pinson and Paul Cheyrou-Lagrèze explain how materials are handled in Sketchfab. They describe the challenges of implementing robust material support for a wide array of 3D model formats, and present Sketchfab's material pipeline and optimizations. They describe how the scene is streamed to the viewer and how shaders are generated at runtime with a shader graph. The shader graph is a graph of nodes representing individual parts of a shader that are compiled together to generate a final shader, allowing developers to develop only individual nodes and still providing flexible materials for artists.

As shaders get larger, they become harder to manage. In Chapter 13, "glslify: A module system for GLSL," Hugh Kennedy, Mikola Lysenko, Matt DesLauriers, and Chris Dickinson present an elegant solution: glslify. They show how modeling after npm, the Node.js package manager, provides a clean and flexible system for reusing and transforming GLSL code on the server or as part of the build process.

A typical frame budget, 16 or 33 milliseconds, can go by quickly, especially if our engine has a large processing job it tries to perform in a single frame. In Chapter 14, "Budgeting Frame Time," Philip Rideout provides WebGL performance tips starting with a design for making all WebGL calls inside `requestAnimationFrame`. To stay within a frame budget, this chapter looks at strategies to amortize work across multiple frames, including using the new `yield` keyword in ECMAScript 6, offloading CPU-intensive tasks to Web Workers, and optimizing fillrate by using a low-resolution canvas.

9
WebGL Engine Design in Babylon.js

David Catuhe

9.1 Introduction

About a year ago, I decided to sacrifice all my spare time for a project I had had in mind for a very long time: creating a pure JavaScript 3D engine using WebGL. IE11 had just shipped with early WebGL support, which meant that all major modern browsers were now able to render accelerated 3D content. I had been writing 3D engines since I was 18; the very first one I wrote in C/C++ did all its rendering on the CPU. Then I switched to the Glide SDK (from 3DFX). It was my very first contact with 3D accelerated rendering and I was absolutely blown away by the raw power that I was able to control!

With Windows 95, I decided to port my engine to DirectX and, using this engine (named Nova), I founded a company in 2002 named Vertice, where I remained for 9 years as CTO and lead developer. Moving to Microsoft in 2012, I created a new engine called Babylon using Silverlight 5 and XNA. This engine was used as the basis of Babylon.js,* a pure JavaScript engine that relies on WebGL for rendering.

---

* http://www.babylonjs.com

Using my experience with 3D rendering, I tried to create an engine built upon two foundations:

- Simplicity
- Performance

For me, a successful framework is not about beauty of the code or heroic stuff that you are doing. It is all about simplicity of use. I consider Babylon.js successful when users can tell me that "this is easy to use and understand." Also, performance matters because we are dealing with real-time rendering.

With these two concepts in mind, I tried to create an architecture that will be both easy to use and performant. I really hope that web developers can use it without having to read documentation (or at least not too much!). This is why the engine is shipped with satellite tools:

- Playground*: The playground is a place where we can try to experiment with Babylon.js directly inside the browser. Examples are provided alongside a complete autocompletion and live documentation system that can help while we're typing (see Figure 9.1).
- CYOS†: CYOS (Create Your Own Shader) is a tool where we can experiment with creating shaders that we can then use with Babylon.js (see Figure 9.2).
- Sandbox‡: The sandbox is a page where we can drag and drop 3D scenes created with Blender 3D or 3DS Max exporters.

In this chapter, I explain how things work under the hood. Writing an engine is not just about rendering triangles and shaders. For the sake of readability and brevity, we focus on parts of the code related to WebGL.

Tip: Feel free to use the playground with code examples used during this chapter to get live results.

9.2 Global Engine Architecture

Before digging into the core of Babylon.js, let's take a step back to see the global picture as shown in Figure 9.3. The engine surface is wide and in order to better understand how everything is linked, it is important to get a vision of all involved actors.

The root of everything in Babylon.js is the `Engine` object. This is the link between the object model and WebGL. All orders sent to WebGL are centralized in the engine. To create it, developers just have to instantiate it with the rendering canvas used as a parameter:

```
var engine = new BABYLON.Engine(canvas);
```

---

* http://www.babylonjs.com/playground
† http://www.babylonjs.com/cyos
‡ http://www.babylonjs.com/sandbox

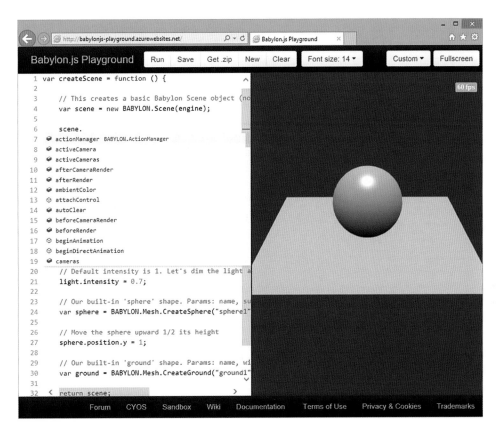

Figure 9.1

Playground with autocompletion screen opened.

The next important object is the Scene object. We can have as many scenes as we want. They are the container for all other objects (meshes, lights, cameras, textures, materials, etc.):

```
var scene = new BABYLON.Scene(engine);
```

Once the scene is created, we can start adding components like a camera (i.e., the user's point of view):

```
var camera = new BABYLON.ArcRotateCamera("Camera", 0, 0, 10, new BABYLON.
  Vector3(0, 0, 0), scene);
```

Camera's constructor needs the scene in order to attach to it. This is true for all actors except Engine.

Meshes contain geometry information (vertex and index buffers) alongside a world matrix:

```
var sphere = BABYLON.Mesh.CreateSphere("sphere1", 16, 2, scene);
```

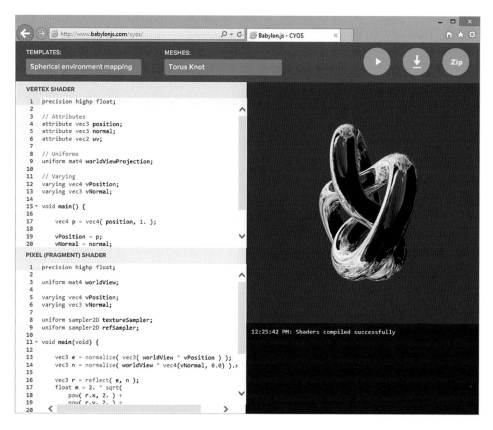

Figure 9.2

CYOS screen.

Shaders are handled by materials. The most general one is the `StandardMaterial`:

```
var materialSphere3 = new BABYLON.StandardMaterial("texture3", scene);
materialSphere3.diffuseTexture = new BABYLON.Texture("textures/misc.jpg",
   scene);
```

In Section 9.4, we'll see how shaders are controlled by materials. Listing 9.1 shows a simple scene involving a small cube and a material.

Figure 9.4 shows the result of executing the code in Listing 9.1

Figure 9.3

Global overview of Babylon.js classes and internal engines.

Listing 9.1 Creating a simple scene.

```
var scene = new BABYLON.Scene(engine);

var camera = new BABYLON.FreeCamera("camera1", new BABYLON.
   Vector3(0, 5, -10), scene);
camera.setTarget(BABYLON.Vector3.Zero());
```

```
camera.attachControl(canvas, false);

var light = new BABYLON.HemisphericLight("light1", new BABYLON.
   Vector3(0, 1, 0), scene);

var sphere = BABYLON.Mesh.CreateSphere("sphere1", 16, 2, scene);

sphere.position.y = 1;

sphere.material = new BABYLON.StandardMaterial("red", scene);
sphere.material.diffuseColor = BABYLON.Color3.Red();
```

These objects are the backbone of the framework and work together with several other objects (e.g., postprocess, shadows, collisions, physics, serialization, etc.) to enable different scenarios.

From the user point of view, the complexities of shaders and WebGL are completely hidden under the API surface.

When we call `scene.render()`, the following process is triggered:

- Using an octree, if activated, and frustum clipping, the scene establishes a list of visible meshes.
- The scene goes through all visible meshes and dispatches them into three groups:
  - Opaque meshes, sorted front to back
  - Alpha tested meshes, sorted front to back
  - Transparent meshes, sorted back to front

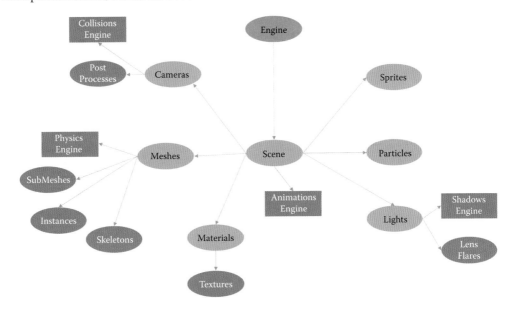

Figure 9.4

Simple scene.

- Each group is rendered using associated engine states, such as alpha blending. To render an individual mesh, the following process is executed:
  - Mesh's material is activated: Inner shader, samplers, and uniforms are transmitted to WebGL.
  - Mesh's index and vertex buffers are sent to WebGL.
- Draw command is executed (`gl.drawElements`).

Now that we have a clearer view over the engine, let's discuss the internal architecture and optimizations.*

9.3 Engine Centralized Access

During the rendering process, everything related to WebGL is handled by the *Engine* object. It was architected this way to centralize cache and states.

The Engine contains a reference to the WebGLContext object (`gl`). If we open the following url in Github, we will see that most of the code is about calling `gl.xxxxx` functions:

https://github.com/BabylonJS/Babylon.js/blob/master/Babylon/babylon.engine.js

The Engine is responsible for supporting the following WebGL features:

- Creating and updating vertex and index buffers
- Creating and updating textures (static, dynamic, video, render target)
- Shader creation, compilation, and linking into programs
- Communication with programs through uniforms and samplers bindings
- Buffers and textures bindings
- State management
- Cache management
- Deleting all WebGL resources
- Full-screen management

It also drives frame rendering with a simple function, called `runRenderLoop`, that uses `requestAnimationFrame` to render frame:

```
engine.runRenderLoop(function () {
    scene.render();
});
```

Capabilities detection is supported by a simple API to query for a specific extension:

```
if (engine.getCaps().s3tc) {...}
```

It keeps track of all compiled programs and all scenes to be able to clean everything when `engine.dispose()` is called.

---

* Complete code is available on Github: https://github.com/BabylonJS/Babylon.js

9.4 Smart Shader Engine

In order to achieve maximum performance, Babylon.js is built upon a system that tries to compile the most efficient shader for a given task.

Let's imagine we need to have a material with only diffuse color. From the user's point of view, the following code will do the job:

```
mesh.material = new BABYLON.StandardMaterial("red", scene);
mesh.material.diffuseColor = new BABYLON.Color3(1.0, 0, 0);
```

Under the hood, the `StandardMaterial` object creates a `BABYLON.Effect` object, which is responsible for everything related to shaders and program. This effect is connected to the engine in order to get vertex and fragment shaders compiled and linked into a program.

The `StandardMaterial` object uses only one big shader—the Uber Shader—that supports all features (e.g., diffuse, emissive, bump, ambient, opacity, fog, bones, alpha, Fresnel, etc.). We made this choice because it is far easier for us to work with one big shader instead of trying to have multiple small shaders for each specific task.

The drawback of using only one piece of code is removing an unwanted feature. In our previous example, we only needed the diffuse part of the code and nothing else. Using conditions with uniforms was not an option because this is not optimal for GPU performance in WebGL.

9.4.1 Removing Conditions from Compiled Programs

We decided to work with the compiler's preprocessors using #define. Listing 9.2 is part of the fragment shader used by `StandardMaterial`:

Listing 9.2 Fragment shader snippet used by `StandardMaterial`.*

```
//Lighting
vec3 diffuseBase = vec3(0., 0., 0.);
vec3 specularBase = vec3(0., 0., 0.);
float shadow = 1.;

#ifdef LIGHT0
#ifdef SPOTLIGHT0
    lightingInfo info = computeSpotLighting(viewDirectionW, normalW,
        vLightData0, vLightDirection0, vLightDiffuse0.rgb,
        vLightSpecular0, vLightDiffuse0.a);
#endif
#ifdef HEMILIGHT0
    lightingInfo info = computeHemisphericLighting(viewDirectionW,
        normalW, vLightData0, vLightDiffuse0.rgb, vLightSpecular0,
        vLightGround0);
```

* Complete shader is available on Github: https://github.com/BabylonJS/Babylon.js/blob/master/Babylon/Shaders/default.fragment.fx

```
#endif
#ifdef POINTDIRLIGHT0
    lightingInfo info = computeLighting(viewDirectionW, normalW,
        vLightData0, vLightDiffuse0.rgb, vLightSpecular0,
        vLightDiffuse0.a);
#endif
#ifdef SHADOW0
#ifdef SHADOWVSM0
    shadow = computeShadowWithVSM(vPositionFromLight0,
        shadowSampler0);
#else
    #ifdef SHADOWPCF0
    shadow = computeShadowWithPCF(vPositionFromLight0,
        shadowSampler0);
    #else
    shadow = computeShadow(vPositionFromLight0, shadowSampler0,
        darkness0);
    #endif
#endif
#else
    shadow = 1.;
#endif
    diffuseBase + = info.diffuse * shadow;
    specularBase + = info.specular * shadow;
#endif
```

Depending on the active options, all defines are gathered and the effect used by the
`StandardMaterial` is compiled. Listing 9.3 shows an example of how the material
configures the list of defines:

Listing 9.3 Preparing a list of defines.

```
if (this.diffuseTexture && BABYLON.StandardMaterial.
  DiffuseTextureEnabled) {
    if (!this.diffuseTexture.isReady()) {
        return false;
    } else {
        defines.push("#define DIFFUSE");
    }
}
```

Before compiling the effects, all defines are gathered and transmitted to the effect constructor (Listing 9.4):

Listing 9.4 Compiling an effect.

```
var join = defines.join("\n");
this._effect = engine.createEffect(shaderName, attribs, […], […],
                join, fallbacks, this.onCompiled, this.onError);
```

The effect communicates with the engine to get the final program using specific defines (Listing 9.5):

Listing 9.5 Compiling and linking a program using defines.

```
var vertexShader = compileShader(this._gl, vertexCode, "vertex",
  defines);
var fragmentShader = compileShader(this._gl, fragmentCode, "fragment",
  defines);

var shaderProgram = this._gl.createProgram();
this._gl.attachShader(shaderProgram, vertexShader);
this._gl.attachShader(shaderProgram, fragmentShader);

this._gl.linkProgram(shaderProgram);
```

Once this task is done, we are sure that the resulting program only contains the necessary code with no extra conditional statements.

9.4.2 Supporting Low-End Devices through Fallbacks

Along with the performance benefits, using defines as a way to control the final program gives us a way to provide fallbacks when shaders are too complex for specific hardware.

Babylon.js is a pure JavaScript library and thus can be used on mobile devices where GPUs are less powerful than on desktops. With simplicity in mind, I wanted to allow the user to get rid of the burden of handling these issues.

This is why the Effect object allows us to define a list of fallbacks. The idea is pretty simple: While preparing the list of defines to be used for the compilation, the developer can also set up a list of optional defines that can be removed if shaders are not successfully compiled.

The main reason why a working shader cannot be compiled on a given hardware is because of complexity: The number of instructions required to execute it is beyond the current limitation of the hardware. Based on this, the fallback system is used to remove some defines in order to produce simpler shaders. We can then automatically degrade the rendering in order to get a working shader.

It is up to the developer to define which options to remove. Since the fallback system is a multistage system, the developer can also define priorities (Listing 9.6):

Listing 9.6 Showing how Fresnel is configured as an optional feature.

```
var fresnelRank = 1;

if (this.diffuseFresnelParameters && this.diffuseFresnelParameters.
  isEnabled) {
    defines.push("#define DIFFUSEFRESNEL");
    fallbacks.addFallback(fresnelRank, "DIFFUSEFRESNEL");
    fresnelRank++;
}
```

```
if (this.opacityFresnelParameters && this.opacityFresnelParameters.
  isEnabled) {
    defines.push("#define OPACITYFRESNEL");
    fallbacks.addFallback(fresnelRank, "OPACITYFRESNEL");
    fresnelRank++;
}

if (this.reflectionFresnelParameters && this.reflectionFresnelParam-
  eters.isEnabled) {
    defines.push("#define REFLECTIONFRESNEL");
    fallbacks.addFallback(fresnelRank, "REFLECTIONFRESNEL");
    fresnelRank++;
}

if (this.emissiveFresnelParameters && this.emissiveFresnelParameters.
  isEnabled) {
    defines.push("#define EMISSIVEFRESNEL");
    fallbacks.addFallback(fresnelRank, "EMISSIVEFRESNEL");
    fresnelRank++;
}
```

Developers can define a rank when creating a fallback. The `Engine` object will try to compile shaders with all options and if this fails, it will try again with some options removed based on ranking (Listing 9.7):

Listing 9.7 Program compilation code with fallback system.*

```
try {
    var engine = this._engine;
    this._program = engine.createShaderProgram(vertexSourceCode,
        fragmentSourceCode, defines);
} catch (e) {
    if (fallbacks && fallbacks.isMoreFallbacks) {
        defines = fallbacks.reduce(defines);
        this._prepareEffect(vertexSourceCode, fragmentSourceCode,
            attributesNames, defines, fallbacks);
    } else {
        Tools.Error("Unable to compile effect: " + this.name);
        Tools.Error("Defines: " + defines);
        Tools.Error("Error: " + e.message);
    }
}
```

* Complete code is available on Github, in the effect and engine classes:
  https://github.com/BabylonJS/Babylon.js/blob/master/Babylon/babylon.engine.js
  https://github.com/BabylonJS/Babylon.js/blob/master/Babylon/Materials/babylon.effect.js

But even with the greatest shaders you can write, you also have to ensure that only required commands are sent to WebGL. This is where caching systems go up on stage.

9.5 Caches

Babylon.js contains many different caching systems. These systems are intended to reduce overhead implied by recreating or resetting something already done. Through our testing, we identified several places where caching could be useful:

- World matrices computation
- WebGL states
- Textures
- Programs
- Uniforms

9.5.1 World Matrices

World matrices are computed per mesh in order to get the current position/orientation/scaling. These matrices not only depend on the mesh itself but also on the hierarchy. Before rendering a mesh, we have to compute this specific matrix. Instead of computing it on every frame, we decided to cache it and only compute it when required. The caching overhead is insignificant compared to the cost of building a world matrix where we have to

- Compute rotation matrix
- Compute scaling matrix
- Compute translation matrix
- Compute pivot matrix
- Get parent's world matrix
- Compute billboarding if activated
- Multiply all these matrices

Tip: You can see the complete code for this here: https://github.com/BabylonJS/Babylon.js/blob/master/Babylon/Mesh/babylon.abstractMesh.js#L321

Instead of this long code, Babylon.js can check properties against cached values (Listing 9.8):

Listing 9.8 Comparing current values against cached values.

```
Mesh.prototype._isSynchronized = function () {
    if (this.billboardMode ! = = AbstractMesh.BILLBOARDMODE_NONE)
        return false;

    if (this._cache.pivotMatrixUpdated) {
        return false;
    }

    if (this.infiniteDistance) {
        return false;
    }

    if (!this._cache.position.equals(this.position))
        return false;
```

```
    if (this.rotationQuaternion) {
        if (!this._cache.rotationQuaternion.equals
            (this.rotationQuaternion))
        return false;
    } else {
        if (!this._cache.rotation.equals(this.rotation))
        return false;
    }

    if (!this._cache.scaling.equals(this.scaling))
        return false;
    return true;
};
```

9.5.2 Caching WebGL States

On low-end devices, changing WebGL states can be extremely expensive due to the inner nature of the WebGL state machine. State objects in Babylon.js are used to centralize state management. Thus, instead of directly changing a WebGL state, the engine changes the corresponding value in a state object. This state object is then used to apply effective updates just before a render. This saves a lot of unrequired changes—for example, when we go through a list of meshes that all require the same value for a specific state. Some browsers can take care of not overwriting an already set value but this behavior is pretty uncommon with, for example, mobile devices.

Listing 9.9 shows how a state object dedicated to alpha management works. An internal dirty flag is kept up to date and is used in order to decide if a specific internal state should be applied:

Listing 9.9 Partial alpha state management.

```
function _AlphaState() {
    this._isAlphaBlendDirty = false;
    this._alphaBlend = false;
}

Object.defineProperty(_AlphaState.prototype, "isDirty", {
    get: function () {
        return this._isAlphaBlendDirty;
    },
    enumerable: true,
    configurable: true
});

Object.defineProperty(_AlphaState.prototype, "alphaBlend", {
    get: function () {
        return this._alphaBlend;
    },
    set: function (value) {
        if (this._alphaBlend = = = value) {
            return;
        }
    }
```

9. WebGL Engine Design in Babylon.js

```
            this._alphaBlend = value;
            this._isAlphaBlendDirty = true;
        },
        enumerable: true,
        configurable: true
});

_AlphaState.prototype.apply = function (gl) {
    if (!this.isDirty) {
        return;
    }

    //Alpha blend
    if (this._isAlphaBlendDirty) {
        if (this._alphaBlend = = = true) {
            gl.enable(gl.BLEND);
        } else if (this._alphaBlend = = = false) {
            gl.disable(gl.BLEND);
        }

        this._isAlphaBlendDirty = false;
    }
};
```

The engine itself uses these states objects when `drawElements` is required as shown in Listing 9.10:

Listing 9.10 States objects applied just before `drawElements`.

```
Engine.prototype.applyStates = function () {
    this._depthCullingState.apply(this._gl);
    this._alphaState.apply(this._gl);
};

Engine.prototype.draw = function (indexStart, indexCount) {
    this.applyStates();

    this._gl.drawElements(this._gl.TRIANGLES, indexCount, this._
      gl.UNSIGNED_SHORT, indexStart * 2);
};
```

9.5.3 Caching Programs, Textures, and Uniforms

Babylon.js also provides a caching system for various data related to WebGL. For instance, each time an `Effect` object wants to compile a program, the `Engine` object checks if the program was not already compiled by keeping a list of active programs.

Textures are handled in the same way. Each time a new `Texture` object is instantiated, the `Engine` checks if the specific url was not loaded before in order to share resources.

We applied the same policy to uniforms as shown in Listing 9.11:

Listing 9.11 Setting a color3 uniform through *Effect* object.

```
Effect.prototype.setColor3 = function (uniformName, color3) {
    if (this._valueCache[uniformName] && this._valueCache[uniformName]
    [0] = = color3.r && this._valueCache[uniformName][1] = =
    color3.g && this._valueCache[uniformName][2] = = color3.b)
        return this;

    this._cacheFloat3(uniformName, color3.r, color3.g, color3.b);
    this._engine.setColor3(this.getUniform(uniformName), color3);

    return this;
};
```

Even this.getUniform is cached. All uniforms' locations are gathered when programs are compiled and linked in order to optimize uniforms' access as well.

9.6 Conclusion

This architecture where we try to automatically optimize for end users allowed us to create great demos like *Assassin's Creed Pirates Race* by UbiSoft (Figure 9.5) and a complete train attraction (Figure 9.6). See a complete list of astonishing demos at www.babylonjs.com.

Figure 9.5

Assassin's Creed Pirates Race by UbiSoft.

9. WebGL Engine Design in Babylon.js

Figure 9.6

Train simulator.

If you want to discuss this chapter with me, feel free to ping me on twitter (@deltakosh). I will be more than pleased to respond. Also, Babylon.js is an open-source engine, so if you want to contribute, please do it! The Github repository can be found here: https://github.com/BabylonJS/Babylon.js.

10

Rendering Optimizations in the Turbulenz Engine

David Galeano

10.1 Introduction

The Turbulenz Engine is a high-performance open source game engine available in JavaScript and TypeScript for building high-quality 2D and 3D games.

In order to extract maximum performance from both JavaScript and WebGL, the Turbulenz Engine needs to reduce waste to the minimum, for the benefit of both the CPU and the GPU. In this chapter, we focus on the rendering loop and the removal of waste originating from redundant and/or useless state changes. In order for this

Figure 10.1

The game *Oort Online.*

removal to be optimal, the engine employs several strategies at different levels, from high-level visibility culling, with grouping and sorting of geometries, to low-level state change filtering.

Throughout the chapter we give specific examples from our game *Oort Online* (Figure 10.1), a massive multiplayer game set in a sandbox universe of connected voxel worlds.

10.2 Waste

JavaScript is now a high-performance language. It is still not as fast as some statically typed languages, but fast enough for using WebGL efficiently in order to achieve high-quality 3D graphics at interactive frame rates. Both JavaScript and WebGL are perfectly capable of saturating a GPU with work to do. See [Echterhoff 14] for an analysis of WebGL and JavaScript performance compared to a native implementation.

However, saturating a CPU or GPU with work that is either redundant or useless would be a waste of resources, and identifying and removing that waste is our focus.

There are several kinds of waste that we classify into:

1. Useless work: doing something that has no visible effect
2. Repetitive work: doing the expensive state changes more than once per frame
3. Redundant work: doing exactly the same thing more than once in a row

We will tackle each one in turn, but first let's discuss the cost of removing unneeded work.

10. Rendering Optimizations in the Turbulenz Engine

10.3 Waste Avoidance

Waste avoidance is simply avoiding the production of waste. It works on the principle that the greatest gains result from actions that remove or reduce resource utilization but deliver the same outcome.

There is always a price to pay for removing waste. In order to save CPU and GPU time, we first need to use some CPU time, and there will always be a trade-off. Too much CPU time spent on filtering out useless work may actually make our application run slower than if there was no filtering at all, but the opposite is usually also true.

The Turbulenz Engine is data-oriented and mostly data-driven; some special rendering effects may be handwritten to call the low-level rendering API directly for performance reasons, but the bulk of the work is defined at runtime based on loaded data.

In general, it is much cheaper to avoid waste at a high level than at a lower level. The higher level has more context information to use and can detect redundancy at a higher scale. The lower levels can only deal with what is known at that instant. For example, the scene hierarchy and the spatial maps available at the high level together allow culling of nonvisible objects in groups, instead of having to check visibility individually for all objects.

The Turbulenz Engine avoids waste at different levels, each one trying to reduce work as much as possible for the lower ones. We will explain the different strategies used at every level.

10.4 High-Level Filtering

This is where the scene is managed, passing information to the middle-level renderer.

At this level, we mostly focus on trying to remove work that provides no usable result. Examples of this kind of waste are geometries that are not visible because they are

- Out of the view frustum
- Fully occluded by other geometries
- Too far away and/or are too small for the target resolution
- Fully transparent

This list also applies to other game elements like, for example, lights affecting the scene, animation of skinned geometries, entities AI, 3D sound emitters, etc.

To optimize away each of these cases, we require extra information that is only available at a high level, for example:

- View frustum
- Geometry AABB
- Dimensions of the rendering target
- Transparency information

The Turbulenz Engine filters out elements that are not visible using a two-step frustum culling system. First we cull the scene nodes and then their contents. Our scene hierarchy has an AABB on each node that contains renderable geometry or lights. Each scene node

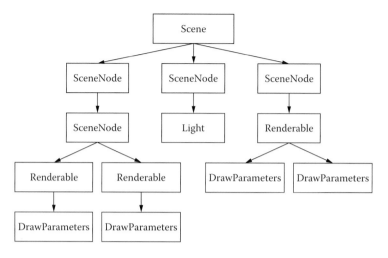

Figure 10.2

Scene example.

can contain an unlimited number of lights or renderable geometries. Figure 10.2 shows an example of a scene. These scene node AABBs are added to spatial maps, one for static objects and a separate one for dynamic objects. They are separate because they are updated with different frequency and we can use different kinds of spatial maps for each case. The spatial maps supported out of the box are

- AABB trees with different top-down building heuristics for static or dynamic objects
- Dense grids
- Sparse grids

For example, our game *Oort Online* uses AABB trees for static objects and dense grids for dynamic ones. Figure 10.3 shows a screen capture showing the renderables bounding boxes.

Once a node is deemed visible then we proceed to check for the visibility of each of the renderables and lights that it contains. When there is a single object on the node, then there is no need to check anything else and it is added to the visible set. When there are multiple objects, we check whether the node was fully visible in the frustum—if so, then all its contained objects are also visible; otherwise, we check visibility for each of them separately. For lights, we may go an extra step by projecting the bounding box into the screen to calculate how many pixels it would actually light, discarding the light or disabling its shadows' maps depending on its contribution to the scene. However, all this work only removes objects outside the view frustum; it does not do anything about occluded objects.

For occlusion culling, the main trade-off is that doing occlusion culling on the CPU is usually very expensive and hence only applied in a limited number of cases. Our game *Oort Online* is made of cubes and hence it is much simpler to calculate occluder frustums

Figure 10.3

Bounding boxes of visible renderables.

than in games using complex nonconvex geometry, but even in this case we only apply occlusion from chunks of $16 \times 16 \times 16$ blocks of opaque cubes closer to the camera. Unfortunately, it only helps in a limited number of situations—for example, looking into a mountain from its base where most of the voxels inside the mountain will be discarded. In other games we employed portals to find the potentially visible set of elements from a particular camera viewpoint.

Once we have removed as much useless work as possible without spending too much time on it, then it is time to pass that information to the next level down. The output from this level is a collection of arrays of visible objects: visible nodes, visible renderables, visible lights, visible entities, etc. We are going to focus just on the list of renderables.

10.5 Middle-Level Rendering Structures

Each visible renderable can represent several rendering geometries. Each rendering geometry is defined by a `DrawParameters` object. Managing the collection of visible rendering geometries for optimal dispatching is the role of the next level down.

`DrawParameters` contains all the information required to render a particular geometry with a particular shading technique:

- Vertex buffers: could be more than one
- Vertex offsets: start of the used region on the vertex buffers
- Vertex semantics: to match a particular vertex component with a particular shader input

- Index buffer: optional, for indexed primitives
- Index buffer offset: start of the used region on the index buffer
- Primitives count: the number of primitives to render
- Primitive type: triangle list, triangle strip, etc.
- Technique: the shading technique to be used to render this geometry
- Technique parameters: a dictionary containing the custom shading parameters for this geometry—for example, the world location, the material color, the diffuse texture, etc.
- User data object: used to group geometries by framebuffer, opacity, etc.
- Sort key: used to sort geometries belonging to the same group

Both user data and sort key are the keys to filtering out waste at this level.

10.5.1 Group Index on the User Data Object

The user data object can be used by the game to store anything about this particular instance of the geometry; the relevant part for this chapter is that the game generally uses this object to tell the renderer what particular group this geometry belongs to.

In hindsight, we should have named this property differently and we should have separated out its different components into different values. The group information is stored as an integer index into the array of groups that the renderer manages; this is game- and renderer-specific.

For example, these are the groups used for our game *Oort Online*:

- Prepass: for smoothing the normals in screen space on geometry close to the camera
- Opaque: for opaque voxels, entities, and opaque vegetation like tree trunks
- Alpha cutouts (or decals): for grass, leaves, etc.
- Transparent: for transparent voxels or entities
- Effects: for particle systems like the smoke or flames from the torches
- Water
- Lava
- Clouds

The different groups for nonopaque geometry that require alpha blending (water, lava, transparent, clouds, etc.) are rendered in different order depending on the camera direction and vertical position in order to get better visual results; otherwise, the different layers may render on top of each other incorrectly.

Other games may have more or fewer groups depending on the renderer. For example, games with dynamic shadows implemented using shadow mapping would have one or more groups for each of the shadow maps.

10.5.2 Sort Key

The `DrawParameters` sort key is a JavaScript number used to sort geometries belonging to the same group. As is often the case in JavaScript, it can be a floating-point value or an integer; it is up to the game renderer to set it correctly. This value can be used to sort the `DrawParameters` objects in ascending or descending order, depending on what makes more sense for its particular render group.

10. Rendering Optimizations in the Turbulenz Engine

For example, in the game *Oort Online*, the keys are calculated as follows:

- For opaque geometry or for additive-only blending:

```
(distance << 24) |
(techniqueIndex << 14) |
(materialIndex << 8) |
(vbIndex & 255)
```

The `distance` value is the distance to the camera near plane quantized into 64 logarithmic-scaled buckets.
The `techniqueIndex` value is the unique shading technique id.
The `materialIndex` value is the unique material id.
The `vbIndex` value is the unique vertex buffer id.

- For transparent geometry:

```
(distance * 1024 * 16) | 0
```

The `distance` value is the distance to the camera near plane in world units.

For opaque geometries, the main reason for sorting is performance. We build the sorting key in such a way that geometries with the same technique, material, or vertex buffer are one after the other in the sorted list. A perfect sorting would require a multilevel bucket hierarchy, which could be too expensive to build, while the sorting key provides a cheap enough solution being fast to generate and fast to sort by. The higher bits of the key are used to sort for the most important state change to optimize by, while the lower bits are left for the cheaper changes. The order will usually depend on the scene complexity. If overdraw is a significant issue, then we will use the distance to the camera on the higher bits, while in some other cases it is the shading technique that requires optimization, either because overdraw is not significant or because the hardware can deal with it efficiently—for example, on tile-based deferred GPUs.

For transparent geometries, the main reason is correctness. Transparent geometries are sorted by distance to get the correct rendering, from back to front. However, if the blending is additive-only, then this is not needed. In the case of our game *Oort Online*, the "effects" pass only contains geometries rendered with additive blending, which is correct no matter the order; this means that we can employ more efficient sorting for this group. Unfortunately many rendering effects do not fall into this category; in our game *Oort Online*, only a handful of particle systems use additive blending—for example, the sparks coming out of a torch.

In general, the Turbulenz Engine tries to use integers smaller than 30 bits in order to force the JavaScript JIT compilers into using integer operations for sorting instead of floating-point ones. Comparing integers is significantly cheaper than comparing floating-point values. The JIT compiler will optimize for either integers or floating-point values, but performance is more predictable if we use the same types everywhere, every time. For more information about low-level JavaScript optimizations, see [Wilson 12].

10.6 Middle-Level Filtering

This is where the renderer groups and sorts `DrawParameters` to be passed down to the low-level rendering functions.

This level focuses on removing repetitive work that is done multiple times but not in a row. Examples of this kind of waste include:

- Bind the same framebuffer more than once.
- Enable blending more than once.

To remove this waste, we rely on the game and the renderer to provide enough information to group and sort `DrawParameters` in the most efficient way to reduce the number of state changes to the minimum.

At this point it is worth explaining the real cost of making WebGL calls.

10.6.1 WebGL Costs

Making a function call is never totally free, but there are some functions that are far more expensive than others. Sometimes the real cost of function calls is not immediately obvious; they may have big performance side effects that only appear in combination with other function calls, and most WebGL calls are in this category. Older versions of rendering APIs like D3D9 or OpenGL ES 2 (which WebGL represents) do not actually represent how modern GPUs work. A lot of state and command translation and validation is lazily evaluated behind the scenes. Changing a state may imply just storing the new value and setting a dirty flag that is only checked and processed when we issue a draw call.

This lazy evaluation cost heavily depends on the hardware and the different software layers that drive it, and there are big differences between them; for example, changing frequently from opaque rendering to transparent rendering and vice versa by changing the blend function is very expensive on deferred tile rendering architectures because they force flushing the rendering pipeline for each change, which could be very wasteful if there is a significant amount of overdraw. Defining the specific costs of each change for each piece of hardware is complicated, but there are some general rules that should work fine in most hardware.

The following are in order of cost of change from high to low:

- Framebuffer
- Blend state
- Shader program
- Depth state
- Other render states
- Texture
- Vertex buffer
- Index buffer
- Shader parameters (uniforms)

This ordering may change significantly from hardware to hardware, but as generic rules they are good enough for a start. For each particular platform we should profile and adapt. For more in-depth information, see [Forsyth 08].

Even if the geometries are rendered in the right order to minimize state changes, there is still one big source of GPU overhead that we are also trying to get rid of at this level: overdraw.

10.6.2 Overdraw

Overdraw is one of the main sources of wasted effort on complex 3D environments; classified as useless work, some pixels are rendered but are never visible because other pixels are rendered on top of them. This is a massive waste in terms of data moving through the app and the graphics pipeline.

In theory this kind of waste is cheaper to be removed early on at a higher level; we should detect occluders and remove all occludes from the rendering list. However, this is a lot of work to do on the CPU and most occludees are only half covered, which means that we should, in theory, split the triangles into visible and not visible parts. And if we go down that route, then most of our CPU would be wasted on occlusion calculations while the GPU sits idle waiting for work to do.

Instead, we rely on grouping and sorting opaque geometry based on the distance to the camera near plane. By first rendering opaque geometry closer to the camera, we can use the depth test to discard shading for occluded pixels. It does not eliminate all the waste from vertex transformations and triangle rasterization, but at least it removes the cost of shading in a relatively cheap way for the CPU. This is one case where we optimize for the CPU, relying on the massive parallel performance of modern GPUs.

This optimization makes sense because of early-z reject hardware implementations that provide fast early-out paths for fragments that are occluded; some hardware even employs hierarchical depth buffers that are able to cull several occluded fragments at the same time. This optimization makes less sense on tile-based deferred architectures that only shade the visible fragment when required at the end of the rendering.

If the game is not CPU limited by the number of draw calls, then a z-only pass could be a good solution to remove most of the overdraw waste. Rendering only to the depth buffer can often be optimized to be significantly faster, but an excessive amount of draw calls is too often the norm for complex games, so this solution is not usually more optimal than just trying to sort opaque geometry front to back. A z-only pass is even more handicapped in JavaScript/WebGL because of the additional overhead compared to native code. And, as before, on tile-based deferred architectures, this solution may actually go against the optimizations done in hardware and result in more GPU overhead.

10.6.3 Benefits of Grouping

There are several reasons why we group `DrawParameters` into different buckets:

1. Because they are rendered into different framebuffers
   - For example, they gather screen depth or view space-per-pixel normal information into textures to be used by subsequent groups.
2. Because they need to be rendered in a specific order for correctness
   - For example, transparent geometry should be rendered after all the opaque geometry; otherwise, they may be blending on top of the wrong pixel.
3. Because they need to be rendered in a specific order for performance
   - For example, opaque geometry closer to the camera could be rendered in its own group before the distant ones.
   - For example, enabling and disabling blending may be quite expensive on some hardware, so we try to do it only once per frame.

Points 2 and 3 require clarification. There is some overlap between the sort key and the group index. In theory we could encode all the required information to sort `DrawParameters` for performance reasons into the sort key, but there is a performance limit on the key size and it is relatively easy to use more than 31 bits and then sorting starts to be expensive for big collections. By moving part of the key information into the group index you can increase the effective key size without penalties; this could be seen as some kind of high-level bucket sorting.

Once all our `DrawParameters` are grouped and sorted efficiently, then they are passed to the low-level API for rendering.

For more on draw calls ordering, see [Ericson 08].

10.7 Low-Level Filtering

This is where we dispatch the `DrawParameters` changes to WebGL. Big arrays of `DrawParameters` objects are passed down to the low-level rendering functions for dispatching.

This level focuses on removing redundant work that is done multiple times in a row. Examples of this kind of waste include:

- Binding the same vertex buffer repeatedly
- Binding the same texture on the same slot repeatedly
- Binding the same shading technique repeatedly

To avoid this redundant work, we shadow all the internal WebGL state, including uniforms for shader programs, avoiding changes that do not alter it. There is overhead for keeping this shadow of the WebGL state in both memory usage and CPU cost, but as the rendering has been sorted already to keep redundant changes close together, most of the time the check for superfluous work saves a lot of work due to the high overhead of the underlying WebGL implementations.

To reduce the cost of state checks as much as possible, we need to store the rendering data in optimal ways. The main rendering structure at this level is the Technique.

10.8 The Technique Object

A Technique object contains the required information for shading a given geometry:

- Vertex and fragment shaders
  - A unique program is linked for each combination and is cached and shared between techniques.
- Render states
  - Contains only the delta from our predefined default render states:
    - DepthTestEnable: true
    - DepthFunc: LEQUAL
    - DepthMask: true
    - BlendEnable: false

- BlendFunc: SRC _ ALPHA, ONE _ MINUS _ SRC _ ALPHA
- CullFaceEnable: true
- CullFace: BACK
- FrontFace: CCW
- ColorMask: 0xffffffff
- StencilTestEnable: false
- StencilFunc: ALWAYS, 0, 0xffffffff
- StencilOp: KEEP, KEEP, KEEP
- PolygonOffsetFillEnable: false
- PolygonOffset: 0, 0
- LineWidth: 1

- Samplers
  - The texture samplers that the shaders would require.
  - They are matched by name to textures from the technique parameters objects passed in by the DrawParameters object.
  - The sampler object contains only the delta from our predefined sampling states:
    - MinFilter: LINEAR _ MIPMAP _ LINEAR
    - MagFilter: LINEAR
    - WrapS: REPEAT
    - WrapT: REPEAT
    - MaxAnisotropy: 1

- Semantics
  - The vertex inputs that the vertex shader would require.
  - These are some of our predefined semantics:
    - POSITION
    - NORMAL
    - BLENDWEIGHT
    - TEXCOORD0

- Uniforms
  - The uniform inputs that the program would require
  - Matched by name to the values from the technique parameters object passed on by the DrawParameters object.

The Technique objects are immutable; if we want to use the same program with a different render state, we need to create a new Technique. This could potentially scale badly, but in practice none of our games use more than a couple dozen techniques. The main reason to have potentially too many techniques would be the need to support too many toggleable rendering configurations—for example, water with or without reflections—but in that case we can just load the techniques that we actually need.

The Technique object contains lots of information; when we change a shading technique, we are actually potentially changing many WebGL states, which is why sorting by technique is so important in many cases, although we still need to optimize it as much as possible.

10.9 Dispatching a Technique

When we have to dispatch a new Technique, we still need to reduce the state changes to the minimum. We keep track of the previously dispatched Technique and we only update the delta between the two.

- Program: only changed if different from the previous one
- Render states: the new state values are only applied if they are different from the values previously set. The render states set by the previous technique that are not present on the current technique are reset to their default values.
- Samplers: the new sampling values are only applied if they are different from the values previously set. The sampling states set by the previous technique that are not present on the current technique are reset to their default values.

Once the main states have been updated, the remaining changes are the buffers and the uniforms.

10.10 Dispatching Buffers

Vertex and index buffers are only dispatched if they differ from the current value set by the previous `DrawParameters` object. The Turbulenz Engine creates big buffers that are shared among different geometries, which helps to reduce the number of buffer changes.

Index buffers are easy; if the current buffer is different from the last one, we change it. If the `DrawParameters` does not contain an index buffer because it is rendering an unindexed primitive, then we just keep the old buffer active because it may be needed again in the future.

Vertex buffers are more complicated. For each vertex buffer, we match each of its vertex components to the relevant vertex shader semantic and then we check if, for that semantic, we already have used the same vertex buffer. If so, we do nothing; otherwise, we need to update the vertex attribute pointer for that semantic.

Shader semantics are hardcoded to match specific vertex attributes; for example, POSITION is always on the attribute zero. This makes dispatching simpler and faster because the same semantic will always be set to the same attribute, which means that in many cases we do not need to set a vertex attribute pointer again when rendering the same vertex buffer with multiple techniques. When the program is linked, we remap its attribute inputs to match our semantics table. This has potential issues when the vertex shader only supports a very limited number of attributes (which happens sometimes on mobile phones) and it means that the mapping between semantics and attributes is calculated right after creating the WebGL context; however, once it is done it never changes.

If vertex array objects (VAOs) are supported, then we build one for each combination of vertex buffers and index buffer present in the `DrawParameters` objects. As we share the buffers between many different geometries, the actual number of combinations is usually quite low. This allows us at dispatch time to simplify all the buffer checks to a single equality comparison between the current VAO and the previous one. Even when the VAOs are different, setting them with WebGL is cheaper on the CPU than setting all the different buffers and vertex pointer attributes, which makes them a big win for complex scenes.

10.11 Dispatching Textures

Textures are nontrivial to set in WebGL because of the stateful nature of WebGL. At loading time, we assign a specific texture unit to each sampler used by a shader. These units are shared by all the shaders; the first sampler on each shader will use unit zero, the second sampler on each shader will use unit one, and so on. When we need to set a texture to a sampler, first we need to activate the required texture unit (but only if it was not already the active one), and then we need to bind the texture for the required target (2D or CUBE _ MAP). This requires a bit of juggling trying to avoid changing the active unit and the bound texture too many times when changing techniques. Our sort key includes a material id, which is derived from the collection of textures applied to a particular renderable; this way, we try to keep groups of textures together and to minimize the number of times a texture is bound.

An alternative system would bind every texture to a different unit at loading time and then we would need to tell the shader which texture unit to use for each renderable at dispatch time. However, the maximum number of texture units supported varies heavily between video cards, some of them limited to 32 or less. This means that, if we are using more textures than the limit, then we still need to constantly bind textures to different units; we found this much slower in practice than our current system of hardcoded texture units at loading time.

10.12 Dispatching Uniforms

Dispatching the uniforms required for each renderable is accumulatively the most expensive thing our games do—not because changing a uniform is too expensive, but rather because we have lots of them. Potentially, we could have a single big uniform array for each renderable in order to reduce the number of WebGL calls, but that would mean that any tiny difference in the array would require dispatching the whole array (which is a waste), so we group uniforms by frequency of change. Shading parameters that change infrequently between renderables are set in the same uniform array, while dynamic parameters are separated out into individual uniforms. This organization is quite effective in order to minimize the number of changes but it means that some techniques have dozens of individual parameters.

Values for each uniform are extracted at dispatch time from the technique parameters dictionaries stored in each DrawParameters object and then compared against the current values set for that particular uniform on the current active program. Only when the values differ do we update the uniform.

To avoid setting the same values twice, we employ a two-level filtering system. We store the JS array that was last set on the uniform; a quick equality check avoids setting the same object twice in a row for the same uniform. This works quite well in practice because of the sorting by material and because we reuse the same typed arrays for the same kinds of data as much as possible. Of course this information is only valid within the function that dispatches an array of DrawParameters and needs to be reset once it finishes. If the uniform consists of a single value, then we skip this level.

Once the JS array is determined to be different from the previous one, then we check each individual value on the array. By default we use equality checks for each value on each uniform, which seems slow but is vastly faster than actually changing the uniform when it is not required. But, when the uniform contains a single floating-point value, we do not use an equality check; we compare whether the absolute difference between the new value and the old value is greater than 0.000001 and only then do we update it. This could be potentially risky if the precision required for that value is higher than the threshold, but we have never found a problem in practice. We could be more aggressive and apply the threshold update requirement for all uniforms, or even use a lower threshold, but we did find rendering issues when doing so. In particular, some rotations encoded in matrices require quite a lot of precision to be correct. However, this is something that could potentially be enabled per project; many users will not notice that an object did not move the tenth of a millimeter that it should have moved.

10.12.1 Performance Evaluation

A typical frame in the game *Oort Online* would have metrics similar to these:

- 9.60 ms total dispatch time
- 3,316 uniform changes
- 2,074 draw calls
- 68 technique changes
- 670 vertex array object changes
- 3 index buffer changes
- 19 vertex buffer changes
- 23 vertex attribute changes
- 78 render state changes
- 181 texture changes
- 32 framebuffer changes

These numbers were captured on a machine with the following specs:

- NVIDIA GeForce GTX 750 Ti
- Intel Core i5-4690 CPU 3.50 GHz
- Ubuntu 14.10
- Google Chrome version 39

After disabling the low-level checks for equality of values, the metrics change to

- 10.53 ms total dispatch time
- 7,477 uniform changes

Disabling also the higher level checks for JS array equality changes the metrics to

- 11.56 ms total dispatch time
- 16,242 uniform changes

These numbers vary a lot depending on the camera position in this heavily dynamic game, but the gains are consistent in any situation. Each of our filtering levels reduces the number of uniform changes by almost 50% and reduces the total dispatch time by about 10%.

10.13 Resources

The Turbulenz Engine is Open Source and can be found at https://github.com/turbulenz/turbulenz_engine. The documentation for the Turbulenz Engine is online and can be found at http://docs.turbulenz.com/

Bibliography

[Echterhoff 14] Jonas Echterhoff. "Benchmarking unity performance in WebGL." http://blogs.unity3d.com/2014/10/07/benchmarking-unity-performance-in-webgl/, 2014.

[Ericson 08] Christer Ericson. "Order your graphics draw calls around!" http://realtime-collisiondetection.net/blog/?p=86, 2008.

[Forsyth 08] Tom Forsyth. "Renderstate change costs." http://home.comcast.net/~tom_forsyth/blog.wiki.html#[[Renderstate%20change%20costs]], 2008.

[Wilson 12] Chris Wilson. "Performance Tips for JavaScript in V8." http://www.html-5rocks.com/en/tutorials/speed/v8/, 2012.

11

Performance and Rendering Algorithms in Blend4Web

Alexander Kovelenov, Evgeny Rodygin, and Ivan Lyubovnikov

11.1 Introduction

Blend4Web is an open-source WebGL framework that uses Blender 3D as its primary authoring tool. It started as an experimental project to replace Adobe Flash and then evolved into a feature-rich platform for any kind of 3D web development. In this chapter, we share insights into the advanced techniques we have implemented in our engine. Among them are prerender optimizations performed on the CPU to increase overall engine performance, discussed in the next section; a worker-based physics engine; a fast shader technique to render realistic oceans; and a feature-rich shader compilation pipeline.

11.2 Prerender Optimizations

Reducing the number of draw calls has always been among the fundamental goals of rendering optimization. The history of the OpenGL API distinctly indicates this tendency, so every consecutive version appends something new, allowing us to do more things in just a single call. In WebGL 1, many such features are unavailable; however, some may be accessed by using extensions, such as `OES_vertex_array_object`,

`ANGLE_instanced_arrays`, `OES_element_index_uint`, and others. However, requiring the engine to use such extensions may have some compatibility costs, so they should be considered only as auxiliary means to increase performance on supported platforms.

11.2.1 Batching

Given that Blend4Web is implemented in JavaScript, we needed a straightforward and fast batching solution. This is especially important when rendering big scenes consisting of thousands of objects. As in the famous presentation by Nvidia, "Batch, Batch, Batch" [Wloka 03], a batch includes geometry from all objects that share the same state of the rendering pipeline. To implement, we extract the state from Blender's materials; combine it with the geometry of separate objects (*meshes* in Blender terminology), forming batches on a per-object basis; and finally merge these per-object batches into the global batches, which will be rendered in the scene.

A first cut may do this at the resource preparation stage and provide such "static" geometry inside exported resource files. Despite the obvious benefit of reducing time needed to process geometry during the scene loading, this has considerable drawbacks. The first is the inability to modify the batches (e.g., upon receiving the information about the target platform or changing the user settings). The second and more essential drawback is the considerable increase in the file size for scenes with instanced objects. For example, if there are five trees and five stubs in the scene, the final geometry will be five times bigger than the original one, resulting in implications such as increasing the loading time, especially for low-bandwidth mobile connections. Due to these drawbacks, we chose to construct batches during scene loading and therefore we required a fast algorithm.

As shown in Figure 11.1, after object batches are constructed, we need a method to compare them in order to make further merging possible. We use a simple hashing algorithm. First, we compute the hashes for each object batch, and then we compare their hashes and merge the batches if the hashes are equal. The key to performance is to keep the static-like typing of JavaScript objects representing the batch so that a modern JavaScript engine can optimize the code. This also allows developing a fast hashing function for every type in the same manner as Java's `hashCode()` function. These hashes can be calculated very fast, but they are not 100% reliable because they are only 32 bits long and prone to hash collisions (i.e., there is a high possibility for two different JavaScript objects to have the same hash). Thus, we needed to compare object batches themselves at the final stage of batch construction. This comparison is made by iteration through all its properties.

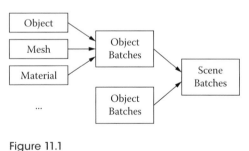

Figure 11.1

The batching scheme.

We measured performance for this on a machine with an i7-3770K 3.50 GHz CPU and GeForce GTX 680 GPU using Chrome 39. The batches were created from low-polygonal objects containing 1,000 triangles each. All the objects have the same material assigned to them, making this test the worst-case scenario (maximum number of intermediate batches joined into the single one). The size and complexity of the material only define the time to perform the hash calculation and this stage does not exceed 10% of

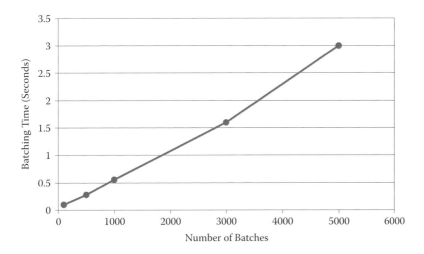

Figure 11.2

Batching performance.

the entire batching time. Thus, 90% of the time is spent for operations such as copying and repositioning the geometry in the final batch. As shown in Figure 11.2, the presented algorithm has a linear time complexity (i.e., the time to complete calculations is proportional to the number of batches). This provides a decent performance for real-world large scenes considering that 5,000 batches consist of five million triangles.

11.2.2 Frustum Culling and LODs

Besides static geometry, there are also dynamic objects that can freely move across the scene. For these, we use view frustum culling and level-of-detail (LOD) algorithms.

In our implementation of frustum culling, we use bounding ellipsoids to represent objects. This compromise solution was chosen because it is fast enough and allows us to effectively designate boundings for objects extended in one or two dimensions. Particularly, a human body is perfectly described by an ellipsoid extended in the up/down dimension. The algorithm itself is just an extension of a typical sphere–plane intersection, where an effective radius is calculated per each direction toward one of six planes making the camera's frustum volume.

The LOD algorithm is fairly straightforward. The artists produce the LODs as separate models and also assign the activation distances to them. Then, just before rendering, the algorithm selects a required LOD by checking the distance between its center and the camera.

The results are shown in Figure 11.3. The screenshot shows a typical scene view, which can be used to analyze overall scene performance and for which the given FPS plot was created. Batching is the most important feature for this scene. Culling has a second priority while LODs produce the best results for hardware, the performance of which is heavily bound by the GPU. The presented algorithms are reasonably inexpensive for the CPU (e.g., for the demo scene presented on the screenshot, which has more than 1,000 objects); frustum culling and LOD calculations take ~2% of all the time available for the browser's main execution thread.

 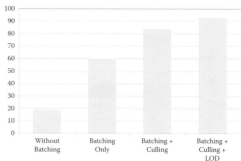

Figure 11.3

Prerender optimization performance: screenshot and FPS on the Farm demo.

11.3 Threaded Physics Simulation

Physics is a vast field that embraces more than simple constrained motions and collision detection algorithms. A typical real-time physics engine implements fast and intricate algorithms to achieve decent quality with minimum computation overhead. We had a strong intention to find an existing solution instead of writing our own. Luckily, there was a great project to compile the Bullet open-source physics engine directly to JavaScript called ammo.js. By using the Emscripten compiler to translate intermediate LLVM bit-code directly to a JavaScript subset known as asm.js (see Chapter 5), some really impressive results were achieved.

However, ammo.js does not include several features we require. First, we needed floating objects and vehicles. Second, we required a fully asynchronous design, which is somewhat incompatible with ammo.js class-based architecture. Third, we needed our own mechanism to interpolate physics simulation results instead of Bullet's built-in implementation based on Motion State. Thus, the uranium.js project was created.

We moved physics calculations to the single independent Web Worker process, freeing the main thread of execution from physics and allowing it to spend more time on rendering, animation, frustum culling etc. The most difficult part to this approach was the mechanism of fast interworker communication (IWC) and time synchronization between the two threads.

Fast IWC is required for scenes with a lot of objects because of the browser's inherent overhead imposed on message passing. Our IWC implementation is based heavily on typed arrays and caching. This was done in such a way that all data frequently passing between the threads of execution are serialized to typed arrays (float or unsigned integer), which are created in advance and stored in cache to minimize the JavaScript garbage collector overhead. This approach is more streamlined than using transferable objects[*] because it is fully supported by all major browsers and there is no need to recreate arrays after each transfer. Also see Chapter 4.

IWC messages are formed by using the following simple approach: the first position of transferred array is a message identifier (message id); the second through last positions

---

[*] http://www.w3.org/html/wg/drafts/html/master/infrastructure.html#transferable-objects

are occupied by the message payload. For example, to set a position of some object on the physics scene we send the message OUT_SET_TRANSFORM to the worker process. OUT_SET_TRANSFORM is a message id, so it will be stored at the first position of transferred array. The second position in the message array is occupied by a number representing a physics body id; next to it X,Y,Z coordinates of the object position follow and then X,Y,Z,W values form a quaternion vector representing the object's rotation. This is explained in Figure 11.4.

Time synchronization is needed because the threads use different clocks. The ticks of the main thread clock are generated by the browser and are linked with the browser's internal rendering process. The worker process clock is free from any constraints and may generate ticks as fast as the performance of the target platform allows.

For example, a helicopter may fly at 250 km/h, or ~70 m/s. At 60 FPS, there will be a displacement of more than a meter per frame. If the time is out of sync or acquired with considerable errors, there will be visual glitches with the flying object. For rapidly moving objects, we need precision in time determination at the level of 1/10 of the frame duration, which is equivalent to several milliseconds. The widely supported method performance.now() returns times as floating-point numbers with up to microsecond precision. Also, with a dedicated Web Worker, synchronization is greatly simplified because it uses the same time origin as the main thread of execution. Not all browsers have built-in support for a high-precision timer in dedicated Web Workers, so for unsupported ones we needed to use the old-fashioned Date object and synchronize time explicitly at the worker startup.

For a simple benchmark, we used a scene with a varying number of cubes colliding with each other and a terrain mesh composed of 8,192 triangles. Each cube has a variety of forces acting upon it: There is a persistent gravity force and multiple transient collision forces per each point of interaction (pressing and friction forces).

The scene is shown in Figure 11.5 along with the performance plot representing FPS for the given number of colliding objects. The performance decreases for many dynamic

MSG ID	BODY ID	X POS	Y POS	Z POS	X QUAT	Y QUAT	Z QUAT	W QUAT

Figure 11.4

OUT_SET_TRANSFORM message.

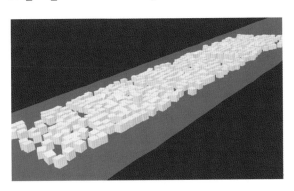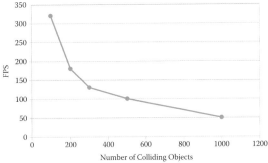

Figure 11.5

Physics performance.

objects as the swarm of IWC messages rises [Priour 14]. However, it suits real-world scenarios well, where a reasonably small number of objects (usually below 100) are calculated simultaneously and where we need polygonal meshes to represent their shape.

11.4 Ocean Rendering

While working on one of our biggest WebGL demos, called "Capri," we encountered the complicated problem of reconstructing the sea surface with a number of effects previously not reproduced in browsers, including:

- LOD system to reduce polygon count
- Single draw-call for water object
- Shoreline influence for rolling waves
- Refraction and caustics on underwater objects
- Dynamic reflections
- Foam
- Fake subsurface scattering
- Realistic floating object behavior

11.4.1 Preparing the Mesh

In WebGL, it is important to move as many calculations as possible to the GPU. However, there are only two types of shaders available: vertex and fragment. This means that we need to prepare all the required geometry on the CPU.

Our approach is based on a geometry clipmap technique that proved to be a good method for rendering static meshes such as terrains (also see Chapter 18) deformed by height-maps but has never been used for dynamic surfaces. The idea is to construct a static mesh from multiple square rings with different polygon densities. The mesh should move with a camera translation. This method achieves a fairly precise simulation in regions close to the camera and delivers a reasonable polygon budget. The main difference, compared to the standard geoclipmapping technique [Hoppe et al. 04], is that there are no seams between the LOD levels. The view of such a mesh is shown in Figure 11.6. The different LOD levels are painted in three different colors.

11.4.2 Shore Parameters

In order to reproduce rolling waves and water color gradient, the distance to the shore and the direction to the nearest shore point are needed. We wrote a Python script for Blender that, for a given terrain mesh and

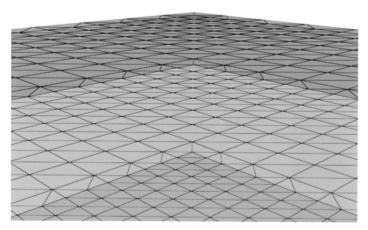

Figure 11.6

Seamless mesh generated on the CPU.

11. Performance and Rendering Algorithms in Blend4Web

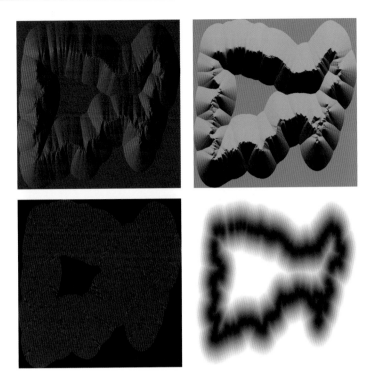

Figure 11.7

Direction and distance to the shore baked into the RGBA image.

a plane representing the water surface, constructs an RGBA image. For every pixel on this image, it finds the nearest terrain vertex and stores the distance and the direction to it. Therefore, all the calculations are done on the CPU before the scene is exported and there is no need for heavy calculations during scene loading. The direction is packed into the red and green channels, and the distance is packed into the blue and alpha channels (Figure 11.7).

11.4.3 Waves

There are three types of waves applied to the water surface:

- High distant waves which have greater influence far from the shore
- Small waves evenly distributed over the surface
- Rolling waves moving toward the shore

11.4.3.1 Distant and Small Waves

Distant and small waves are generated by mixing several noise functions (Figure 11.8). For each noise,

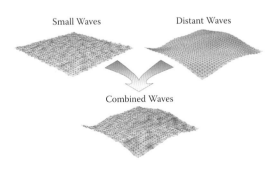

Figure 11.8

Combination of small and distant waves.

a different movement speed and scale are chosen. Distant waves use two simplex noises [McEwan 11]. Small waves are a combination of two cellular noise functions [Gustavson 11]. GLSL calculations are shown in Listing 11.1.

Listing 11.1 Waves generation.

```
//waves far from the shore
float dist_waves =
  snoise(DST_NOISE_SCALE_0 * (pos.xz + DST_NOISE_FREQ_0 * time))*
  snoise(DST_NOISE_SCALE_1 * (pos.zx - DST_NOISE_FREQ_1 * time));

//high resolution geometric noise waves
float cel_coord1 = 20.0/WAVES_LENGTH * (pos.xz - 0.25 * time);
float cel_coord2 = 17.0/WAVES_LENGTH * (pos.zx + 0.1 * time);
float small_waves = cellular2x2(cel_coord1).x +
                    cellular2x2(cel_coord2).x - 1.0;
```

Summing up these functions we get dynamic waves that are well suited for rendering the open ocean [Mátyás 06].

11.4.3.2 Rolling Waves

Closer to the shore the length and amplitude of the wave should decrease. Also, as the underwater current is moving in the opposite direction, the wave receives a little incline toward the shore. Approximate shape is shown in Figure 11.9.

To generate rolling waves the parameters of the shore are extracted from the prepared RGBA texture. Because the distance is packed into the blue and alpha channels, we need to divide the texture's blue channel by 255 and add it to the value of the alpha channel. In other words, we reconstruct the original value from two parts: the bigger and less precise one—alpha channel—and the lesser but more precise one—blue channel. It is done with a bit_shift constant vector. Actually, this is a common technique for packing precise values in several channels of the image. Calculations are shown in Listing 11.2.

Listing 11.2 Shore parameters extraction.

```
//shore coordinates from world position
vec2 shore_coords = 0.5 +
  vec2((pos.x - SHORE_MAP_CENTER_X)/SHORE_MAP_SIZE_X,
       -(pos.y + SHORE_MAP_CENTER_Y)/SHORE_MAP_SIZE_Y);

//unpack shore parameters from texture
vec4 shore_params = texture2D(u_shore_dist_map,shore_coords);

const vec2 bit_shift = vec2(1.0/255.0, 1.0);
float shore_dist = dot(shore_params.ba, bit_shift);
vec2 dir_to_shore = normalize(shore_params.rg * 2.0 - 1.0);
```

11. Performance and Rendering Algorithms in Blend4Web

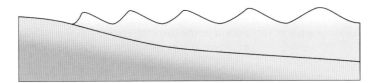

Figure 11.9

Rolling waves.

Given the distance to the shore and the direction to it, we can calculate the sinusoidal waves moving toward the shore as shown in Listing 11.3. The `shore_waves_length` variable is divided by PI because the `WAVES_LENGTH` constant is defined in meters and needs to be converted to properly calculate the wave length.

Listing 11.3 Waves moving toward the shore.

```
float shore_waves_length = WAVES_LENGTH/MAX_SHORE_DIST/M_PI;
float dist_fact = sqrt(shore_dist);

float shore_dir_waves = max(shore_dist, DIR_MIN_SHR_FAC)
  * sin(dist_fact/shore_waves_length + DIR_FREQ*time)
  * max(snoise(DIR_NOISE_SCALE*(pos.xz + DIR_NOISE_FREQ*time)),
      DIR_MIN_NOISE_FAC);
```

To make rolling waves more natural, the result is multiplied by one more `snoise` function value.

Listing 11.4 Waves moving toward the shore.

```
float dir_noise =
    max(snoise(DIR_NOISE_SCALE*(pos.xz + DIR_NOISE_FREQ*time)),
        DIR_MIN_NOISE_FAC);
shore_dir_waves *= dir_noise;
```

11.4.3.3 Combining Waves

All wave types are mixed to get the final vertical offset as shown in Listing 11.5.

Listing 11.5 Mixing all waves together.

```
float waves_height = WAVES_HEIGHT *
    mix(shore_dir_waves, dist_waves, max(dist_fact, DST_MIN_FAC));
waves_height += SMALL_WAVES_FAC * small_waves;
```

11.4.3.4 Waves Inclination

Higher vertices have greater inclination in the direction to the nearest shore point.

Listing 11.6 Horizontal offset for waves inclination.

```
float wave_factor = WAVES_HOR_FAC * shore_dir_waves
        * max(MAX_SHORE_DIST/35.0 * (0.05 - shore_dist), 0.0);
vec2 hor_offset = wave_factor * dir_to_shore;
```

11.4.3.5 Normal Calculation

In order to perform further shading, the normals need to be calculated in the vertex shader. Therefore, the "offset" function is called for three adjacent vertices by stepping with cascade steps in x and y directions. This step is stored in the vertex's y coordinate. By using such information, we can calculate bitangent, tangent, and normal vectors.

Listing 11.7 Normal calculation.

```
vec3 bitangent = normalize(neighbour1 - world.position);
vec3 tangent = normalize(neighbour2 - world.position);
vec3 normal = normalize(cross(tangent, bitangent));
```

11.4.4 Material Shading

Water color consists of the components displayed in Figure 11.10(a–d).

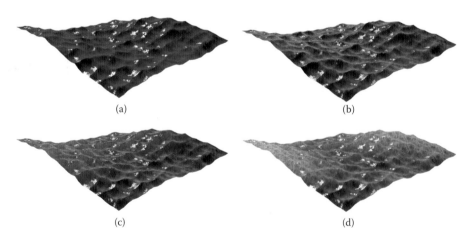

(a)

(b)

(c)

(d)

Figure 11.10

Rolling waves. (a) Lambert shading, Wardiso specular; (b) + Fresnel reflection; (c) + Subsurface scattering (SSS); and (d) + Foam.

11. Performance and Rendering Algorithms in Blend4Web

Reflections are produced by flipping the original camera vertically and rendering the needed objects to a new framebuffer, which has a lower resolution for greater performance. *SSS* simulation is based on light and camera directions and surface normals [Seymour 12]. These simple calculations give us surprisingly realistic results.

Foam is influenced by three major factors used for mixing with the original color:

1. High waves
2. Rolling waves with a normal close to the direction toward the shore
3. Depth-based foam on objects close to the surface of the water

11.4.4.1 High Waves Foam

When water mesh vertices reach a specific height, foam starts to be mixed in, influencing the resulting color. The mask for this type of wave should look like the one in Figure 11.11. White areas have more foam:

11.4.4.2 Rolling Waves Foam

Based on the wave normal vector and the direction to the shore, the foam mix factor increases. So we can see more realistic rolling waves, which generate foam in areas facing toward the shore. The GLSL equation is as follows:

```
float foam_shore = 1.25 * dot(normal, shore_dir) - 0.1;
foam_shore = max(foam_shore, 0.0);
```

As seen in Figure 11.12, the mask has much sharper edges.

Figure 11.11

High waves foam factor.

Figure 11.12

Rolling waves foam factor.

11.4.4.3 Depth-Based Foam

The last component of the resulting foam is a depth-based foam. Mathematically, a mask for such an effect is calculated by subtracting the depth of the underwater object's pixel from the depth of the water surface pixel. If the result is close to zero, the foam factor has maximum value (Figure 11.13).

11.4.4.4 Final Look of the Foam

By combining these three types of foam we achieve plausible waves. The result is shown in the Figure 11.14.

11.4.5 Refractions and Caustics

11.4.5.1 Refractions

Refraction uses previously rendered underwater objects stored in a specially created framebuffer and distorts them in accordance with normal vectors of the water's surface (Figure 11.15a, b).

11.4.5.2 Caustics

In order to reduce texture memory usage, caustics are approximated by cellular noise with deformed texture coordinates (Figure 11.16a, b).

11.4.6 Floating Objects Physics

In Blend4Web, every object that can float requires several points to be defined. They are described by metaobjects called "bobs." When a bob goes underwater, the buoyancy force

Figure 11.13

Depth-based foam factor.

Figure 11.14

Final look of the foam.

(a)   (b)

Figure 11.15

Water refractions: (a) without refraction; (b) with refraction.

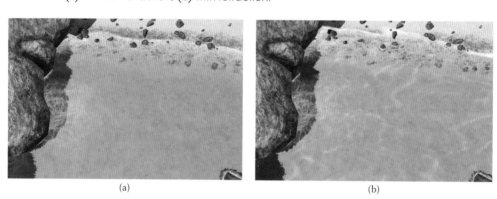

(a)   (b)

Figure 11.16

Water refractions: (a) without caustics; (b) with caustics.

is applied to its center and affects the floating object. In Figure 11.17, bobs are visualized as yellow spheres. In order to calculate the height of waves at a specific point, the equations mentioned in Section 11.4.3 must be implemented in the physics thread but with some simplifications.

11.5 Shaders in Blend4Web

Modern graphics applications may use a large number of shaders for creating different effects and improving the overall rendering quality. There is no exception for WebGL applications despite the specifics imposed by the web platform (e.g., limitations on the loading time and the need to compile the shaders upon application startup).

GLSL code can be stored inside an HTML page using, for example, `<script>`; however, this approach contradicts modularity and leads to bulky code. Instead, Blend4Web uses a shader processing system. A shader's source code is stored in separate text files and passed through our custom preprocessor for error checking, optimization, and obfuscation.

11. Performance and Rendering Algorithms in Blend4Web

Figure 11.17

Bobs attached to the yacht object to simulate floating.

11.5.1 Directives

The standard GLSL preprocessor controls the shader compilation using different directives. By using branching macros code, reuse of code becomes possible—for example:

Listing 11.8 Branching using directives.

```
# if TEXTURE_COLOR
float alpha = (texture2D(u_sampler, v_texcoord)).a;
# else
float alpha = u_diffuse_color.a;
# endif
```

This way, different combinations of macros can significantly reduce the source code volume. However, the WebGL API does not allow compiler directives and therefore additional manipulations with the shader code are required, which is essentially improvised preprocessing.

The first and rather simple solution is to add #define directives containing macro values to shader code before the compilation, in order to entirely rely on the standard GLSL preprocessor. While this solution was quite efficient, it was not flexible enough for our needs. As shaders grew, the need arose to separate specific functionalities, such as lighting or refraction implementation, using include files to improve the structure and code reuse. The absence of the #include directive in the standard GLSL preprocessor was the reason for creating the more advanced shader processing utility, which could support not only macros but also some new and specific directives.

We needed a parser, which was generated using the PEG.js* generator on the basis of the preprocessor directives grammar. This parser produces an abstract syntax tree (AST)—a JavaScript object containing GLSL and preprocessor tokens. Then, the analysis of this object is performed in order to substitute the macro values, which saves us the necessity to specify the #define directives. At the next stage, the shader text code is created from the AST. When an #include directive is met, the AST from the corresponding include file is simply inserted into the current tree. Finally, the shader is compiled using the standard WebGL API. As an optimization, ASTs can also be generated for all shaders offline and loaded at application startup as a JSON object.

11.5.2 Node Materials

Node materials are a simple yet effective method to visually program shaders, which allows for creating rich and realistic surfaces. This can be found in numerous 3D modeling and developer tools such as Blender, Autodesk Maya, UnrealEd, and others. Developed on the basis of Blender, Blend4Web inherits its node material system.

Blender's node material is a directed graph; the nodes denote certain transformations for every pixel. Therefore, the graph itself represents a shader that performs such transformations (Figure 11.18).

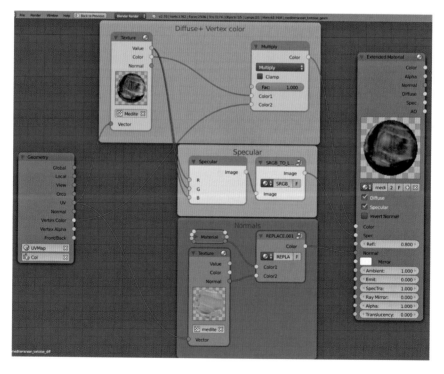

Figure 11.18

Complex node structures being processed and transformed to GLSL.

---

* http://pegjs.majda.cz/

In Blend4Web, these shaders are created using a pair of special fragment and vertex shaders containing code blocks according to the nodes used by a material. This toolkit mirrors Blender's visual shader programming system.

Using our own preprocessor, we introduced new directives specific for this node system—for example:

Listing 11.9 Node example.

```
#node TEXTURE_COLOR
    #node_in vec3 uv
    #node_out vec3 color
    #node_out float value
    #node_param uniform sampler2D texture
{
    vec4 texval = texture2D(texture, vec_to_uv(uv));
    color = texval.xyz;
    srgb_to_lin(color);
    value = texval.w;
}
#endnode
```

The code for a node is located between the #node and #endnode directives. The required parameters are identified by the corresponding directives:

- #node_in—node input parameters
- #node_out—node output parameters
- #node_param—additional parameters such as attributes, uniforms, etc.

Shaders are constructed as follows. First, the Blender material data are parsed and loaded into a node graph object. Next, preprocessing is performed using the previously generated AST. During preprocessing, we traverse the node graph and create code blocks according to its nodes, which we insert into the target shader code in the corresponding places. The shader will contain only the commands required for rendering the source material. Finally, the shader is passed to WebGL to be compiled and linked. With this approach, it is possible to build an individual shader for every node material of the scene.

11.5.3 Obfuscation and Validation

Given our preprocessor, we can go a step further and implement additional tools for obfuscation (minifying) and validation of the shader code.

11.5.3.1 Obfuscation

The obfuscator substitutes most identifiers in the AST with generated short names similar to other popular obfuscation tools. Also, redundant spaces and comments are removed, resulting in a more compact code.

The most difficult part is the need to process a combination of GLSL code and directives instead of just the GLSL code. In practice, the GLSL parser ignores preprocessor directives as if they were nonexistent. As a result, a code, which could be invalid either semantically or syntactically, can be processed—for example:

Listing 11.10 Invalid code example from the obfuscator's point of view.

```
main_bend(vertex_position, object_center, au_wind_bending_amp,
    au_wind_bending_freq, u_time, wind,
#if MAIN_BEND_COL
    a_bending_col_main);
#else
    1.0);
#endif
```

When the AST is built, semantics are not taken into account. However, the syntax analysis is performed by the parser and an additional limitation is imposed for input data—namely, the GLSL code has to be syntactically valid. After obfuscation, AST is translated into the shader code, which is used at the next stage of preprocessing.

A benefit of obfuscation is that the volume of code is reduced. For typical shaders used in Blend4Web, the reduction is about 32%–49% of their original size. Obfuscation also significantly decreases readability and therefore it can be useful for creating proprietary Blend4Web modifications.

11.5.3.2 Validation

The GLSL parser also allowed us to implement a utility to validate shaders. This utility checks for unused variables and functions and for cases where reserved or undeclared identifiers are used. The validator is also able to reveal certain types of errors that are hidden by some WebGL implementations, and thereby serves as an auxiliary shader debugger.

Validation allowed us to improve the overall structure of shader code. Particularly, for `include` files, `#import` and `#export` directives were introduced to improve the code readability in cases where an `include` file still retains global dependencies:

- `#import`—specifies which external variables and functions are used inside the `include` file and enforces their outside declaration
- `#export`—specifies which internal variables and functions, which are declared inside the `include` file, can be used outside it

It can be noted that further development of the GLSL parser can have other useful applications, such as debugging, shader complexity analysis, and additional code optimizations.

11.6 Resources

All the source code is available in the Blend4Web GitHub repository at https://github.com/TriumphLLC/Blend4Web.

Bibliography

[Gustavson 11] Stefan Gustavson. "Cellular Noise in GLSL." http://webstaff.itn.liu.
se/~stegu/GLSL-cellular/, 2011.

[Hoppe et al. 04] *GPU Gems 2* Chapter 2, "Terrain Rendering Using GPU-Based Geometry
Clipmaps." http://http.developer.nvidia.com/GPUGems2/gpugems2_chapter02.html,
2004.

[Mátyás 06] Zsolt Mátyás. "Water Surface Rendering Using Shader Technologies. http://
shiba.hpe.sh.cn/jiaoyanzu/wuli/soft/Water/thesis.pdf, 2006

[McEwan 11] Ian McEwan. "Ashima Arts. Array and textureless GLSL 2D simplex noise
function." https://github.com/ashima/webgl-noise/blob/master/src/noise2D.glsl, 2011.

[Priour 14] Matt Priour. "Making Workers Work for You." http://mpriour.github.io/
workers-f4g14.

[Seymour 12] Mike Seymour. Assassin's Creed III: The Tech Behind (or Beneath) the
Action." http://www.fxguide.com/featured/assassins-creed-iii-the-tech-behind-or-
beneath-the-action/, 2012.

[Wloka 03] Matthias Wloka. "Batch, Batch, Batch: Presentation at Game Developers
Conference 2003." http://developer.nvidia.com/docs/IO/8230/BatchBatchBatch.pdf.

12

Sketchfab Material Pipeline
From File Variations to Shader Generation

Cedric Pinson and Paul Cheyrou-Lagrèze

12.1 Introduction

Sketchfab is the online platform to publish, share, and embed interactive 3D content. With Sketchfab's real-time viewer (Figure 12.1), users can upload and interact with almost any type of 3D model. Consequently, our architecture demands handling a wide variety of file types—formats like FBX, COLLADA, Blender, PLY, STL, and OBJ. Each format describes 3D data, such as geometry and material properties, with diverse techniques. Geometry is easy to handle. With reasonably standardized component definitions (position, texture coordinates, normals, and tangents), vertex information is easily deciphered across different file formats and export variations.

Materials, on the other hand, are significantly more problematic. Even if a format is well documented, 3D authoring tools interpret and export material descriptions in an array of different ways. The variations between exported files, even files of the same format, create unique challenges when handling material data.

In this chapter, we give an overview of the material obstacle Sketchfab faced. We then describe the development of our materials pipeline and the pipeline's approach to shader generation. Concluding with the pros and cons of our system, we discuss potential future approaches to tackle the materials issue with even greater efficiency.

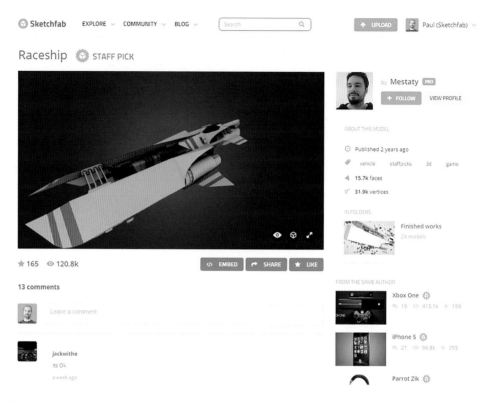

Figure 12.1

Sketchfab viewer.

12.1.1 The Materials Nightmare

In the 3D industry materials can be a nightmare; the disparity between 3D formats creates a maze of material data. These format deviations pose a particular challenge for platforms like Sketchfab that aim to support a broad range of 3D creators—from architects and 3D scanners to video game designers and character modelers. Even if an individual format is well documented, maneuvering variations between files of the same format can be arduous.

Next we outline a few examples of file formats and their variations.

12.1.1.1 COLLADA

COLLADA is well known for its extensibility. While authoring tools enjoy the convenience of adding custom data by the extension of existing COLLADA elements, this flexibility may create complications during import. Typically, importers ignore the customizations and apply only the standard data set.

12.1.1.2 STL

STL (the native CAD format) is commonly employed by rapid prototyping and computer-aided manufacturing software. Although it is one of the oldest 3D formats,

STL files are numerous due to the popularity of consumer 3D printing. Binary STL has at least two nonstandard deviations for the addition of color information. The format contains an 80-character header generally overlooked by most importers. Yet, some 3D software applications utilize the header to insert their own custom data. *Materialize Magics* software, for example, exploits the header to store both color (a default overall color) and materials data (with channels for diffuse, specular highlight, and ambient light).

12.1.1.3 WAVEFRONT OBJ

One of the most common formats our platform encounters is OBJ, a simple, mature, and well-documented type. Despite the fact that almost every authoring tool imports and exports to OBJ, individual handling of materials data (and often geometry data) diverge. For example, certain software applications export a diffuse channel in both `ka_map` and `kd_map` slots, while others employ only the `kd_map` slot.

In addition, several 3D tools extend OBJ with vertex colors, placing RGB color values after XYZ position values. Although vertex color is not standard for OBJ, the abundance of users implementing it (plus the popularity of the format) forces other authoring tools and importers to support the additional component. Thus, a mutant 3D file format (and a new hurdle for importers) is born.

As Sketchfab currently supports over 20 file formats, the preceding examples are only a sample of the complications we encounter when parsing different types of 3D files.

12.2 Our Materials Pipeline

Sketchfab allows users to upload models in a couple of different ways: either directly through our website or with one of our custom exporters. Our software exporters allow the inclusion of more file information, like the creation tool and software version, which is tremendously helpful when interpreting materials data.

The OSG* framework is the foundation for Sketchfab's WebGL viewer. OSG.JS is a WebGL framework based on `OpenSceneGraph` concepts that allows developers to interact with WebGL via JavaScript with an "`OpenSceneGraph`-like" toolbox. Generic shader generation code is open source in OSG.JS. The material pipeline described next and our specific implementation, on the other hand, are particular to Sketchfab's code base.

Our pipeline first processes the 3D model on our server, using different filters to clean up, optimize, and generate a generic 3D scene file (Figure 12.2). The resulting OSG. JS file contains both the geometry and the materials metadata, which were extracted during the initial processing step. Next, after

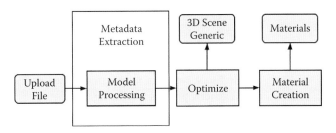

Figure 12.2

Creation of materials and the generic 3D scene.

---

* OSG is the C++ version of OpenSceneGraph.

determining which processor to apply, the original materials data are translated into our own materials system.

Listing 12.1 Example metadata (for each corresponding material) stored in the generic 3D scene.

```
[
  {"Name": "source", "Value": "wavefront"},
  {"Name": "ka", "Value": "[0, 0, 0]"},
  {"Name": "kd", "Value": "[0.5, 0.5, 0.5]"},
  {"Name": "ks", "Value": "[0, 0, 0]"},
  {"Name": "ns", "Value": "0"},
  {"Name": "sharpness", "Value": "0"},
  {"Name": "emissive", "Value": "[0, 0, 0]"},
  {"Name": "tr", "Value": "1"},
  {"Name": "UniqueID", "Value": "1"}
]
```

Since Sketchfab users can edit their 3D models in the material editor, we do not fully optimize the generic 3D scene file. A more optimized 3D scene is generated posteditor for rendering purposes. The final optimization of models includes vertex unification, vertex cache optimization, the generation of tristrips, index reordering, and, finally, mesh compression.

12.2.1 Handling Material Variations

When mediating variations between formats, the original file's metadata are invaluable. Therefore, a copy of the originating metadata is extracted and copied into the OSG.JS file during the first processing step. Using this information, we can identify the original file format (and often the corresponding authoring tool) to help select the optimal processor (Figure 12.3).

With a format-specific processor, we can more accurately implement all necessary changes when transforming the 3D scene into our native format. Even when handling identical file formats, different authoring tools express material data by varying methods. In COLLADA files, for instance, interpreting transparency requires contextual information. SketchUp and Google Earth invert the A_ONE blend transparency property. Adobe Photoshop adds its own "alpha_mask" metadata on top, making transparency interpretation even more complex. Our distinct processors, dependent on the file type and authoring tool, result in more streamlined code. If no format metadata are available, however, a generic processor must be used.

12.2.2 Our Material Model

Our material system is composed of the following channels:

- Diffuse
- Lightmap

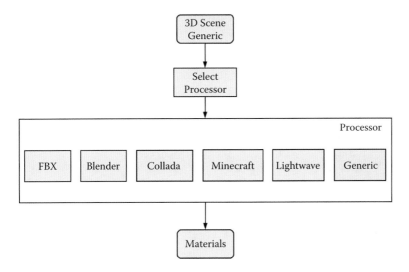

Figure 12.3

A format-specific processor more accurately converts materials.

- Specular
- Specular hardness
- Normal map
- Bump map
- Opacity
- Emission

Additionally, an HDR environment can be utilized for reflections and irradiance. For more on HDR, see Chapter 16.

Each channel consists of a texture (or a color) and a corresponding factor for scaling. An individual channel might additionally exhibit a specific attribute or behavior, which the other channels do not inherit. For example, the opacity channel cannot implement a color; the factor is used to scale either the level of opacity or the texture.

12.2.3 User Custom Materials

Using the Sketchfab material editor (Figure 12.4), users can edit and save material changes to their uploaded 3D model. When the materials are saved, the server is notified and a new 3D scene is generated specifically for the modified materials (Figure 12.5).

Using the generic 3D scene file and the saved material edits, a new 3D scene file is checked against a list of possible optimizations for faster rendering. More aggressive optimizations are now possible, including

- Duplicate materials removal
- Geometry mergers (resulting in a reduction of draw calls)
- Scene graph simplification and flattening of transformations
- Unused vertex attribute disposal

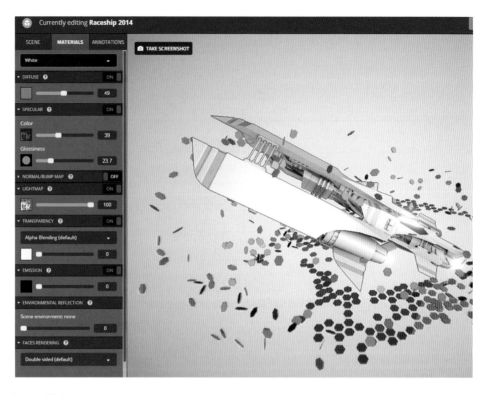

Figure 12.4

Sketchfab material editor.

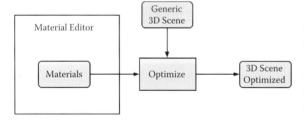

Figure 12.5

Work flow when users modify materials.

12.2.4 WebGL Viewer

Our WebGL viewer loads a 3D scene in the order described by Figure 12.6.

First, a resource manager (first in/first out [FIFO] queue simultaneously downloading resources) schedules the assets to load. We limit[*] the number of threads loading data to create independence from the browser implementation limit. After a resource is loaded, a slot is freed to download the subsequent resource. At time zero, the materials data and scene information are available to the viewer. Data are then pushed in the following order:

1. 3D geometry (binary array) using a device-dependent Level of Detail
2. Low-resolution textures[†]
3. High-resolution textures[‡]

---

[*] We use five simultaneous download requests, after testing different configurations.

[†] Low-texture resolution is 16×16.

[‡] High-texture resolution depends on device (mobile/desktop) and WebGL implementation.

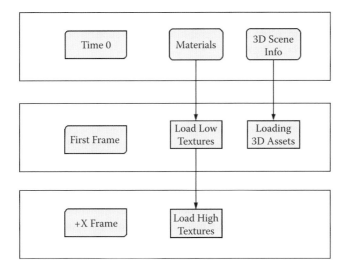

Figure 12.6

Order of data loading for scene rendering.

To draw the first frame, a promise indicates when all 3D geometry and low-resolution textures are ready. Once the first frame is drawn, high-resolution textures are downloaded to progressively replace low-resolution textures.

The viewer excludes ineffective textures—a material with an incorrectly configured channel that does not contribute to rendering. For example, diffuse time 0.0 does not require downloading a diffuse texture. These exclusions reduce shader complexity and prevent downloading superfluous data.

Finally, to improve user experience, we designed a texture load order that prevents distracting viewer flicker. Textures are loaded in the following order: opacity, diffuse, specular, light map, normal map, bump map, and emissive.

12.3 Shader Generation

The first Sketchfab iteration used an ubershader, which quickly created a jumble of spaghetti code and a massive number of compilation permutations. An *ubershader* is a giant, fixed shader containing all the material's characteristics, such as environment map, normal map, specular, bump map, etc. To configure the shader, some preprocessor symbols are specified with #defines, which enable and disable features as needed.

With the ubershader, numerous possible channel combinations (coupled with the fact we incrementally created shader code) forced our team to regularly debug and add code for new materials/metadata combinations. Clearly, our shader architecture was insufficient.

In comparison to the ubershader, a *shader graph* is a system that constructs shaders dynamically, using building blocks of shader code snippets.* Combinatorial material

---

* http://www.cs.cmu.edu/afs/cs/academic/class/15869-f11/www/readings/cook84_shadetrees.pdf

inputs are transformed into a shader graph, which is ultimately compiled into shader code. Shader graphs are often used in graphical editors such as Maya HyperShade and the Unreal Engine Material Editor.

The shader graph system addressed all the limitations of Sketchfab's original architecture. For instance, adding shader effects (such as the transition between two materials via transitional nodes) would be challenging inside an ubershader.

Before being fed to our shader generator system, materials and channels are translated into low-level primitives. These primitives inherit from OSG.JS's `StateAttribute`. `StateSets` represent WebGL graphics states and correspond to nodes in the scene graph. A `StateSet` encapsulates a series of `StateAttributes`, which describe the scene/material properties. Sketchfab applies the following OSG.JS `StateAttributes`: `TextureMaterial`, `Environment`, and `Light`.

The `Environment` state attribute contains environmental scene data, including specular, background, and irradiance map. `Light` contains lighting information (point, spot, or directional) and is read either from the scene or by using a default lighting scheme. And lastly, for material texture data, `TextureMaterial` encapsulates the material channel information—like the type of the texture (diffuse, specular, etc.), as described in Section 12.2.

Material channels contain a texture (or a color) and a scaling factor. To simplify low-level code, color data are converted to a one-pixel texture and also encapsulated into the `TextureMaterial` state attribute. We now recognize, however, that our approach to color handling might have been unwise. While performance is a nonissue, WebGL only guarantees eight texture units, and our method does not maximize the WebGL limitation.

When changes are detected in `StateAttributes`, the shader generator compiles a new shader. However, during loading, our ultimate goal is to display the first frame as quickly as possible. Waiting for a texture to create `TextureMaterials` and generate shaders slows the initial frame display. Therefore, prior to fetching 3D assets and textures, we prepare the needed elements as much as possible. Shaders are precomputed during asset download. For instance, rather than waiting on the real texture, placeholder textures of one pixel are fed to the shader generator. When the textures are received, they supersede the placeholder. This method allows the first frame to be visible more quickly.

12.3.1 From Inputs to Shader: An Outline

Figure 12.7 outlines shader code generation from inputs:

1. State attributes (`TextureMaterial`, `Environment`, and `Light`) are converted into shader nodes.
2. The resulting shader graph is traversed multiple times. Each traversal requests shader nodes to output different shader code blocks:
   a. Global variables definition: uniforms, varyings, etc.
   b. Functions
   c. Code used inside the main function
3. The preceding code outputs are concatenated into the final text of the shader code.

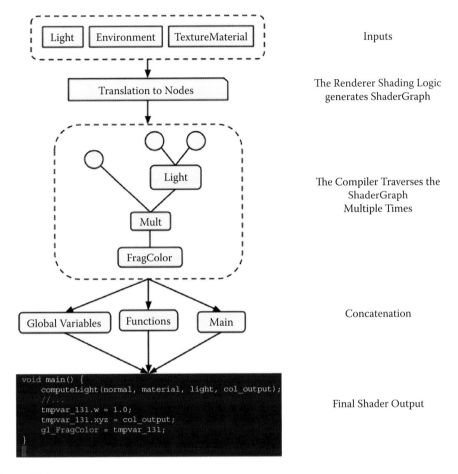

Figure 12.7

Transformation of input into final shader code.

12.3.2 Shader Node Allocation: Choices

When we developed our shader graph system, the goal was to maximize graph flexibility with renderer-independent, reusable shader nodes. To accomplish this, material inputs (TextureMaterial, Environment, and Light state attributes) and shader nodes do not directly correspond. Rather, the correspondences are created by the shader graph generation code (via the compiler). All shading logic decisions are made during shader graph generation. This approach allows us to retain a library of reusable shader nodes (see Table 12.1), which are independent of the final renderer type.

For example, a Light state could be fed into a *cook torrance* or *phong* light node. The compiler builds the shader graph, allocating and linking code from inputs. The orientation of the renderer (e.g., custom shader logic in the case of postprocess effects or the type of shading to attach) is defined for each chosen node from the corresponding input.

Table 12.1 Shader Nodes Used in Sketchfab

Data	Textures	Operations	Functions	Lights
Sampler	TextureRGB	Blend	sRGB2Linear	Light
Variable	TextureRGBA	MultVector	Linear2sRGB	Lambert
Varying	TextureAlpha	AddVector	DotClamp	SpheremapReflection
Uniform	TextureCubemapRGB	InlineCode	NormalTangentSpace	CookTorrance
InlineConstant	TextureSpheremap	ReflectionVector	EnvironmentTransform	
FragColor	TextureSpheremapHDR	SetAlpha	TonemapHDR	
	TextureTranslucency	PassValue	Bumpmap	
	TextureIntensity	Vec3ToVec4	NormalAndEyeVector	
	TextureNormal	Vec4ToVec3	NormalMatcap	
	TextureGradient	DotVector	FrontNormal	
		PreMultAlpha		

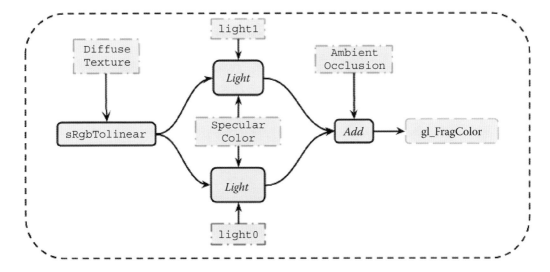

Figure 12.8

Shader graph. Red dotted boxes are inputs. gl _ FragColor is the end of the graph.

12.3.3 Shader Graph Traversal and Compilation

A shader graph is a directed acyclic graph and has a straightforward internal compilation mechanism. As the path is traversed, each node's output is followed. The output from the previous node becomes the input for the next node in the graph. For a shader node to be taken into account during compilation, it must ultimately traverse to gl_FragColor (see Figure 12.8). During each traversal, different fragment shader parts are gathered. The final *node output* is the fragment shader gl_FragColor, and the final *traversal* produces the main() function. This ensures that main() always ends with the correct output.

The graph is traversed multiple times, once for each part of the resulting shader code (see Listing 12.2), thereby outputting an "ordered" shader source. Global variables (like varyings and uniforms) are declared first. Global functions are declared next. Lastly, the main() code is produced.

Listing 12.2 Compiler multiple traversal.

```
//collect all active attributes from state for the current geometry
   and filter
//Material Attributes we are interested in

var attributes = collectMaterialAttributes();

//get a hash from all active Material Attributes to check if in cache
var hash = attributes.getHash();
```

```
if (caches[hash]) return caches[hash];

//Shader Generation Starts now

//generate a shader graph from shader node
var shaderGraph = createShaderGraph(attributes);
var fragmentShader = '';

//evaluate the shader graph to define global variables
fragmentShader + = traverseGraph(shaderGraph,
    defineGlobalVariableFunctor);

//evaluate the shader graph to declare functions
fragmentShader + = traverseGraph(shaderGraph, defineFunctionFunctor);

//evaluate the shader graph to declare main() variables and code
fragmentShader + = [
    "void main()",
        traverseGraph(shaderGraph, declareVariableFunctor),
        traverseGraph(shaderGraph, computeNodeFragmentFunctor),
    "}"].join('\n');

//post process the content (precision, files includes, and other
    custom post process)
var finalFragment = shaderProcessor(fragmentShader);
cache[hash] = finalFragment;
return finalFragment;
```

Both simple shader nodes (those outputting only GLSL code in the main function) as well as more complex shader nodes (like "light," which can add new varyings/uniforms/ functions, etc.) are allowed using the preceding shader graph implementation. Identical declarations by multiple nodes are easily handled, and shader code is constructed without duplication. An output link, needed for compiler traversal, is the only mandatory code for a shader node. The implementation's simplicity builds smaller code and improves customization[*] opportunities.

12.3.4 Shader Graph Custom Usage: A Code Sample

To demonstrate our system, let's look at an example.[†] In the following, we implemented two basic custom shader nodes and a custom shader compiler. The two nodes—RampNode (using a different color for some range of lighting) and a NegatifNode (which changes the color to 1.0)—are activated/disabled by custom StateAttributes. These state attributes are declared as RampAttribute[‡] and NegatifAttribute.[§]

---

[*] https://github.com/cedricpinson/osgjs/blob/develop/sources/osgShader/Compiler.js
[†] https://github.com/cedricpinson/osgjs/tree/develop/examples/shader-generator
[‡] https://github.com/cedricpinson/osgjs/tree/develop/examples/shader-generator/ramp.js
[§] https://github.com/cedricpinson/osgjs/tree/develop/examples/shader-generator/negatif.js

The `StateAttributes` and shader nodes are the two basic ingredients feeding the shader compiler. We also implemented a custom `createFragmentSha-derGraph` function that handles shader logic and builds shader code from inputs (see Listing 12.3) in exactly the same manner as the original OSG.JS `createShad-erGraph` method.[*]

Listing 12.3 Custom compiler pseudocode demonstrating usage of custom `StateAttributes` and shader nodes.

```
createFragmentShaderGraph: function() {

  if (!MaterialAttribute) return;

  var fragColor = createFragColorAsRootNode();
  var diffuse = Material.diffuse || firstTextureFound() * Material.
     diffuse;

  if (LightsAttributes) {
    var color = createLights(diffuse);//diffuse as input

    if (RampAttribute && RampAttribute.isEnable())
        color = createRampNode(color);//connect color as input

    if (NegatifAttribute)
        color = createNegatifNode(color);//connect color as input

    fragColor = emission + color;
  } else {
      fragColor = diffuse;
  }
}
```

12.3.5 Shader Code Output: A Comparison

After all compiler traversals are complete, the final code compilation is easily readable and debuggable. The only flaws in the code's readability are the temporary variables created as "links" between shader nodes. Unfortunately, the links spawn a long list of `tmp_X` variables (see Listing 12.4) on the main level. Depending on GLSL compiler vendors, a potential side effect is register spilling. A "hand-coded" shader would undoubtedly create fewer temporary variables and therefore fewer registers.

Nonetheless, thanks to complex nodes that use functions rather than inline code (see `computeSunLightShading` in Listing 12.4), the resulting code is still surprisingly straightforward and readable.

---

[*] https://github.com/cedricpinson/osgjs/blob/develop/examples/shader-generator/CustomCompiler.js

Listing 12.4 Sample output from shader graph compilation.

```
... define, global vars, func, etc...
void main() {
  vec3 frontNormal; vec3 normal; vec3 eyeVector; vec4 tmp_22; float
    tmp_23; vec4 lightOutput; vec4 lightTempOutput; vec4 tmp_26; vec4
    tmp_27;

  tmp_22.rgb = MaterialDiffuse.rgb;
  if (ArrayColorEnabled = = 1.0) {
    tmp_22 * = VertexColor.rgba;
  }
  frontNormal = gl_FrontFacing ? FragNormal : -FragNormal ;
  normalizeNormalAndEyeVector(frontNormal, FragEyeVector, normal,
    eyeVector);

  lightTempOutput = computeSunLightShading(normal,   eyeVector,
                                           MaterialAmbient,
                                           tmp_22, MaterialSpecular,
                                           MaterialShininess,
                                           Light0_uniform_ambient,
                                           Light0_uniform_diffuse,
                                           Light0_uniform_specular,
                                           Light0_uniform_position,
                                           Light0_uniform_matrix,
                                           Light0_uniform_invMatrix);

  lightOutput = lightTempOutput.rgba;
  tmp_26 = MaterialEmission.rgba+lightOutput.rgba;
  tmp_23 = MaterialDiffuse.a;
  tmp_27.rgb = tmp_26.rgb * tmp_23;

  gl_FragColor = vec4(tmp_27.rgb, tmp_23);
}
```

12.4 Conclusion

An absolutely essential (albeit complicated) requirement for Sketchfab is seamlessly supporting an array of 3D file formats and variations. In particular, we sought to minimize the effort necessary when handling the numerous combinatorial inputs of material data, each with its own peculiar parameterization. Our materials pipeline required continuity when handling currently supported materials, while simultaneously easing the introduction of new ones.

To allow maximum flexibility for any material input, Sketchfab adopted a shader graph architecture. In the first iteration of Sketchfab, a huge fixed shader created a significant amount of unused code. With the shader graph approach, however, the code is created by simple, isolated node operations. Overall code complexity is reduced, resulting in reusable and more easily debuggable code.

The shader graph system is easily extensible and efficiently handles the addition of new (or multiple) inputs. Customization only requires including new nodes and adjusting the shader generation code (gathering the tailored input and linking new nodes into shaders). The primary goals for our system are achieved: maximum flexibility for material inputs, minimal impact on the current code base, and the final production of easily debuggable shader code.

As for future improvements to our materials system, an optimization traversal could improve shader code performance based on GPU constraints (counting registers, uniforms, texture fetches, etc.). The introduction of these elements, however, would require that code complexity be minimally impacted.

As it stands, the JavaScript code is still quite verbose. We would like to streamline the API to add custom behavior to the system. Also, greater simplification of shader graph debugging (by using real-time shader graph display, incorrect node usage detection, or input/output connection shortening) might also further improve our shader development process.

13

glslify
A Module System for GLSL

*Hugh Kennedy, Mikola Lysenko, Matt DesLauriers,
and Chris Dickinson*

13.1 Introduction

In 2009, Ryan Dahl released Node.js,* a platform for writing applications with JavaScript. Shortly thereafter Mikeal Rogers released npm,† Node's officially endorsed package manager and registry. npm came with features and ideals that have since become central to supporting the Node.js ecosystem. These include Semantic Versioning, nested dependencies, and, in particular, the encouragement of "small modules": granular dependencies that do one thing well, making them easier to combine and reuse.‡

   glslify is an effort to apply some of the same functionality and ideals to GLSL by introducing a simple module system inspired by Node.js, taking advantage of the existing npm ecosystem for module distribution. In doing so, we can improve the initial learning curve

---

* http://nodejs.org/
† https://www.npmjs.com/
‡ http://substack.net/node_aesthetic

of GLSL and WebGL, as well as improve our productivity by sharing shader code between projects for straightforward consumption.

In this chapter, we introduce the glslify work flow by covering the fundamentals of using CommonJS, Semantic Versioning, npm, and, of course, glslify itself.

13.2 Modular Programming in Node.js

The main idea behind glslify is to apply the modular programming practices of Node.js to GLSL shaders. By analogy to manufacturing, the goal of modular programming is to reduce programs to the composition of interchangeable parts. These parts themselves are recursively divided into simpler components until, ultimately, the final components are small generic structures (like screws or washers in manufacturing). These components need not be built by even the same author and can be shared and reused across many projects.

In the Node.js community there is a strong culture of modular programming that has been fostered by npm. The npm approach has proven popular and has well over 80,000 public packages available.[*] Other tools, such as browserify,[†] enable the use of these packages within the browser. As the name may now imply, glslify allows us to use npm to distribute and consume shader code.

npm relies on a small number of conventions that have contributed to its success. These conventions have made using large dependency trees pleasant—and even preferable in many cases—when working with JavaScript.

13.2.1 CommonJS

Node.js uses a variant of CommonJS[‡] for handling imports and exports between modules, and its implementation is simple.

In Node, each module is represented as a single JavaScript file with a `module.exports` object, the value of which is "exported" for use in other modules. This exported value may be of any type. Objects, functions, numbers, and strings are all acceptable, though a single function or class is generally preferred.

```
//sum.js
module.exports = function sum(a, b) {
  return a + b;
};
```

To then import that value elsewhere, one may use the global `require()` function, specifying the path of the module:

```
//index.js
var sum = require('./sum.js');

sum(1, 2);//3
```

---

[*] http://alexandros.resin.io/npm-now-the-largest-module-repository/
[†] http://browserify.org/
[‡] http://wiki.commonjs.org/wiki/CommonJS

Scope within a module is local to that module, and variables declared in one file will not be accessible elsewhere unless made available through `module.exports`.

Tip: For more detail on the intricacies of Node's CommonJS implementation, refer to the documentation: http://nodejs.org/api/modules.html

13.2.2 node_modules

Node's module resolution algorithm makes a special case for directories within the `node_modules` directory: Each of these is a *package* installed from npm and may be required without the need for specifying a relative path.

For example, after installing the browserify and gulp packages, one should be able to require them as in the following:

```
var browserify = require('browserify');
var gulp = require('gulp');
```

We can use the `npm install` command from our terminal to install one or more packages from npm. To install `browserify` and `gulp` simply run:

```
npm install browserify gulp
```

These will be downloaded from the npm registry along with their own dependencies and extracted into the `node_modules` directory.

Importantly, `node_modules` is not a flat list of dependencies. Every Node package has a dependency tree, where each package may have its own locally scoped dependencies in its own `node_modules` directory, as shown in Figure 13.1. This is handled by npm when installing dependencies, so we only need to concern ourselves with top-level dependencies for each project.

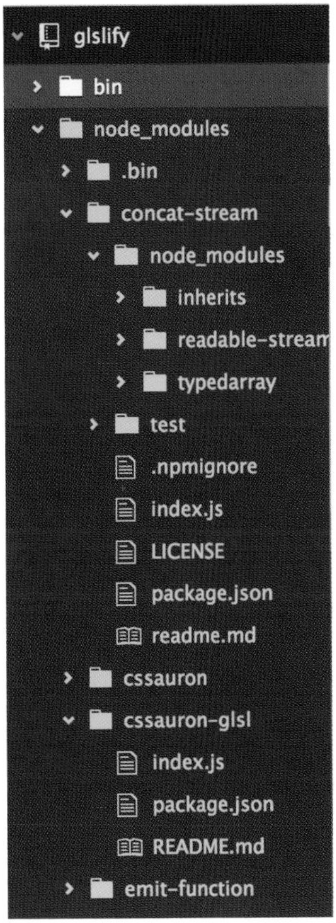

13.2.3 Semantic Versioning

With a greater number of dependencies comes a greater deal of complexity in managing their versions, fondly referred to as *dependency hell*.

npm minimizes this issue by taking advantage of Semantic Versioning[*] (SemVer). Each time a package is published, it is given a unique version number incremented from the previous one. The version is incremented according to the following rules [Preston-Werner 11]:

Given a version number MAJOR.MINOR.PATCH, increment the:

1. MAJOR version when you make incompatible API changes
2. MINOR version when you add functionality in a backward-compatible manner

Figure 13.1

Example directory structure after performing `npm install` on a project with nested dependencies.

---

[*] http://semver.org/

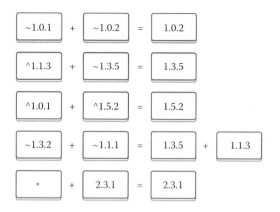

Figure 13.2

Some example SemVer combinations for a single package required by two different dependent packages. This figure demonstrates the resulting versions shared between them, where possible.

3. PATCH version when you make backward-compatible bug fixes

When dependencies are specified, they are not given a static version, but rather a version range. This way, one may safely update a package deep within the tree and its dependent packages will install the most up-to-date version that their authors are comfortable with.

npm uses several shorthand "operators" for specifying version ranges:

- `*` is shorthand for `>=0.0.0` or any available version.
- ~a.b.c is shorthand for > = a.b.c <a.(b+1).0
- ^a.b.c is shorthand for > = a.b.c <(a+1).0.0

Some examples are shown in Figure 13.2. Most often, package dependencies are specified using either the ~ or ^ operators, based on the apparent stability of the package and the preference of the package author depending on them. The full semantics of npm SemVer ranges are covered in detail in the `npm/node-semver`[*] repository on GitHub.

If two packages use a dependency with compatible version ranges, they will find the best available common version and share a single copy between them. Otherwise, each package receives a copy of its best available version to use separately.

The end result is that we get transparent updates where possible, and also the penalty for having two conflicting versions within our dependency graph is minimized.

13.2.4 package.json

Every package published to npm and every project using npm dependencies has a package.json file. This is used to specify dependencies and any additional metadata for Node or other tools to take advantage of. At its simplest, a `package.json` file has two fields: a `name` and a `version`. Generally, we also want to include a `main` field, which specifies which file to load from the package when imported using `require`, and a `description` field.

```
{
  "name": "my-package",
  "version": "1.0.0",
  "main": "lib/my-package.js",
  "description": "My first npm package"
}
```

There are a number of other fields that may be included in a `package.json` file. These fields are documented in full on the npm website.[†]

---

[*] https://github.com/npm/node-semver
[†] https://docs.npmjs.com/files/package.json

13.3 What Is glslify?

glslify is a tool built using Node.js that aims to statically analyze, transform, and write GLSL shader code. At its core, it enables developers to modularize and author shader programs using the same practices used in Node modules on npm. Reusable parts can be decoupled from the application-specific code surrounding them, and these parts can be published with semantic versioning to npm. Using Node's file lookup algorithm for `require` statements, it becomes possible to safely depend on reusable snippets of GLSL and even receive patches upstream without the need to modify our shaders.

This is achieved by parsing the shader source, applying transformations to it, and then rewriting it as runnable and conformant GLSL code that can be run in WebGL, as shown in Figure 13.3. Since the tooling has been built on top of Node.js, glslify can be used on the server or during the build step of client-side JavaScript source using a tool such as browserify or require.js. A command-line interface is also provided for more generic uses, enabling GLSL source transformations outside the scope of JavaScript and WebGL.

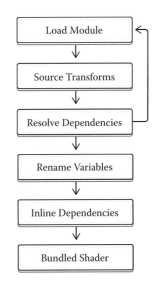

Figure 13.3

glslify recursively loads, transforms, and inlines shaders throughout the dependency tree, resulting in a single shader source as the final output.

13.3.1 Syntax: Exporting Modules

glslify gives us a new #pragma command that we can use to "import" and "export" symbols to and from modules. Much like Node, in glslify, modules and files have a one-to-one correspondence: Each file is also a module that exports a single symbol. At a basic level, this allows us to split up shaders into multiple files for better organization.

As an example, suppose that we would like to separate the implementation of the lighting computation from the rest of the shader. For simplicity we will use the Blinn–Phong lighting model. Let us begin by defining a light data type:

```
//File: light.glsl
struct PointLight {
  vec3 position;
  vec3 diffuse;
  vec3 ambient;
};

#pragma glslify: export(PointLight)
```

The line #pragma glslify: export(PointLight) tells glslify to export the struct PointLight to whatever module requires this file. We can use the same technique to define a material data type as well:

```
//File: material.glsl
struct Material {
```

```
  vec3 color;
  float kAmbient, kDiffuse, kSpecular;
  float phongExponent;
};

#pragma glslify: export(Material)
```

Next, we will create a subroutine that takes these two data types as well as some information about the surface/viewing geometry as input and computes the intensity of the light in each channel:

```
#pragma glslify: Material = require(./material.glsl)
#pragma glslify: PointLight = require(./light.glsl)

vec3 blinnPhong(
  Material material,
  PointLight light,
  vec3 surfaceNormal,
  vec3 surfacePosition,
  vec3 viewDirection) {

  vec3 L = normalize(light.position - surfacePosition);
  vec3 H = normalize(viewDirection + surfaceNormal);

  vec3 ambient = light.ambient * material.color;
  vec3 diffuse = max(0.0, dot(L, surfaceNormal)) * light.diffuse *
      material.color;
  float specular = pow(max(0.0, dot(L, H)), material.phongExponent);

  return material.kAmbient * ambient +
      material.kDiffuse * diffuse +
      material.kSpecular * specular;
}

#pragma glslify: export(blinnPhong)
```

13.3.2 Syntax: Requiring Modules

Once we have declared and exported a module, we can use it by requiring it. For example, we can combine the preceding modules for the Blinn–Phong lighting model to create the following fragment shader (see Figure 13.4):

```
precision mediump float;

#pragma glslify: Material = require(./material.glsl)
#pragma glslify: Light = require(./light.glsl)
#pragma glslify: computeColor = require(./blinnPhong.glsl)

#define NUM_LIGHTS 16

varying vec3 normal, position, viewDirection;

uniform Material material;
uniform Light lights[NUM_LIGHTS];

void main() {
  vec3 color = vec3(0,0,0);
```

```
  for(int i = 0; i<NUM_LIGHTS; ++i) {
    color += computeColor(
      material,
      lights[i],
      normal,
      position,
      viewDirection);
  }

  gl_FragColor = vec4(color, 1.0);
}
```

Here, we are using `require` to bring in the `blinnPhong` function and assign it the local name `computeIntensity`. The require statement uses the same algorithm as Node's require lookup:

- A require path starting with a forward slash (/) is assumed to be an absolute path to a glslify module.
- A require path starting with a dot and forward slash (./) is assumed to point to a glslify module relative to the path of the file calling `require()`.
- A require path without a starting slash or dot-slash will look for modules by name inside a `node_modules`, recursively moving up directories until it's found.

Unlike Node's require, quotes are optional in glslify's require statements. This is because GLSL itself does not have the concept of strings, and some parsers will crash on handling a shader with quotes.

Figure 13.4

An example rendering from the Blinn–Phong shader.

13.3.3 Using Modules from npm

One of the most important features of glslify is that it allows users to install and use modules. Any user can author a module and publish it using the npm package manager. Dependencies of modules are installed using npm and may be loaded easily due to glslify's use of Node's module resolution algorithm.

For example, in the following fragment shader, we import a function from a module that has been authored with glslify and distributed on npm. The glsl-noise[*] module provides us with the popular simplex and classic noise functions by Ashima Arts and Stefan Gustavson.[†]

```
varying vec2 vUv;

#pragma glslify: snoise2d = require(glsl-noise/simplex/2d.glsl)

void main(void) {
  float n = snoise2d(vUv);
  gl_FragColor = vec4(vec3(n), 1.0);
}
```

In the previous snippet, the algorithm finds the `glsl-noise` folder that has been installed into `node_modules` and then looks inside it to find the `simplex/2d.glsl` file. We can choose to omit the file extension if we're using .glsl, so the preceding require statement could have also been written in the following way for brevity:

```
#pragma glslify: snoise2d = require(glsl-noise/simplex/2d)
```

If the require statement points to a directory, it will instead check the directory's contents for GLSL files. It will first look for a `package.json` file in that directory and select the file defined in that package's `main` field. If no `package.json` exists, or the `main` field is missing, it will use that directory's `index.glsl` file.

Files are always imported in preference to directories. In the preceding example, `glsl-noise/simplex/2d.glsl` will always be used instead of `glsl-noise/simplex/2d/index.glsl` if both files exist.

Some examples of valid require statements include:

```
//a module installed by npm into node_modules
#pragma glslify: random = require(glsl-random)

//a relative file
#pragma glslify: bar = require(../foo/bar.glsl)

//a relative directory with an index.glsl entry point
#pragma glslify: blend = require(./blend-color-dodge)

//a relative path within a module
#pragma glslify: snoise2d = require(glsl-noise/simplex/2d)
```

---

[*] https://github.com/hughsk/glsl-noise
[†] https://github.com/ashima/webgl-noise

13.3.4 Syntax: Local Values

glslify also provides a means of passing names across modules. One use case for this behavior is creating higher-order functions. For example, suppose that we want to implement a derivative calculation. Then we could write a module to do this as follows:

```
//File: derviative.glsl
float func(float);

float derivative(float t, float epsilon) {
  return 0.5 * (func(t+epsilon) - func(t-epsilon));
}

#pragma export(derivative)
```

To use this module, we then require as before, only we replace the local value func with whatever custom function we want to differentiate:

```
float position(float t) {
  return 0.5 * t * t - t + 1.0;
}

#pragma glslify: velocity = require(./derivative, func = position)
```

Other uses for this feature include ray marching and constructive solid geometry (CSG) using signed distance fields.

13.3.5 Installing Modules

Firstly, ensure that a package.json file has been created in the project's root directory. The file must be valid JSON and contain both a name and version field. Here's an example we can use:

```
{
  "name": "glsl-project",
  "version": "1.0.0"
}
```

Once we've discovered a package and its name on npm, we should be able to install it locally to use in our project as in the following:

```
npm install --save <package-names...>
```

If we wanted to install glsl-noise, for example, we could use the following:

```
npm install --save glsl-noise
```

The --save flag used here will automatically add the packages to the closest available package.json file, which in most Node projects should be the root project directory. Take a look at our package.json file again. It should look similar to the following:

```
{
  "name": "glsl-project",
  "version": "1.0.0",
```

```
  "dependencies": {
    "glsl-noise": "0.0.0"
  }
}
```

Now if somebody else were to use our project, they could reinstall our dependencies by simply running:

```
npm install
```

We can try it ourselves now by removing our `node_modules` directory and installing our dependencies again:

```
rm -rf node_modules
npm install
```

13.3.6 Using the Command-Line Interface (CLI)

At its simplest, one can build a shader with glslify as in the following:

```
glslify input.glsl --output output.glsl
```

This will take the `input.glsl` file, build it using glslify, and write the output to `output.glsl`. There are also additional flags:

- `— transform <tr>` applies a local transform (see Section 13.4) to the included shaders.
- `— global-transform <tr>` applies a global transform (see Section 13.4) to the included shaders.
- `— list` lists all of the files in the dependency graph and is useful for Makefiles.
- `— deps` lists out the dependency graph as JSON with additional metadata.

To install glslify for use from the command-line, use npm to install it globally instead of locally:

```
sudo npm install --global glslify
```

13.3.7 Publishing Shaders

Once we have a project set up, it's not too difficult to then publish it to npm.

Before getting started, we need to create an account by logging into npm from the CLI:

```
$ npm login
Username: your-username
Password: **********
Email: your@email.com
```

Next, we want to create a new package.json file for the project:

```
{
  "name": "glsl-example-sum",
  "version": "1.0.0",
  "description": "An example glslify shader: sums two values"
}
```

The name field must be unique, as npm has a single namespace for published packages. If unsure, check http://npmjs.org/package/<name> to make sure the name isn't already taken. The version field should be familiar, and description is an additional field used by npm for searching for and displaying the package.

Let's create an index.glsl file: This file will be the default file loaded when users of our package use require(glsl-example-sum):

```
float sum(float a, float b) {return a + b}
float sum(vec2 a, vec2 b) {return a + b}
float sum(vec3 a, vec3 b) {return a + b}
float sum(vec4 a, vec4 b) {return a + b}

#pragma export(sum)
```

If desired, we can include additional files of any kind in the project (e.g., a runnable example demonstrating how to use the package). It's also advised to include a Markdown-formatted readme file, generally named README.md, with any relevant information one might need before using the package.

Otherwise, this should be all that's required to publish a package to npm. From the project directory, run:

```
npm publish
```

This will publicly publish the package, immediately available for installation through npm. There are no rules on what may or may not be published to npm, though more information on what is considered "best practice" can be found at Matt DesLaurier's module-best-practices repository on GitHub: https://github.com/mattdesl/module-best-practices.

13.4 Source Transforms

GLSL is a simple language that works very well within its intended domain. However, much like any other language, there is always room for improvement when dealing with specific problems—be it to make our shader code more concise, more performant, or less logically complex.

This is evidenced in the JavaScript community, for example, with the increasingly large number of "compile to JS" languages, which add features for clarity, performance, and developer productivity. Much like JavaScript is for our browsers, GLSL is the only available shader language when working with WebGL and, as such, there's room for new variants of the language that could potentially provide the same benefits.

glslify has support for *source transforms* to allow us to easily extend or modify the language either at the application level or for each shader module individually. Transforms are written in JavaScript and made available as Node.js modules that can be published to and downloaded from npm.

Each transform may be applied in one of three ways, each with a different use case:

1. **Local transforms** are the default and are applied only to application-level shader files: Shaders inside node_modules are ignored.

2. **Global transforms** ignore this restriction and are applied to every shader file.
3. **Post transforms** are applied when the build is complete on the entire resulting shader and are particularly useful for applying shader optimizations.

13.4.1 glslify-hex

glslify-hex is a simple, practical example of a source transform: It allows us to specify color vectors as hex values, making it easier as a web developer to share colors between our CSS and our shaders.

By enabling the transform, we can use the following in our shader file:

```
void main() {
  gl_FragColor = vec4(#FF0066, 1);
}
```

During the process of building the shader, the preceding code will be converted into this:

```
void main() {
  gl_FragColor = vec4(vec3(1.0, 0.0, 0.4), 1);
}
```

The code required for creating this transform is reasonably simple:

```
var hexArray = require('hex-array');
var hexFloat = require('hex-float');

module.exports = transform;

function transform(file, source, options, done) {
 //Updates the original file's shader source
 //to replace hex values with vec3s and vec4s.
 var hexExpression =/\#([0-9a-f]{6}|[0-9a-f]{8}?)/gi;

source = source.replace(hexExpression, function(hex) {
  var channels = hexArray(hex).map(function(n) {
   return hexFloat(n);
  });

  return channels.length === 3
   ? 'vec3(' + channels.join(',') + ')'
   : 'vec4(' + channels.join(',') + ')';
 });

    //Sends the updated source back to glslify to
    //continue building the shader.
    done(null, source);
}
```

In many other situations we might prefer to fully parse the shader into an abstract syntax tree (AST) before making any modifications, but thankfully hex values are unlikely to appear unintentionally in shader code. As such it's relatively safe in this case to simply perform a find/replace operation on the raw source. In most other cases, we could use the glsl-parser[*] and glsl-deparser[†] libraries available on npm to assist in parsing and safely modifying GLSL code: glslify uses these to perform its own source transformations.

13.4.2 Using Transforms

Adding a transform is similar to installing a shader module: Start by installing the transform as a dependency:

```
npm install --save glslify-hex
```

Once installed, we can tell glslify to use it by adding it to the `glslify.transforms` array in our `package.json` file:

```
"glslify": {
  "transforms": ["glslify-hex"]
}
```

After performing this, our `package.json` file should look something like the following:

```
{
  "name": "glsl-project",
  "version": "1.0.0",
  "dependencies": {
    "glsl-noise": "0.0.0",
    "glslify-hex": "^2.0.0"
  },
  "glslify": {
    "transforms": ["glslify-hex"]
  }
}
```

We should now be able to create a shader file in the same directory called index.glsl:

```
precision mediump float;

void main() {
   gl_FragColor = vec4(#FF0000, 1.0);
}
```

And when building it with glslify, the hex transform will be applied automatically:

```
$ glslify index.glsl

precision mediump float;

void main() {
   gl_FragColor = vec4(vec3(1.0, 0.0, 0.0), 1.0);
}
```

---

[*] https://github.com/stackgl/glsl-parser
[†] https://github.com/stackgl/glsl-deparser

13.5 Shortcomings

Because glslify relies on npm and the filesystem, some form of preprocessing step is required to make use of it. Efforts have been made to remove this step from similar tools,[*] but the implementation is currently nontrivial and introduces a network dependency on the npm registry. Whether or not this is an issue for a project depends on our current work flow for deploying WebGL applications. In many cases it should be negligible for projects already using a build tool such as browserify or WebPack, as is becoming an increasingly common practice in the frontend web development community; however, some teams prefer to avoid the overhead of such a tool.

Because glslify requires a preprocessing step, it also limits our ability to generate shaders at runtime in the browser, which may be desirable for improving developer productivity (as explored in Chapter 12, "Sketchfab Material Pipeline: From File Variations to Shader Generation"). Tools such as `glslify-deps`[†] allow us to preprocess our shader sources and update shader modules on the fly, though this is less flexible than manipulating strings directly.

At the time of writing, glslify officially supports only WebGL-compatible variants of GLSL. This will likely change in the future to accommodate OpenGL and OpenGL ES projects. Despite variations between the myriad versions of GLSL, the core syntax does not differ greatly. Often, support would only be a matter of considering new keywords and core functions, and support for WebGL 2 should already be available.

13.6 Contributing

glslify is still under development and new contributors are more than welcome to help out where they see fit—be it in the form of documentation, bug fixes, or additional features. Another area of interest is in framework interoperability: making glslify work well alongside a range of 3D frameworks, WebGL or otherwise.

Outside of contributing to the core codebase, you can, of course, also contribute by creating glslify packages and publishing them to npm!

The project is hosted at http://github.com/stackgl/glslify/ and is available under the MIT license. It's maintained by the stack.gl community, an "Open Open Source"[‡] project with the aim of producing small, independent packages for working with WebGL without the need for a large framework.

Bibliography

[Preston-Werner 11] "Semantic Versioning 2.0.0" https://github.com/mojombo/sem ver/ blob/a2769c455b57e7e5e7e9 1d81cc0a61acad68122a/semver.m d#summary, 2013.

---

[*] https://github.com/mafintosh/browserify-browserify and http://requirebin.com

[†] https://github.com/stackgl/glslify-deps

[‡] http://openopensource.org/

14

Budgeting Frame Time

Philip Rideout

14.1 Introduction

In this chapter, we cover some common techniques for optimizing WebGL graphics engines. At a high level, these techniques all focus on one of two questions: *when* and *how long*. *When* is the best time to make WebGL calls? *How long* is the budget for a sequence of WebGL-related tasks?

First, we'll discuss *when* it's appropriate to make GL calls. Calling any GL function outside the `requestAnimationFrame` entry point can hurt performance. Even if we make an innocent-looking call to `uniform1f` while `onload` is at the top of the stack, we've violated this requirement. We'll discuss a design technique that makes it easy to satisfy this requirement in a natural way.

In the second half of the chapter, we'll focus on *how long* the graphics work takes in each animation frame. Ideally, this never exceeds 16 ms, because Chrome and other modern browsers render DOM elements at around 60 frames per second (fps). We'll cover several techniques that can help us to stay in budget.

Tip: The 60 fps limit can be lifted during development to help with performance measurements, by passing disable-gpu-vsync to Chrome. See Chapter 5 for more information, including a similar tip for Firefox.

14.2 Deferring until the Draw Cycle

The RequestAnimationFrame (RAF) API is the recommended way of triggering a redraw of our WebGL scene; it kicks off the game loop in almost every WebGL application. Unlike `setInterval` and `setTimeout`, RAF is meant specifically for animation, so the browser can do smart things with it. For example, RAF isn't called when the app is backgrounded or overloaded.

You're already using RAF instead of `setInterval`? That's great, but don't stop there! For example, if a certain mouse action causes the engine to call `bufferSubData` to modify a mesh, it's usually better to defer this work until the next RAF, rather than performing it immediately. Several mouse events might occur between each RAF event, and you often only care about the last one. Moreover, on some browsers, performing GL calls outside the RAF entry point can incur a context switch. In the next section, we'll illustrate a common convention for deferring GL-related work.

14.2.1 Example: Texture Wrapper

Suppose our graphics engine proffers a simple `Texture` class to clients. The `Texture` class doesn't do much; it simply loads an image from a URL and creates a WebGL texture object from it. The class could be used like this:

```
var mytexture = new MyEngine.Texture();
mytexture.load("http://lorempixel.com/g/1920/1080/");
//then, some time later:
gl.bindTexture(gl.TEXTURE_2D, mytexture.webgl_texture);
```

This sounds simple enough; a naive implementation for this class is shown in Listing 14.1:

Listing 14.1 Naive texture wrapper class.

```
MyEngine.Texture = function() {
    this.webgl_texture = gl.createTexture();
    this.image_element = new Image();
    this.image_element.addEventListener('load',
        this.loaded.bind(this));
};

MyEngine.Texture.prototype.load = function(url) {
    this.image_element.src = url;
};

MyEngine.Texture.prototype.loaded = function() {
    gl.bindTexture(gl.TEXTURE_2D, this.webgl_texture);
    gl.texImage2D(gl.TEXTURE_2D, 0, gl.RGBA, gl.RGBA,
        gl.UNSIGNED_BYTE, this.image_element);
    gl.texParameteri(gl.TEXTURE_2D, gl.TEXTURE_MIN_FILTER,
        gl.LINEAR);
    gl.bindTexture(gl.TEXTURE_2D, null);
};
```

Setting aside the fact that it doesn't have an error handler, the problem with Listing 14.1 is that several WebGL calls occur outside the RAF entry point. Instead they occur in the `loaded` handler. Listing 14.2 remedies this by moving all WebGL calls into an `update` method, which `MyEngine` calls during the RAF cycle.

Listing 14.2 Improved texture wrapper class.

```
MyEngine.Texture = function() {
    this.webgl_texture = null;
    this.image_element = null;
};

MyEngine.Texture.prototype.load = function(url) {
    this.image_element = new Image();
    this.image_element.src = url;
};

MyEngine.Texture.prototype.update = function() {
    var img = this.image_element;
    if (img && img.src && img.complete) {
        this.webgl_texture = gl.createTexture();
        gl.bindTexture(gl.TEXTURE_2D, this.webgl_texture);
        gl.texImage2D(gl.TEXTURE_2D, 0, gl.RGBA, gl.RGBA,
            gl.UNSIGNED_BYTE, img);
        gl.texParameteri(gl.TEXTURE_2D, gl.TEXTURE_MIN_FILTER,
            gl.LINEAR);
        gl.bindTexture(gl.TEXTURE_2D, null);
        this.image_element = null;
    }
};
```

In the new implementation in Listing 14.2, not only do we avoid WebGL work during the load event, but we also avoid it inside the class constructor. This gives clients the freedom to instantiate the object whenever they want.

By restricting all WebGL calls to a small handful of well-defined methods, our graphics engine deals with RAF in a consistent way. It also makes it easier to debug and interoperate with other graphics engines.

Aside: We should mention that the normative GL state of our hypothetical graphics engine has the active stage set to TEXTURE0, and that there is no texture bound to that stage. By "normative state," we mean that all `update` methods in the engine are expected to leave the GL state machine in much the same state that they found it, to be polite to other entities.

14.3 Amortizing Work across Frames

We've discussed *when* to make WebGL calls; now let's discuss *how long*. If an application spends more than 16 ms inside a given animation tick, then it's blocking the browser from its usual upkeep, such as garbage collection and processing events. Even though the browser's refresh rate is around 16.6 ms, in practice our budget should be closer to 12 ms, because we need to give the browser 2–4 ms for its own work [Thompson 12].

Aside: One tool that shows great potential in the analysis of the WebGL frame is Google's open-source tracing framework. See their website for more: http://google.github.io/tracing-framework/index.html

If we have some frames that spill over the budget, the frame rate can appear even more un-smooth with WebGL than with other graphics environments, because the browser's scheduler is continuously trying to adjust. In the best case, the browser's scheduler calls RAF at 60 fps, but if it sees that frames are taking longer, it adjusts to a lower frequency. This is why staying in budget is crucial.

If an application occasionally exceeds 16 ms even after employing the techniques in this chapter, one alternative is to clamp the rate to a solid, modest 30 fps, as suggested in [Thompson 12]. This can be done by rendering only the even-numbered frames. The scheduler can adapt to this frequency, and at least the frame rate will not be jaggy.

14.3.1 Amortization Technique 1: Queue of Updateable Objects

It's common for graphics engines to define a base class—say, Entity—that proffers polymorphic methods for drawing, updating, responding to certain events, and so on. By making a strong distinction between updating (or initialization) and actual rendering, we can impose a time limit only on the update tasks.

For example, WebGL calls that involve transferring data to the GPU, such as texImage2D and bufferSubData, should occur at "update time" when following this engine design, because they are relatively slow operations. Calls like bindVertexArrayOES and drawElements should occur at "draw time" because they are necessary for rendering the frame. If we follow this rule of thumb, we can amortize the update tasks but not the draw tasks. Listing 14.3 is our first attempt at designing a game loop with distinct update and draw passes. It has some major problems that we'll fix later:

Listing 14.3 Top-level game loop that needs improvement.

```
MyEngine = function() {
    this.entities = [];//drawable objects
};

MyEngine.prototype.tick = function() {

    var MAX_FRAME_TIME = 12,//in milliseconds
        i, ilen = this.entities.length,
        now = window.performance.now,
        start = now(), elapsed = 0;

    //First, allow objects to perform update or init tasks.
    for (i = 0; i < ilen && elapsed < MAX_FRAME_TIME; i++) {
```

```
        this.entities[i].update();
        elapsed = now() - start;
    }

    //Next, render the scene.
    for (i = 0; i < ilen; i++) {
        this.entities[i].draw();
    }
};
```

These are some of the issues with Listing 14.3:

- Objects near the start of the list are more likely to be updated than objects near the end.
- If many objects do nothing in their update method, performing two passes is wasteful.
- If a single task incurs enough work, we'll exceed MAX_FRAME_TIME anyway.

Let's try to address some of these points by creating a first in/first out (FIFO) of entities that need an update, as in Listing 14.4:

Listing 14.4 An improved game loop.

```
MyEngine = function() {
    this.entities = [];//drawable objects
    this.queue = []; //objects that need updating
};

MyEngine.prototype.requestUpdate = function(entity) {
    this.queue.push(entity);
};

MyEngine.prototype.tick = function() {
    var MAX_FRAME_TIME = 12,//in milliseconds
        i, ilen = this.entities.length,
        now = window.performance.now,
        start = now(), elapsed = 0;

    //Process the update FIFO.
    while (this.queue.length && elapsed < MAX_FRAME_TIME) {
        this.queue.shift().update();
        elapsed = now() - start;
    }

    //Render the scene.
    for (i = 0; i < ilen; i++) {
        this.entities[i].draw();
    }
}
```

In Listing 14.4, the list of drawables (`this.entities`) is kept separate from the FIFO of updateables (`this.queue`). If an entity needs to be updated—for example, to have its texture replaced or VBO rewritten—then it needs to be explicitly added to the FIFO.

Aside: Sort your drawables! Order matters. Semitransparent objects should be at the end of the list for proper blending. Large occluders should be at the beginning of the list, so that subsequent entities that are occluded will fail the depth test right away, rather than going through fragment shading.

14.3.2 Amortization Technique 2: Task Manager

One variation on the preceding technique is to generalize the notion of updateable objects into a FIFO of tasks [Olmstead 14]. Here's one possible implementation for a simple task manager:

```
//Returns a list of remaining tasks for the next frame.
MyEngine.prototype.runTasks = function(tasks) {
    var now = window.performance.now,
        start = now(), elapsed = 0;
    while (tasks.length && elapsed < MAX_FRAME_TIME) {
        this.runTask(tasks.shift());
        elapsed = now() - start;
    }
    return tasks;
};
```

The engine should call `runTasks` once per invocation of `requestAnimation-Frame` and stash any tasks that didn't have a chance to run, so that they can be passed back to the function in the following frame.

14.3.3 Amortization Technique 3: Split Tasks with Yield

Recall one of the criticisms of Listing 14.3:

- If a single task incurs enough work, we'll exceed MAX_FRAME_TIME anyway.

For example, suppose our application needs to run a task that involves a lengthy image-processing sequence. Maybe we're using a pair of ping-pong framebuffer objects (FBOs) to generate a distance field [Tan 06]. What should we do if all this image processing work is best represented with a single task, but takes much longer than 16 ms?

The new `yield` keyword in ECMAScript 6 comes in handy for these situations. The first step is to define a *generator* function. Note the asterisk:

```
function* imageProcessingTask() {
    //bind Texture B, render to FBO A
    yield;
    //bind Texture A, render to FBO B
    yield;
    repeat...
}
```

Generator functions are actually factories for *iterator objects*, which proffer a `next` method. Let's go ahead and modify our task scheduler to work with iterators. See Listing 14.5 and note the usage of `next` inside the `if` statement:

Listing 14.5 Using ECMAScript 6 iterators to amortize tasks.

```
//Returns a list of remaining tasks for the next frame.
MyEngine.prototype.runTasks = function(tasks) {
    var now = window.performance.now, elapsed = 0,
        start = now(), task, incompleteTasks = [];

    //Execute as many tasks as possible in this frame.
    while (tasks.length && elapsed < MAX_FRAME_TIME) {
        task = tasks.shift();

        //Execute the task up until the next "yield".
        if (!task.next().done) {
            incompleteTasks.push(task);
        }
        elapsed = now() - start;
    }
    return tasks.concat(incompleteTasks);
};
```

Of course, using yield in the preceding way favors making *some* progress on *many* tasks, but each task might take longer to complete overall. It's the classic amortization trade-off and should only be used after careful consideration.

For more information about generators, take a look at the article "Generators: The Gnarly Bits" [Posnick 2014].

14.4 Threading with Web Workers

The preceding sections dealt with relatively small-grained tasks. Long CPU-intensive jobs that can be decoupled from the rest of your application can benefit greatly from parallelism. This is where Web Workers come to the rescue.

For more about Web Workers, see Chapter 4. They're especially useful for tessellation of procedural geometry, since typed arrays can be passed back from workers without incurring any copies.

For example, consider the knot-browsing application shown in Figure 14.1, which allows users to browse mathematical knot shapes (left) and zoom in on particular knots to see them in high detail (right). Since there are virtually an infinite number of knots, it's crucial to make the load time short. We can optimize the load time by sending only the spline control points from the server, rather than full-fledged triangle meshes. Then it becomes the client's job to tessellate the tubes into triangles. Since tessellation is a math-heavy process that occurs on the CPU, it's well suited for Web Workers.

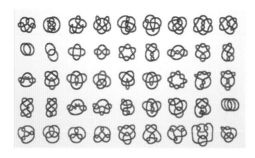

Figure 14.1

Mathematical knot viewer; tessellation is done in-browser.

14.4.1 Example 1: CPU Bottleneck: Knot Tessellation

Web Workers only allow strings, typed arrays, and simple dictionaries to go over the wire between workers. For tessellation, we definitely want to send typed arrays over the wire, since they're the most efficient way of representing a large swath of data. Let's design a simple worker messaging protocol based on small JSON dictionaries that have a string property called command. Listing 14.6 shows the design for a simple protocol with three commands: do-tessellation, mesh-data, and console-message. The latter is useful for diagnostic purposes, since workers cannot directly write to the console.

Listing **14.6** Simple JSON message protocol for tessellation workers.

```
//- - - - - - - - - - - - -
// command: 'do-tessellation'
// type: client to worker
// url: string
//- - - - - - - - - - - - -
// command: 'mesh-data'
// type: worker to client
// points: Float32Array
//triangles: Uint16Array
//- - - - - - - - - - - - -
// command: 'console-message'
// type: worker to client
// text: string
//- - - - - - - - - - - - -
```

The worker thread doesn't make any actual GL calls; it simply does math. Upon receiving the do-tessellation command, it downloads the centerline data, tessellates the tube shapes, and sends a mesh-data command back to the client. The client then stashes the typed arrays and uploads them into vertex buffer objects during the next animation cycle.

14.5 Performing Work When Idle

If we absolutely cannot achieve the level of visual quality that we need within the frame budget, one possible recourse is to switch into a high-fidelity "slow mode" only when the camera is idle, or when user interactivity has paused momentarily. This technique is especially useful in nongaming applications and platforms that have trouble with fill rate. Fill rate is an increasingly important consideration with the advent of high DPI displays, such as 4K monitors and Apple Retina displays.

14.5.1 Example 2: GPU Bottleneck: Glass Effect

Consider the glass effect shown in Figure 14.2, which is rendered to a floating-point FBO with additive blending. The result is drawn to the canvas using a full-screen triangle. When the user spins the model around its turntable using a touch interface, a smooth frame rate is achieved by rendering to a low-resolution framebuffer (left). When the user lifts her fingers, we redraw the model in full resolution (right).

On a 13-inch, 2014 MacBook Pro with Retina display, achieving this effect on a full-screen canvas at native resolution exceeds the frame budget, causing a frame rate of around 40 fps. The half-resolution image is well within budget and can be drawn at 60 fps.

At a high level, the draw loop for this demo looks like Listing 14.7. The passed-in `turntable` object has a `state` property, which tells us if the user is currently tumbling the model or resting.

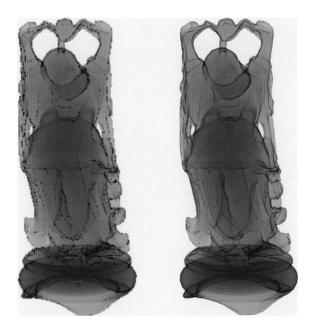

Figure 14.2

Fill-rate intensive glass effect: interactive (left) and at rest (right).

Listing 14.7 Low-fidelity/high-fidelity modes for glass effect.

```javascript
MyEngine.Buddha = function() {
    this.framebuffers = {lo: null, hi: null};
    this.textures = {lo: null, hi: null};
};

MyEngine.Buddha.draw = function(gl, turntable) {
    var texture, canvas = gl.canvas;

    if (turntable.state = = turntable.states.Resting) {
        //Use a full-resolution framebuffer:
        texture = this.textures.hi;
        gl.bindFramebuffer(gl.FRAMEBUFFER, this.framebuffers.hi);
        gl.viewport(0, 0, canvas.width, canvas.height);

    } else {
        //Use a half-resolution framebuffer:
        texture = this.textures.lo;
        gl.bindFramebuffer(gl.FRAMEBUFFER, this.framebuffers.lo);
        gl.viewport(0, 0, canvas.width/2, canvas.height/2);
    }

    //...draw Buddha to offscreen surface...

    gl.viewport(0, 0, canvas.width, canvas.height);
    gl.bindFramebuffer(gl.FRAMEBUFFER, null);
    gl.bindTexture(gl.TEXTURE_2D, texture);

    //...draw fullscreen triangle to canvas...

    gl.bindTexture(gl.TEXTURE_2D, null);
};
```

Bibliography

[Olmstead 14] Don Olmstead. "Optimizing WebGL Applications." SFHTML5 2014.

[Posnick 14] Jeff Posnick. "Generators: The Gnarly Bits." http://updates.html5rocks.com/2014/10/Generators-the-Gnarly-Bits, 2014.

[Tan 06] Guodong Rong and Tiow-Seng Tan. "Jump Flooding in GPU with Applications to Voronoi Diagram and Distance Transform." I3D '06 Proceedings of the 2006 Symposium on Interactive 3D graphics and games.

[Thompson 12] Lilli Thompson. "From Console to Chrome." GDC 2012. (49:30)

Section V
Rendering

Rendering is the generation of an image, given a scene description of geometry, materials, lights, and a camera. It is made up of two areas: finding visible surfaces and shading. Shading is the simulation of material and light to produce a color and is the primary topic of chapters in this section.

A common theme throughout this section is WebGL implementations of algorithms traditionally implemented in more feature-rich desktop graphics APIs—and how to get the best possible performance out of them. This can be something as simple as how to implement packing in GLSL, despite its lack of bitwise operators, to how to do volumetric rendering without 3D textures to simulating `EXT_shader_texture_lod` with octahedral environment mapping.

Deferred shading is a technique that decouples lighting from geometric complexity by performing lighting in a postprocessing step. This allows a large number of dynamic lights and simplifies engine design by only requiring one shader per material and per light type. Deferred shading became popular using desktop graphics APIs in 2008. We are now starting to see deferred shading with WebGL. In Chapter 15, "Deferred Shading in Luma," Nicholas Brancaccio discusses Luma, a physically based renderer for interior architectural spaces. He explores how to perform deferred rendering without multiple render targets through chroma subsampled lighting and creative packing of g-buffer parameters in a conventional floating-point render target.

Image-based lighting (IBL) uses processed image data, stored with high dynamic range (HDR), to properly represent a wide range of intensities to compute lighting. In Chapter 16, "HDR Image-Based Lighting on the Web," Jeff Russell explains how to implement IBL within the limitations of WebGL. This includes memory, performance, and visual-quality trade-offs of HDR decoding in GLSL and HDR transmission from the server to the client. For environment maps, octahedral environment mapping is used to store a cube map in a 2D texture to allow mipmap level of detail (LOD) selection without `EXT_shader_texture_lod`.

Many data sets, especially those in biomedical imaging, are naturally represented as 3D volumes. In Chapter 17, "Real-Time Volumetric Lighting for WebGL," Muhammad Mobeen Movania and Feng Lin discuss how volume rendering with half-angle slicing can be implemented within the constraints of WebGL using `OES_texture_float`,

`OES_texture_float_linear`, and `WEBGL_draw_buffers`. This chapter includes an overview of the theory, a walk-through of the JavaScript and GLSL code, a detailed performance analysis of direct volume rendering and half-angle slicing, and CPU versus GPU approaches.

Terrain rendering is a popular area since it has so many real-world and game-uses cases. Terrain presents geometric LOD challenges and opportunities for clever shading. In Chapter 18, "Terrain Geometry—LOD Adapting Concentric Rings," Florian Bösch presents a terrain rendering approach that is well suited to WebGL since it is very light on the CPU, pushing most of the work to the GPU. Topics include LOD and geomorphing with nested grids, and shading by combining a derivative map and detail mapping. Check out the online demo as you read.

15

Deferred Shading in Luma

Nicholas Brancaccio

15.1 Introduction

Luma is Floored's in-house, browser-based solution for rendering interior architectural spaces. Interior architectural visualization brings with it the challenge of illuminating a variety of clean, real-world materials with many distinct, localized light sources. Like a number of its contemporaries, Luma tackles these challenges with deferred shading [Geldreich 04] complemented by physically based shading [Hill 13]. Luma's implementation is flexible enough to be used for both traditional photorealistic rendering, as well as stylized rendering [Wei 14].

Deferred shading decouples shape rasterization from surface illumination by caching geometric and material parameters per pixel into offscreen memory. This geometry buffer (g-buffer) is typically implemented with multiple draw buffers in order to handle the many parameters necessary for shading. Physically based shading specifies a parameterization for materials that appear plausible and consistent in a wide variety of lighting conditions. Encoding physically based shading parameters in a g-buffer is not a perfect fit for all rendering challenges, but it serves as an excellent starting point for accelerating shading in everyday materials.

Deferred shading techniques were originally developed in the real-time community for more feature-complete APIs such as modern OpenGL and Direct3D. In this chapter, we explore implementing these techniques in WebGL for broad consumption.

Luma is implemented in JavaScript and WebGL. One limitation of WebGL, whose goal is to be a universal baseline API, is that it does not support multiple draw buffers without an extension. At the time of writing, the `WEBGL_draw_buffers` extension exposing such functionality is not widely supported enough for Floored's use case.* Further, few texture data types are available to assist in the tight quantization necessary for efficient bandwidth usage. Luma manages to perform deferred shading without multiple draw buffers through creative packing of g-buffer parameters in a conventional floating-point draw buffer.

15.2 Packing

While GLSL in WebGL has no support for the bitwise operators, we can simulate bit manipulation through integer arithmetic. Bit shift left and right operators are simulated through multiply and divide, respectively. The AND and OR operators, useful for returning specific bits, are simulated through multiplies, moduli, and additions. In practice, we avoid returning individual bits, as simulation becomes arithmetically impractical for real-time performance constraints. In other words, storing 32 unique bit flags in an integer has a high decode cost.

Because WebGL only supports narrow integer formats, we must use floating-point storage. Fortunately, the IEEE-754 32-bit floating point [Goldberg 91] exactly represents all integers from -2^{24} to 2^{24}, inclusive. Larger magnitude values are representable, but the step size between them is greater than one, making these ranges unappealing for integer packing. A 16-bit half-precision floating point has similar properties with a more limited range of -2^{11} to 2^{11}.

For simplicity, we pack bytes in the unsigned 24-bit range of 0 to $2^{24} - 1$. The sign bit could be easily accessed as a flag through a `sign()` call. See Listings 15.1 and 15.2.

Listing 15.1 Example of packing code. Storing 24-bpp color data in a floating point value. Pay careful attention to where inclusive versus exclusive integer ranges are used.

```
float normalizedFloat_to_uint8(const in float raw) {
  return floor(raw * 255.0);
}

float uint8_8_8_to_uint24(const in vec3 raw) {
  const float SHIFT_LEFT_16 = 256.0 * 256.0;
  const float SHIFT_LEFT_8 = 256.0;

  return raw.x * SHIFT_LEFT_16 + (raw.y * SHIFT_LEFT_8 + raw.z);
}

float packcolor(const in vec3 color) {
  vec3 color888;
  color888.r = normalizedFloat_to_uint8(color.r);
  color888.g = normalizedFloat_to_uint8(color.g);
  color888.b = normalizedFloat_to_uint8(color.b);

  return uint8_8_8_to_uint24(color888);
}
```

---

* WebGL Stats www.webglstats.com

```
float uint8_to_normalizedFloat(const in float uint8) {
  return uint8/255.0;
}

vec3 uint24_to_uint8_8_8(const in float raw) {
  const float SHIFT_RIGHT_16 = 1.0/(256.0 * 256.0);
  const float SHIFT_RIGHT_8 = 1.0/256.0;
  const float SHIFT_LEFT_8 = 256.0;

  vec3 res;
  res.x = floor(raw * SHIFT_RIGHT_16);
  float temp = floor(raw * SHIFT_RIGHT_8);
  res.y = -res.x * SHIFT_LEFT_8 + temp;
  res.z = -temp * SHIFT_LEFT_8 + raw;
  return res;
}

float unpackcolor(const in float colorPacked) {
  vec3 color888 = uint24_to_uint8_8_8(colorPacked);

  return vec3(
    uint8_to_normalizedFloat(color888.r),
    uint8_to_normalizedFloat(color888.g),
    uint8_to_normalizedFloat(color888.b)
  );
}
```

15.2.1 Unit Testing

Unit testing is a critical step in developing packing functions. It is easy to miss precision issues, value collisions, and other subtleties when simply evaluating new packing functions directly in your application. Further, while device support should be unified, and pollable through extensions, you may find contract breaches (Figure 15.1). Having experienced such pain directly, we highly recommend implementing unit testing for packing functions from day 1.

Luckily, 2^{24} is a fairly small number. We can exhaustively evaluate all unsigned integers in this domain using a single 4096×4096 draw buffer. In Listing 15.3, we see an example fragment shader of such a unit test:

Listing 15.3 Fragment shader of single-pass packing unit test.

```
void main() {
  vec2 pixelCoord = floor(vUV * pass_uViewportResolution);
  float expected = pixelCoord.y * pass_uViewportResolution.x +
    pixelCoord.x;
```

```
vec3 encoded = uint8_8_8_to_sample(uint24_to_uint8_8_8(expected));
float decoded = uint8_8_8_to_uint24(sample_to_uint8_8_8(encoded));

if (decoded = = expected) {
  //Packing successful. Display as green.
  gl_Fragcolor = vec4(0.0, 1.0, 0.0, 1.0);
} else {
  //Packing failed. Display as red.
  gl_Fragcolor = vec4(1.0, 0.0, 0.0, 1.0);
  }
}
```

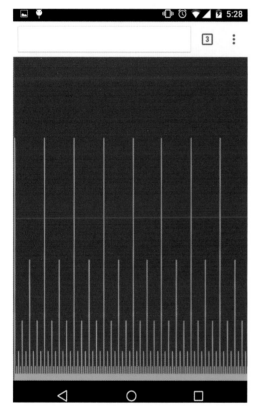

Figure 15.1

Unit tests are useful for validating device support. Here, an Android smartphone fails to store all integers from 0 to $2^{24} - 1$ to a floating-point render target, even after successfully getting the oes_texture_float extension. Numbers in the 0 to 2^{11} range appear green. Larger numbers appear green at diminishing multiples of 2. This output leads us to believe the device is supplying half-float precision under the hood.

Tip: Avoid gl_FragCoord use when rendering to large textures. gl_FragCoord is specified as mediump. Platforms that invoke precision qualifiers in hardware will not uniquely identify pixels at coordinates greater than 2048.

In the first example, data are packed, unpacked, and then compared to an expected value, all within a single shader. This unit test verified that our packing functions are mathematically correct and can be evaluated successfully when full floating-point precision is maintained.

In an application, packing is a means to encode multiple values in a texture. The discretization of these values is dictated by the packing function invoked, as well as by the precision of the texture written to. Full floating-point precision is not maintained when rendering to half-float or unsigned-byte textures, for instance.

A thoroughly realistic test, then, includes texture discretization. Listings 15.4 and 15.5 demonstrate the fragment shaders of such a unit test.

Listing 15.4 Fragment shader of pass 1 of a 2-pass packing unit test.

```
void main() {
  vec2 pixelCoord = floor(vUV *
      pass_uViewportResolution);
  float expected = pixelCoord.y * pass_
      uViewportResolution.x + pixelCoord.x;

  gl_Fragcolor.rgb =
    uint8_8_8_to_sample(uint24_to_
    uint8_8_8(expected));
}
```

Listing 15.5 Fragment shader of pass 2 of a 2-pass packing unit test.

```
void main() {
  vec2 pixelCoord = floor(vUV * pass_uViewportResolution);
  float expected = pixelCoord.y * pass_uViewportResolution.x +
      pixelCoord.x;

  vec3 encoded = texture2D(encodedSampler, vUV).xyz;
  float decoded = uint8_8_8_to_uint24(sample_to_uint8_8_8(encoded));

  if (decoded = = expected) {
    //Packing Successful
    gl_Fragcolor = vec4(0.0, 1.0, 0.0, 1.0);
  } else {
    //Packing Failed
    gl_Fragcolor = vec4(1.0, 0.0, 0.0, 1.0);
  }
}
```

This test now requires two passes. In the first pass, data are packed and written to a draw buffer of the desired type. In the following pass, the target is sampled, decoded, and compared to the expected value. Finally, a status color is drawn to our onscreen canvas, indicating the result of the comparison.

Imprecise packing functions, such as packing a floating-point value across a red, blue, green, alpha (RGBA) unsigned-byte texture [Pranckevičius 09], can find use as well. Unfortunately, a binary unit test is not indicative of the relative success of such packing functions. In these cases, the status color can be trivially modified to illustrate error as in the following:

```
gl_Fragcolor.rgb = abs(decoded - expected) * SCALE;
```

To further enhance the visualization, we can go so far as to employ a color gradient lookup texture for heat-mapped output.

```
gl_Fragcolor.rgb = texture2D(colorLUTSampler, vec2(abs(decoded - expected)
    * SCALE, 0.5)).rgb;
```

Now that we have validated the robustness of our packing functions, they are suitable for encoding our g-buffer.

15.3 G-Buffer Parameters

In Luma, unit-length normal vectors are encoded in two-dimensional octahedral space [Cigolle 14], giving fairly uniform discretization across the unit sphere at a reasonably low encode and decode cost.

Following Burley [Burley 12] and Karis [Karis 13], we transform albedo and specular color into a color and metallic parameter. In real-world materials, albedo and specular are coupled. Nonmetallic surfaces tend to have an achromatic Fresnel response, all within a narrow range of plausible functions. Metallic surfaces have varying chromatic Fresnel responses. Further, fully metallic surfaces transform all refracted energy into nonvisible waveforms such as heat, resulting in no diffuse response. Taking advantage of this parametric coupling, color is conditionally interpreted as albedo or specular color, based on the metallic Boolean. In the case of metallic surfaces, albedo is hard-coded as zero. In the case of nonmetallic surfaces, specular color is hard-coded at (0.04, 0.04, 0.04), a common value in the plausible range [Hoffman 10; Lagarde 11]. We find the slight loss in control well worth the savings in g-buffer storage. Further, this coupling makes it challenging for artists to create implausible architectural materials: a fundamental principle of physically based shading.

Similar to [Sousa 13], our color parameter is transformed to a perceptual basis, separating luminance from chrominance. Following Mavridis [Mavridis 12], we exploit the human visual system's low sensitivity to chroma variance, writing only a single chroma coefficient per pixel and alternating between bases in a checkerboard pattern to store data at a lower frequency (see Figures 15.2 and 15.3).

15.3.1 G-Buffer Format

RGBA float 128-bpp consists of:

    R: ColorY 8 bits, ColorC 8 bits, gloss 8 bits
    G: VelocityX 14 bits, NormalX 10 bits
    B: VelocityY 14 bits, NormalY 10 bits
    A: Depth 31 bits, metallic 1 bit

In fully deferred shading, we must reconstruct our color parameter's missing chroma coefficient for each pixel for each light source, during g-buffer decoding. This procedure incurs

(a)

Figure 15.2

(a) Fully lit scene in Luma.

(b)

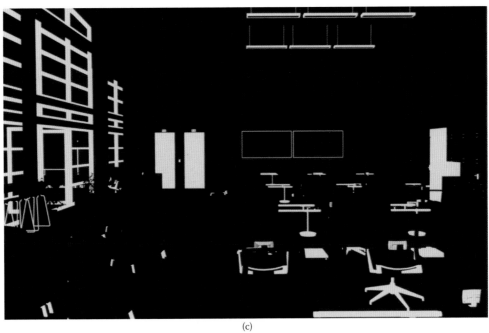

(c)

Figure 15.3

(b) g-buffer color; (c) g-buffer metallic. In Luma, we find use for Velocity in motion blur and temporal re-projection technique. *(Continued)*

(d)

(e)

Figure 15.3 (Continued)

(d) g-buffer gloss; (e) g-buffer normal. In Luma, we find use for Velocity in motion blur and temporal re-projection techniques. *(Continued)*

(f)

(g)

Figure 15.3 (Continued)

(f) g-buffer depth; (g) g-buffer velocity. In Luma, we find use for Velocity in motion blur and temporal re-projection techniques.

an arithmetic cost from our weighted reconstruction function, as well as a potential bandwidth cost, from sampling the g-buffer pixel's cross neighborhood, if samples fall outside the texture cache. In the following sections, we explore two alternative techniques to remove this cost by deferring this chroma reconstruction cost later down the rendering pipeline.

15.4 Light Prepass

Light prepass, also known as deferred lighting [Geldreich 04], is the first production proven variant of deferred shading, designed to reduce read bandwidth usage during light accumulation.

Bandwidth reduction is accomplished through a refactoring of the rendering equation, which pulls albedo and specular color outside the summation [Hoffman 09], allowing these values to be sampled once, after light accumulation.

Many implementations take a forward approach to sampling albedo and specular color, reducing g-buffer size, at the expense of a second scene rasterization pass [Shishkovtsov 05; Mittring 09]. Others move to storing these data in the g-buffer, and sample postlight accumulation in a full-screen viewport-aligned quad (billboard) pass [Lobanchikov 09; Sousa 13].

Light prepass's optimization comes at the cost of shading accuracy.

Diffuse and specular illumination must now be accumulated separately, as two RGB values. Ideally we accumulate to a six-channel render target or 2 three-channel render targets. In practice a four-channel render target is used, where only specular luminance is stored, and specular chroma is approximated from diffuse chroma. Both Mittring [Mittring 09] and Mavridis [Mavridis 12] alternatively propose accumulating luminance and a single chroma component for both diffuse and specular illumination. The chroma basis stored is alternated per pixel in a checkerboard pattern, allowing approximate chroma reconstruction from neighboring samples, in the same fashion as our g-buffer color component encoded.

With specular color no longer available during light accumulation, Fresnel must be evaluated outside the summation, driven by surface normal and view vector as opposed to the surface normal and half-vector [Hoffman 09]. This coarse approximation makes light prepass less appealing for a physically based pipeline [Pranckevičius 13] (Figure 15.4).

We may be able to assume a singular approximate light vector. In interior spaces, surfaces are primarily illuminated from above. Therefore, it may be reasonable to approximate the light vector, outside the summation, with a hard-coded up vector, as seen in Figure 15.5. Unfortunately, this approach increases error on downward-facing surfaces such as ceilings. An adaptive approach toggles between up or down as light vectors as follows:

```
vec3 lightVector = vec3(0.0, 0.0, sign(surfaceNormal.z));
```

Conveniently, for nonmetallic surfaces, Luma's hard-coded specular color of (0.04, 0.04, 0.04) allows Fresnel to be evaluated during light accumulation, parameterized with the more accurate half-vector. Metallic surfaces, which require a texture lookup to retrieve their specular color, still suffer an approximation, though the lower contrast of physically plausible metallic Fresnel curves [Hoffman 10] means lower perceptual error.

Unfortunately, with a g-buffer packed into a single draw buffer, we pay its full bandwidth cost even when sampling a subset of parameters. Transcoding our g-buffer into

Figure 15.4

Gold metallic spheres illuminated by single punctual light source, and image-based lighting. Top row: vertical split-screen comparison of Fresnel parameterized with nDotH, left, and nDotV, right. Bottom row: vertical split-screen comparison of Fresnel parameterized with nDotH, left, and false nDotH, computed with hardcoded upward facing light vector. NDotV exhibits a subtle overestimation of reflectance, particularly visible at the rim of the sphere, where nDotV approaches 1.0.

Figure 15.5

Direct light, evaluated in RGB space at 100% resolution.

multiple draw buffers of narrower bit width over multiple billboard passes is a possibility, though this creates additional memory and bandwidth use regardless of the scene's light rig. Mileage varies on application.

In Luma, with many overlapping light sources, we find transcoding well worth the baseline cost. Additionally, it accelerates other screen space techniques existent in our pipeline, such as ambient occlusion.

Example outputs of such a transcoding follow:

Type	R	G	B	A
RGBA unsigned byte	NormalX (8 bits)	NormalX (4 bits), NormalY (4 bits)	NormalY (8 bits)	Gloss (8 bits)
RGBA unsigned byte	Depth (8 bits)	Depth (8 bits)	Depth (8 bits)	Metallic (1 bit)

15.4.1 Chroma Subsampled Deferred Shading

Mavridis [Mavridis 12] proposed chroma subsampled lighting as a means to accumulate diffuse and specular light in a single four-channel draw buffer, for use in a light prepass pipeline. We repurpose this idea for use in a fully deferred pipeline, with a chroma subsampled g-buffer.

Chroma subsampled lighting factors the g-buffer's color reconstruction out of the inner light accumulation summation, saving significant decode, sampling, and weighting logic per pixel per light. As the name suggests, this factoring is accomplished by computing outgoing radiance in the same chroma subsampled space that our g-buffer's color is compressed in. Leaning on deficiencies of our perceptual system, we avoid direct computation of radiance for the chroma component we did not store in our g-buffer. This missing chroma component is approximated, after light accumulation, from the pixel neighborhood. At a high level the procedure is as follows:

1. Compute and accumulate outgoing radiance in YCoCg space, subsampling chroma by alternatively storing the coefficient of a single chroma basis in a checkerboard pattern.
2. In a fullscreen billboard pass, approximately reconstruct the missing chroma component through a similarity weighting of each pixel's cross neighborhood.

As much of our lighting and brdf code deals in scalar luminance quantities, few changes must be made to support computing outgoing radiance in YC space. One notable modification occurs in the Fresnel equation, inside the brdf. See Listings 15.6 and 15.7.

Listing 15.6 Schlick's approximation of Fresnel in RGB space.

```
vec3 fresnelSchlickRGB(const in float vDotH, const in vec3
  specularcolor) {
  float power = pow(1.0 - vDotH, 5.0);

  return (1.0 - specularcolor) * power + specularcolor;
}
```

Listing 15.7 Schlick's approximation of Fresnel in YC space.

```
vec2 fresnelSchlickYC(const in float vDotH, const in vec2
  specularcolor) {
  float power = pow(1.0 - vDotH, 5.0);

  return vec2(
    (1.0 - specularcolor.x) * power + specularcolor.x,
    specularcolor.y * -power + specularcolor.y
  );
}
```

The luminance function remains identical to Schlick's original approximation as scaling R, G, and B components can be understood as an indirect modulation of luminance. Conversely, the chroma function becomes inverted, approaching zero when perpendicular. Schlick Fresnel is simply a function that interpolates between specular color and white. In RGB space, white is represented by (1.0, 1.0, 1.0), but in YC space, white is represented by (1.0, 0.0).

Conveniently, our new Fresnel function for YC space is cheaper than the original RGB function. While we have devectorized the computation, making the source longer, modern GPUs are primarily single instruction, multiple data (SIMD) across fragment neighborhoods rather than vector lanes. We have saved three scalar instructions. Dropping a channel skips an ADD and MADD; by skipping the "1.0 -" the inverted chroma function skips an ADD. Negation of the power variable should be free on assignment.

Finally, our chroma reconstruction filter is a fairly trivial modification to Mavridis's [Mavridis 12] edge directed filter, seen in Listing 15.8. For tunability, we prefer a narrow Gaussian function over a step function. Additionally, we return both chroma components, to support fallback to full desaturation in the case of poor weights.

Listing 15.8 Filter to approximately reconstruct missing chroma component from cross neighborhood. Sensitivity controls luminance difference penalty. Arguments a1–a4 are cross neighborhood luminance and chroma pairs. (See Figures 15.6-15.9.)

```
vec2 reconstructChromaHDR(const in float sensitivity, const in vec2
  center, const in vec2 a1, const in vec2 a2, const in vec2 a3, const
  in vec2 a4) {
  vec4 luminance = vec4(a1.x, a2.x, a3.x, a4.x);
  vec4 chroma = vec4(a1.y, a2.y, a3.y, a4.y);

  vec4 lumaDelta = abs(luminance - vec4(center.x));

  vec4 weight = exp2(-sensitivity * lumaDelta);

  //Guard the case where sample is black.
  weight * = step(1e-5, luminance);

  float totalWeight = weight.x + weight.y + weight.z + weight.w;

  //Guard the case where all weights are 0.
  return totalWeight > 1e-5 ? vec2(center.y, dot(chroma, weight)/
    totalWeight) : vec2(0.0);
}
```

Figure 15.6
Direct light, evaluated in chroma-subsampled YC space at 100% resolution.

Figure 15.7
Direct light, evaluated in RGB space at 25% resolution.

Figure 15.8

Direct light, evaluated in chroma-subsampled YC space at 25% resolution.

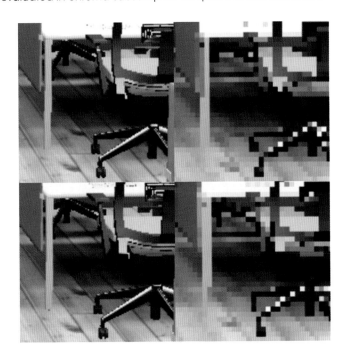

Figure 15.9

Top-left: RGB 100%; top-right: RGB 25%; bottom-left: YC 100%; bottom-right: YC 25%.

(Continued)

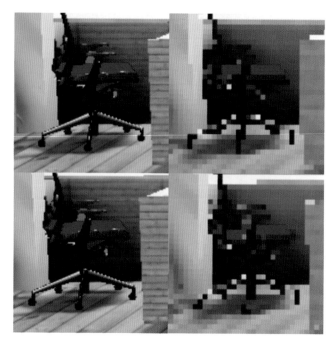

Figure 15.9 (Continued)

Top-left: RGB 100%; top-right: RGB 25%; bottom-left: YC 100%; bottom-right: YC 25%.

Bibliography

[Burley 12] Brent Burley. "Physically Based Shading at Disney." http://disney-animation. s3.amazonaws.com/library/s2012_pbs_disney_brdf_notes_v2.pdf, 2012.

[Cigolle 14] Zina H. Cigolle, Sam Donow, Daniel Evangelakos, Michael Mara, Morgan McGuire, and Quirin Meyer. "A Survey of Efficient Representations for Independent Unit Vectors." http://jcgt.org/published/ 0003/02/01/, 2014.

[Geldreich 04] Rich Geldreich, Matt Pritchard, and John Brooks. "Deferred Lighting and Shading." https://sites.google.com/site/richgel99/home, 2004.

[Goldberg 91] David Goldberg. "What Every Computer Scientist Should Know about Floating-Point Arithmetic." http://docs.oracle.com/cd/E19957-01/806-3568/ncg_goldberg.html, 1991.

[Hill 13] Stephen Hill, Naty Hoffman, Dimitar Lazarov, Brian Karis, David Neubelt, Matt Pettineo, Zap Andersson, Adam Martinez, Christophe Hery, and Ryusuke Villemin. "Siggraph 2013 Course: Physically Based Shading in Theory and Practice." http://blog.selfshadow.com/publications/s2013-shading-course/, 2013.

[Hoffman 09] Naty Hoffman. "Deferred Lighting Approaches." http://www.realtimerendering.com/blog/deferred-lighting-approaches, 2009.

[Hoffman 10] Naty Hoffman. "Physically Based Shading Models in Film and Game Production." http://renderwonk.com/publications/s2010-shading-course/hoffman/s2010_physically_based_shading_hoffman_a_notes.pdf, Siggraph, 2010.

15. Deferred Shading in Luma

[Karis 13] Brian Karis. "Real Shading in Unreal Engine 4." http://blog.selfshadow.com/publications/s2013-shading-course/karis/s2013_pbs_epic_notes_v2.pdf, 2013.

[Lagarde 11] Sébastien Lagarde. "Feeding a Physically Based Shading Model." http://seblagarde.wordpress.com/2011/08/17/feeding-a-physical-based-lighting-mode/, 2011.

[Lobanchikov 09] Igor A. Lobanchikov and Holger Gruen. "GSC Game World's S.T.A.L.K.E.R: Clear Sky—A Showcase for Direct3D 10.0/1." http://amd-dev.wpengine.netdna-cdn.com/wordpress/media/2012/10/01GDC09AD3DDStalkerClearSky210309.ppt, Game Developers Conference, 2009.

[Mavridis 12] Pavlos Mavridis and Georgios Papaioannou. "The Compact YCoCg Frame Buffer." http://jcgt.org/published/0001/01/02/, *Journal of Computer Graphics Techniques*, 2012.

[Mittring 09] Martin Mittring. "A Bit More Deferred—CryEngine 3." http://www.crytek.com/cryengine/cryengine3/presentations/a-bit-more-deferred-cryengine3, 2009.

[Pranckevičius 09] Aras Pranckevičius. "Encoding Floats to RGBA—The final?" http://aras-p.info/blog/2009/07/30/encoding-floats-to-rgba-the-final, 2009.

[Pranckevičius 13] Aras Pranckevičius. "Physically Based Shading in Unity." http://aras-p.info/texts/files/201403-GDC_UnityPhysicallyBasedShading_notes.pdf, Game Developers Conference, 2013.

[Shishkovtsov 05] Oles Shishkovtsov. "Deferred Shading in S.T.A.L.K.E.R." http://http.developer.nvidia.com/GPUGems2/gpugems2_chapter09.html, 2005.

[Sousa 13] Tiago Sousa. "The Rendering Technologies of Crysis 3." http://www.crytek.com/cryengine/presentations/the-rendering-technologies-of-crysis-3, 2013.

[Walter 07] Bruce Walter, Stephen R. Marschner, Hongsong Li, and Kenneth E. Torrance. "Microfacet Models for Refraction through Rough Surfaces." http://www.cs.cornell.edu/~srm/publications/egsr07-btdf.pdf, 2007.

[Waveren 07] J. M. P. van Waveren and Ignacio Castaño. "Real-Time YCoCg-DXT Compression." http://developer.download.nvidia.com/whitepapers/2007/Real-Time-YCoCg-DXT-Compression/Real-Time%20YCoCg-DXT%20Compression.pdf, 2007.

[Wei 14] Implementing a "Sketch" Style of Rendering in WebGL. http://www.floored.com/blog/2014/sketch-rendering, Angela Wei, Emma Carlson, Nicholas Brancaccio, Won Chun, 2014.

16

HDR Image-Based Lighting on the Web

Jeff Russell

16.1 Introduction

Image-based lighting (IBL) is a family of techniques for illuminating surfaces using processed image data. Such images can be prerendered, captured on the fly, or obtained through photography. Regardless of their source, the use of these images for lighting has several advantages for render quality, not least of which is the inclusion of both direct and indirect illumination from a surrounding scene.

Image-based lighting is not new and has in fact become a prevalent rendering technique in games and visual simulations today. As we turn our attention to graphics on the web, and WebGL in particular, it becomes clear that most if not all IBL techniques should be feasible with some modification on this new platform.

This chapter will focus on the aspects of a WebGL implementation that may differ from those of other platforms. Concerns specific to compatibility and performance will be addressed in order to better reflect the broad hardware demographic of the web today, with a special emphasis on mobile devices. This assumes that the reader has a basic familiarity with image-based lighting; for a broader introduction to the topic, see Debevec [Debevec 02].

16.2 High Dynamic Range Encoding

Image-based lighting requires use of high dynamic range (HDR) texture data to properly represent the full range of luminosities present in a given image. Such values often span ranges far outside those of typical display technology, requiring both increased range and precision. Several encodings exist to address these needs; however, at the time of this writing few are well suited for web deployment.

Many WebGL implementations expose extensions such as `OES_texture_float` and `OES_texture_half_float`, providing support for 32-bit and 16-bit floating point values, respectively. At first glance, these formats would seem to be an ideal means for storing HDR data, and indeed they are in large part made available for exactly that purpose. There are, however, several drawbacks to their use. Because they are optional extensions, many implementations do not provide them, and when others do, they often do not supply functionality for linear filtering. On top of this, these formats double or quadruple memory and bandwidth requirements compared to a typical 8-bit format, which can create performance problems on mobile devices and increase page load times.

Many solutions exist for packing HDR data into smaller memory footprints. The Red/Green/Blue/Exponent (RGBE) encoding stores a shared exponent byte in the fourth color channel of a 32-bit image [Ward 97]. This allows for representation of a wide range of luminosities in a compact form, though it does require additional instructions in a shader to unpack and apply the exponent. A similar encoding that uses the CIE LUV color space is "LogLUV" [Ward 98].

A simpler option available to us involves storing a separate scale value in the alpha channel, and decoding these Red/Green/Blue/Multiplier (RGBM) values through simple multiplication in the shader [Karis 09]. This has several advantages. First, like RGBE and LogLUV, it does not require use of any extensions, which makes it workable on all platforms that support the base WebGL specification. Second, it uses much less memory and hence bandwidth during render time than most floating point formats. Finally, it is very simple to encode and decode, requiring fairly little ALU workload.

Listing 16.1 Basic RGBM decoding.

```
mediump vec3 decodeRGBM(mediump vec4 rgbm) {
  return rgbm.rgb * rgbm.a * maxRange;
}
```

Linear interpolation of such RGBM values during texture sampling is technically incorrect, and a naive decoding, as shown in Listing 16.1, will often produce banding artifacts. Filtering precision can be a significant factor in this; some devices perform texture filtering at 8-bit precision, which can worsen banding significantly. The performance advantage of using built-in texture filtering is significant enough that we need to explore ways of minimizing its side effects on RGBM textures.

One method of correcting banding is reducing the range of our "M" multiplier. An 8-bit value could theoretically be interpreted as a linear multiplier on a range as wide as [0, 255]. While this is feasible on some devices, it tends to stretch the limits of texture filtering

Linear, Maximum Range of 255   Nonlinear, Maximum Range of 49

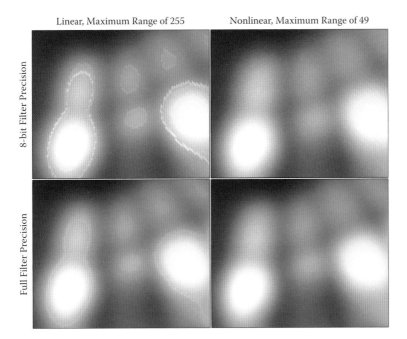

Figure 16.1

Comparison of texture filtering results with RGBM encoding. Linear RGBM with a maximum range of 255 (left column) compared with nonlinear RGBM with a maximum range of 49 (right column). The top row has been rendered on a mobile GPU with 8-bit filter precision.

precision on others. We find that reducing the mapped range of M to [0,7] greatly reduces the appearance of filtering artifacts.

We additionally adopt a nonlinear transformation akin to a gamma curve in order to reduce color banding in the color components. The rationale is analogous to that behind gamma compression, though in our case it will have the additional benefit of using nonlinearity to greatly expand dynamic range. sRGB color space conversion suits this purpose well and is available on many GPUs; however, WebGL does not expose this, so we instead adopt a simpler exponential curve. By applying an exponent of two, we both increase the upper limit on our range from 7 to 49 and provide better precision for values closer to zero, where banding is most apparent (Figure 16.1).

Listing 16.2 RGBM encoding and decoding, with nonlinearity and range reduction to reduce filtering artifacts.

```
highp vec4 encodeRGBM(highp vec3 rgb) {
  highp vec4 r;
  r.xyz = (1.0/7.0) * sqrt(rgb);
  r.a = max(max(r.x, r.y), r.z);
  r.a = clamp(r.a, 1.0/255.0, 1.0);
  r.a = ceil(r.a * 255.0)/255.0;
```

```
    r.xyz/= r.a;
    return r;
}


mediump vec3 decodeRGBM(mediump vec4 rgbm) {
    mediump vec3 r = rgbm.rgb * (7.0 * rgbm.a);
    return r * r;
}
```

As is evident in Listing 16.2, encoding RGBM values require several more instructions than decoding, although both operations are fairly brief. In many cases this is not a problem, as IBL image data are prepared infrequently during runtime, if at all. However, in situations where RGBM encoding is to be used for render targets, the use of floating-point texture formats may be preferable when they are available, particularly if blending is required.

The relatively limited range of RGBM values is another possible drawback. For most scenarios we have found that preconvolved, pre-exposed light data fit well within this modest span. Preconvolved images, such as those used for diffuse and specular lighting, tend to fit well, as any strong highlights in the original image tend to "spread out" their energy as a result of convolution. The use of tone mapping filters may in some cases also hide a lack of range. However, applications wishing to faithfully represent very high luminance values may need to switch to a higher dynamic range format such as RGBE or floating point textures.

RGBM texture data are well suited for compact transmission over the web. In theory, the 32-bit color values map well into the portable network graphics (PNG) image format for which all browsers provide support. PNG images use lossless compression, which is a necessity for RGBM data; use of JPEG or other lossy formats results in severe reconstruction errors.

In practice, several popular browsers have been found to premultiply alpha values after PNG decoding, which introduces pronounced color banding in RGBM color data (this occurs regardless of the "UNPACK_PREMULTIPLY_ALPHA_WEBGL" pixel storage setting). Our solution is to compress the RGBM data ourselves. If we store each color plane separately, readily available content encodings such as gzip can provide compression ratios comparable to PNG. This requires client-side decompression and re-interleaving of color channel data, but is well worth the savings in transmission times. Any convenient image container format may be used for transmission, including DirectDraw Surface (DDS) or custom layouts.

16.3 Environment Mapping

Key to any image-based lighting system is the layout of the environment images themselves. For any image to be usable for lighting, it must cover the entire sphere of possible directions. There are, of course, many layouts with this property, but far and away the most popular is cube mapping. WebGL has good support for cube maps in the base specification, supporting all texture formats, filtering, and more.

It is also important for an IBL system to provide the ability for texture artists to specify, usually through a grayscale mask, differing roughness values. This gives control over the apparent "shininess" of a surface, affecting the size of specular highlights and reflections. This is often known as "gloss mapping" or "roughness mapping" and is a nearly indispensable tool for creating believable surfaces.

Roughness mapping typically interacts with cube map reflections through the manual selection by the shader of different mipmap levels according to roughness values. Environment cube maps are specially prepared such that they contain different convolutions of the base image in each mipmap level—typically becoming "blurrier" as the mip dimensions reduce. In this way the shader can easily control the appearance of environment reflections in much the way it would for analytical light sources (Figure 16.2). Third-party tools for preparing convolutions of this sort are readily available; see www.hdrshop.com, www.knaldtech.com/lys/ or code.google.com/p/cubemapgen/.

Herein lies a difficulty for implementation in WebGL, as mipmap level of detail (LOD) selection is not supported in the core specification. The EXT_shader_texture_lod extension seeks to address this shortcoming; however, at the time of this writing it is not widely supported. In order to make universal use of roughness mapping possible in WebGL today, an alternative means of storage is needed for preconvolved environment maps.

By packing multiple maps into a two-dimensional texture atlas, we can access them with simple texture coordinate transformations, requiring no mipmap selection. However, since cube maps are not easily packed in this way, this again poses our earlier problem of mapping the environment sphere onto a flat surface. The puzzle is as old as map making, and we come to it with the added restriction of requiring a fast, simple formulation with minimal distortion.

Spherical projections making direct use of latitude and longitude provide conceptually straightforward mappings; however, they come with the significant drawback of having polar discontinuities. This results in a highly uneven distribution of texels at poles,

Figure 16.2

Roughness mapping used in conjunction with cube-mapped environment reflections.

as well as filtering artifacts, which can be difficult to fully mask. Dual paraboloid mapping [Heidrich 98] avoids this polar distortion, having only modest discontinuities. It does, however, provoke somewhat uneven texel distribution and does not make efficient use of texture space, as much goes unused around its circular borders.

Octahedral environment mapping [Engelhardt 08] provides a good compromise between simplicity, distortion, and speed, with performance and appearance very similar to cube mapping. Additionally, its two-dimensional surface lies in a perfect square, which means no space goes unused in an atlas layout. The technique maps three-dimensional vectors onto the surface of an octahedron, or "double pyramid" (Figure 16.3).

The three-dimensional surface of the unit octahedron is defined by $|x| + |y| + |z| = 1$. From this relation, unnormalized vectors can be quickly projected into the octahedron's two-dimensional surface. Points with positive Y values are simply "flattened" onto the XZ-plane, and those in the opposite hemisphere are unfolded to fill the corners (see [Engelhardt 08] for a more detailed description). Listing 16.3 displays code to perform this mapping.

Listing 16.3 Octahedral projection from unnormalized 3D vector to octahedral UV coordinates.

```
mediump vec2 octahedralProjection(mediump vec3 dir) {
    dir/= dot(vec3(1.0), abs(dir));
    mediump vec2 rev = abs(dir.zx) - vec2(1.0,1.0);
    mediump vec2 neg = vec2(dir.x < 0.0 ? rev.x : -rev.x,
                            dir.z < 0.0 ? rev.y : -rev.y);
    mediump vec2 uv = dir.y < 0.0 ? neg : dir.xz;
    return 0.5*uv + vec2(0.5,0.5);
}
```

With this environment projection in hand, we are now able to build a full atlas of as many convolutions as we wish. An atlas comprising a vertical column of a few images makes for easy construction through simple memory concatenation and allows simple shader logic for selecting convolutions. A shader making use of such an atlas is also free to

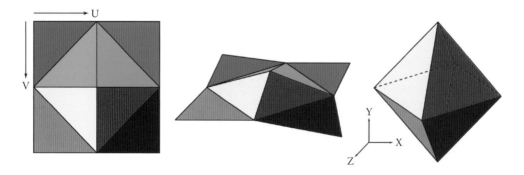

Figure 16.3

Octahedral environment map layout in 2D coordinates (left), and folded into three dimensions (middle and right).

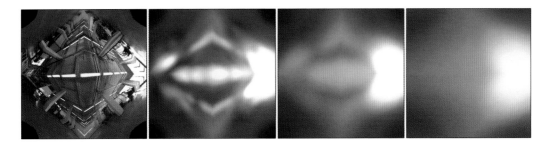

Figure 16.4

Example of octahedral environment maps, convolved for the Phong BRDF with varying levels of surface roughness.

take samples from different convolutions and blend between them, achieving a smoother transition between roughnesses.

Octahedral maps do have some difficulties with texture filtering. Specifically, they do not tile: Their edges will filter improperly if the texture wrap parameter is set to "REPEAT." Use of octahedral maps in texture atlases poses similar problems, causing neighboring maps to filter with one another at map boundaries. This is best resolved by adding a single pixel of padding around each map in the atlas, and altering the shader code to adjust the sample coordinates accordingly.

Octahedral environment maps (Figure 16.4) incur a small performance penalty in exchange for the added compatibility they provide. We have measured a difference of roughly six additional instructions as compared to traditional cube mapping, though this varies with hardware characteristics and compilers. This cost is incurred for each unique direction sampled, but can be amortized for multiple samples with the same direction vector (as in the case of gloss mapping).

Use of an octahedral layout directly for render targets can prove challenging. Engelhardt [Engelhardt 08] provides a brief discussion of a method based on splitting rendered triangles across the eight faces to ensure proper perspective. In practice we have found it preferable to simply remap a cube render target into the octahedral layout as a postprocess. This can be quickly performed by rendering flat the octahedral geometry itself, with cube map texture coordinates assigned to each vertex.

Diffuse IBL data sets generally do not have need of multiple convolutions, freeing them of the restrictions that motivate the use of octahedral maps for reflections. Diffuse convolutions can therefore still make use of cube maps or any other convenient mapping. Such convolutions are also amenable to representation with the spherical harmonic basis functions [Ramamoorthi 01], which provide an extremely compact reproduction of low-frequency image data. For diffuse lighting we have found the spherical harmonic representation preferable to images, due to its flexibility, simplicity, and small memory footprint.

16.4 Conclusion

With the modifications discussed here, WebGL is well suited today for the use of image-based lighting across the panoply of devices and platforms that is the web. Through careful selection of HDR texture encodings, many of the drawbacks of floating point formats can be eliminated while retaining a useful dynamic range; by adopting octahedral environment mapping, we can create a fully featured IBL renderer without relying on device-specific extensions.

It is our hope that as the WebGL standard progresses, some of the limitations outlined in this chapter will be overcome in the same ways they have been in other environments, eliding the need for much special treatment. Image-based lighting is likely to remain a relevant rendering technique for years to come, and the web, being no exception, will see the benefit of solid implementations.

Bibliography

[Debevec 02] Paul Debevec. "Image-Based Lighting." USC Institute for Creative Technologies, http://ict.usc.edu/pubs/Image-Based%20Lighting.pdf, 2002.

[Engelhardt 08] Thomas Engelhardt and Carsten Dachsbacher. "Octahedron Environment Maps." Vision, Modeling and Visualization, http://www.vis.uni-stuttgart.de/~dachsbcn/download/vmvOctaMaps.pdf, 2008.

[Heidrich 98] Wolfgang Heidrich and Hans-Peter Seidel. "View-Independent Environment Maps." Eurographics Workshop on Graphics Hardware. http://www.cs.ubc.ca/~heidrich/Papers/GH.98.pdf, 1998.

[Karis 09] Brian Karis. "RGBM Color Encoding." http://graphicrants.blogspot.com/2009/04/rgbm-color-encoding.html, 2009.

[Ramamoorthi 01] Ravi Ramamoorthi and Pat Hanrahan. "An Efficient Representation for Irradiance Environment Maps." http://graphics.stanford.edu/papers/envmap/envmap.pdf, 2001.

[Ward 97] Gregory Ward Larson. "Radiance File Formats." http://radsite.lbl.gov/radiance/refer/filefmts.pdf, 1997.

[Ward 98] Gregory Ward Larson. "The LogLuv Encoding for Full Gamut, High Dynamic Range Images." *Journal of Graphics Tools*, 3(1):15–31, 1998.

16. HDR Image-Based Lighting on the Web

17

Real-Time Volumetric Lighting for WebGL

Muhammad Mobeen Movania and Feng Lin

17.1 Introduction

Volume rendering is used in several disciplines like biomedical imaging and computational fluid dynamics. It allows us to visualize a large amount of data using specialized algorithms. Several methods have been proposed for visualization of volumetric data sets including texture slicing, shear warp, splatting, polygonization, and cell projection [Engel 06; Preim and Botha 13].

Meanwhile, visualization experts have always aspired to realistic or physically plausible rendering. To achieve this, physically based methods have been proposed recently [Kroes 12] and an open-source project, Exposure Render [Exposure 14], does a good job for that. However, due to the steep computational demands of such physically based methods, these algorithms have not so far been implemented on computationally limited platforms such as WebGL. The fairly simple unlit volume rendering typically implemented for WebGL lacks spatial information, which makes differentiation between near and far areas impossible. High-quality physically plausible volume rendering remains a challenge for WebGL.

To circumvent these shortcomings, we propose to use a well-known visualization method called "half-angle slicing" [Ikits 04]. Our preliminary study shows that this method can be easily implemented on a resource-limited platform. Moreover, it can achieve real-time performance, which makes it an ideal candidate for interactive WebGL applications including games and scientific applications. The strategies and experiences from our implementation, along with the shader details are elaborated in this chapter. Snapshots from the technique are given in Figure 17.1.

The discussion is organized as follows. We start with an introduction of the algorithm and the rendering technique. Next, mathematical underpinnings required to understand the algorithm are elaborated. Following this discussion, the algorithm is presented along with some optimizations that are required for implementation in WebGL, including client-side setup code and shader details. Finally, experimental results are discussed and conclusions are derived.

17.2 Why WebGL?

Prior to the introduction of WebGL, conventional methods for 3D volume rendering in a web browser relied on specialized browser plug-ins. These had compatibility issues, and development outputs varied across different hardware platforms as well as across different web browsers.

Moreover, before WebGL, mobile platforms had to be handled specifically. With the introduction of WebGL, we are relieved as WebGL provides an implementation for a specific OpenGL ES profile. If the given hardware supports that profile, the implementation works subject to the capabilities of the mobile device (see Chapter 8).

In addition to the standard WebGL benefits of being cross-platform, plug-in free, and having built-in debugging tools, WebGL also has specific benefits for visualization:

- **Support of WebGL on Mobile Platforms:** Support for WebGL is available in mobile platforms through mobile versions of Mozilla®, Firefox®, and Google® Chrome® web browsers. This makes the deployment of developed visualization easier on mobile platforms like tablets and smartphones, which are turning out to be the preferred deployment platforms for modern visualization applications.
- **User interfaces can be created quickly:** Using HTML5 form elements like sliders, buttons, text boxes, etc., we can generate simple UI elements to provide interactivity to WebGL content. This enables creation of specialized widgets (e.g., for transfer function editors, color map, lighting parameter widgets, etc.) that are an inherent part of any visualization application.

17.3 WebGL and Volume Rendering

For a general overview of volume rendering, we refer the readers to the following texts: Engel [2006] and Preim and Botha [2013]. In the context of WebGL, volume visualization is not new. Several implementations have been proposed. Initial works are reported

(a)

(b)

(c)

Figure 17.1

Rendering results using the technique detailed in this chapter showing (a) skull data set, (b) engine block data set, and (c) CTHead data set.

by Congote [2011]. As WebGL 1.0 did not support 3D textures, Congote [2011] used flat 2D textures to store the 3D volume. The volume was then reconstructed using modulo arithmetic in the fragment shader. A similar approach was also detailed for large 3D volumes.

There are two basic approaches to volume rendering: multipass [Kruger 03] and single-pass [Stegmaier 05] rasterization. Congote [2011] used the former approach, whereas Movania and Lin [2013a] used the latter approach, which provides better performance on mobile platforms like smartphones and tablets.

17.4 The Volumetric Lighting Model

Volume rendering tries to approximate the emission and absorption characteristics of a volume data set using an optical model. The resultant equation is known as the volume rendering integral, which is defined as follows:

$$L(x_1,\omega) = T(0,l)L(x_0,\omega) + \int_0^l T(s,l)\,R\big(x(s)\big)f_s\big(x(s)\big)T_L(s,t)L_l ds \qquad (17.1)$$

where

$x(l)$ is the current sample position that, for a parametric ray, is given as $x(t) = x_0 + t\omega$
$R(x)$ is the surface reflectivity color
f_s is the BRDF that is the shading model used, which for volume rendering relies on a phase function
L_l is the light source intensity
$T(a,b)$ is the attenuation along the ray from the sample point $x(a)$ to $x(b)$
$T_L(a,b)$ is an additional light attenuation term

The attenuation function $T(a,b)$ is defined as

$$T(a,b) = exp\left(-\int_a^b \tau(s')ds' \right)$$

where τ is the attenuation coefficient at the current sample position s'. The reflectivity term (R) is well defined for conventional polygonal surface rendering. However, it is not well suited for fuzzy phenomena like volumes. The light attenuation term $T_L(a,b)$ is defined as follows:

$$T_L(t,t') = exp\left(-\int_0^{t'} \tau\big(x(t)+\omega_l s\big)ds \right)$$

where ω_l is the light direction from the current sample position $x(t)$. The new light attenuation term captures light attenuation as a ray travels from the current sample point $x(t)$ in the light source's direction ω_l.

17.5 Volumetric Lighting Techniques

Lighting and shadows in volume rendering are dependent on the volume rendering algorithm used. In ray casting, we create another sampling ray that starts at the current sample point $x(a)$ and continues in the light source's direction ω_l. The obtained samples are attenuated using the light's color. In 3D texture slicing, half-angle slicing [Ikits 04] can be used. In this algorithm, a pair of draw buffers is used: one eye buffer and one light buffer. Texture slicing is then carried out in the half-angle direction to ensure that there are no slicing polygons parallel to the light or eye direction. In WebGL, framebuffers are used to provide render-to-texture functionality.

17.5.1 Determining Half-Angle Direction

If we have a view vector $\mathbf{V}$ and a light vector $\mathbf{L}$, we can determine the half-angle direction by calculating the dot product between the view vector $\mathbf{V}$ and the light vector $\mathbf{L}$. If the dot product is less than 0, we invert the view vector ($\mathbf{V} = -\mathbf{V}$) as shown in Figure 17.2(a). Otherwise, we can then determine the half-angle direction vector ($\mathbf{H}$) by simply normalizing the sum of the two vectors $\mathbf{V}$ and $\mathbf{L}$ as shown in Figure 17.2(b).

$$\mathbf{H} = normalize\,(\mathbf{V} + \mathbf{L})$$

17.5.2 Dynamic Texture Slicing in the Half-Angle Direction

Once the half-angle direction ($\mathbf{H}$) is known, we can start the dynamic slicing process as shown in Figure 17.3 by first calculating the intersection of a unit cube with a set of planes perpendicular to the half-angle direction ($\mathbf{H}$). The half-angle direction ($\mathbf{H}$) becomes the plane normal. There are six possible cases for intersection between the plane and the unit cube. The intersection configurations are stored in an edge table [Engel 06]. These are obtained by looping through all three edges of the cube's vertex when moving from the nearest to the farthest vertex for the current view and finding the matching case from the edge table to get the ray intersection point.

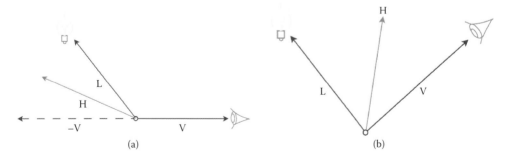

Figure 17.2

Determining the half-angle direction vector (**H**): (a) when dot product between the light vector (**L**) and the view vector (**V**) is less than 0; (b) when the dot product is greater than or equal to 0.

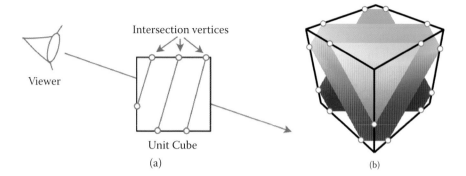

Figure 17.3

Dynamic 3D slicing of a unit cube showing the side on view in (a) and the view from the viewer in (b). The slicing polygons are color coded with their interpolated texture coordinates. Intersection points along each edge for each slicing polygon are shown as points.

Considering two points, P_0 and P_1, any point on the edge $(P_1 - P_0)$ may be given using the following parametric equation:

$$P(t) = P_0 + t * (P_1 - P_0)$$

where $t = [0,1]$.

We can solve this equation for (t), for a plane with normal ($\mathbf{H}$) and distance (d) from origin by plugging in the plane equation

$$H \cdot P(t) + d = 0$$

and then solving for parameter (t), which gives

$$t = \frac{-d - H \cdot P_0}{(P_1 - P_0)} \tag{17.2}$$

This parameter (t) gives the intersection point for one edge $(P_1 - P_0)$. We iterate over all edges from the current vertex by using the edge table and obtain all intersection points. These points are then used to generate the slicing polygon. This process is then repeated for all eight cube vertices going front to back in the half-angle direction. In each iteration, the plane distance is incremented to get a set of planes until the whole unit cube is traversed.

17.5.3 Dynamic Slicing on the GPU

To implement dynamic slicing on the GPU, we first prepare the necessary data structure for the GPU, which includes preparing the vertex/edge table for appropriate edge lookup for intersection calculation in the vertex shader and generating the dynamic geometry. Based on the number of slices required, the plane increment is calculated. Next, the nearest intersection to the unit cube is calculated and becomes the plane start position. The plane start position, the vertex/edge tables, the half-angle direction vector, and the current modelview/projection matrices are passed as vertex shader uniforms. At this stage, we initiate the slicing loop for the desired number of slices. In each iteration of the loop, we pass the current plane position to the vertex shader. Next, we generate a six-sided polygon with dummy positions to just enable calls to our shaders. Finally, we increment the plane position by using the plane increment amount.

In the vertex shader, we calculate the current plane distance (d) by using the plane start and plane increment values. Next, we run a loop for all edges. We obtain the current edge vertices by using the passed vertex and edge lists. The edge direction vector is obtained and it is a dot product with the half-angle direction vector. Then, using Equation (17.2), we obtain the parameter (t) for the current edge's intersection point (provided that the value of the parameter t is between 0 and 1). The intersection point is then multiplied with the current modelview/projection matrix to get the clip space position.

The fragment shader obtains the density value from the volume sampler by using the interpolated intersection point from the vertex shader. Since WebGL 1.0 does not support 3D textures, we implement a custom texture sampling function that takes a given 3D texture coordinate and a volume texture (provided in a flat 2D layout) and returns a density value by using modulo arithmetic [Congote 11]. If a transfer function sampler is available, a dependent texture lookup is carried out by using the volume density value at the lookup value. This lookup returns the color corresponding to the current transfer function sampler for the given density value.

In WebGL v 2.0, support of 3D textures will be available. Our resampling function will then be replaced by a hardware-accelerated 3D texture sampling function, which would further improve performance and simplify the code.

17.6 WebGL Implementation

Now that we understand how the dynamic slicing technique can be supported on the GPU, we can commence the half-angle slicing algorithm. The exact implementation steps for a two-pass half-angle slicing algorithm are detailed in Ikits [2004].

17.6.1 Required WebGL Extensions

For implementing this technique in WebGL, we need support for three extensions: `OES_texture_float`, `OES_texture_float_linear`, and `WEBGL_draw_buffers` extensions.

17.6.2 Set Up Offscreen Rendering Support in WebGL

Next, we generate a framebuffer object (FBO) with two color attachments: a light buffer, attached to color attachment 0, and an eye buffer, attached to color attachment 1. The FBO is bound and the required textures are created. These are bound to texture unit 2 as we reserve texture unit 0 for volume data set (laid in a flat 2D texture) and texture unit 1 for transfer function texture. We then test for FBO completeness to ensure that our FBO setup is correct. After testing for FBO completeness, the FBO and texture are unbound. This is achieved by using the code snippet shown in Listing 17.1.

```
fboID = gl.createFBO();
gl.bindFramebuffer(gl.FRAMEBUFFER, fboID);
gl.activeTexture(gl.TEXTURE2);
lightBufferID = createTexture(gl.viewportWidth, gl.viewportHeight);
eyeBufferID = createTexture(gl.viewportWidth, gl.viewportHeight);
gl.framebufferTexture2D(gl.FRAMEBUFFER, ext.COLOR_ATTACHMENT0_WEBGL,
                        gl.TEXTURE_2D, lightBufferID, 0);
gl.framebufferTexture2D(gl.FRAMEBUFFER, ext.COLOR_ATTACHMENT1_WEBGL,
                        gl.TEXTURE_2D, eyeBufferID, 0);

var status = gl.checkFramebufferStatus(gl.FRAMEBUFFER);
if(status ! = gl.FRAMEBUFFER_COMPLETE)
    alert("Error setting up fbo: Framebuffer not complete");

gl.bindFramebuffer(gl.FRAMEBUFFER, null);
gl.bindTexture(gl.TEXTURE_2D, null);
```

The `createTexture` utility function creates a new WebGL texture with floating point texture format given the width and height as parameters. This function is defined as shown in Listing 17.2.

Listing 17.2 The createTexture utility function.

```
function createTexture(width, height) {
    var id = gl.createTexture();
    gl.bindTexture(gl.TEXTURE_2D, id);
    gl.texParameteri(gl.TEXTURE_2D, gl.TEXTURE_MAG_FILTER, gl.LINEAR);
    gl.texParameteri(gl.TEXTURE_2D, gl.TEXTURE_MIN_FILTER, gl.LINEAR);
    gl.texParameteri(gl.TEXTURE_2D, gl.TEXTURE_WRAP_S,
        gl.CLAMP_TO_EDGE);
    gl.texParameteri(gl.TEXTURE_2D, gl.TEXTURE_WRAP_T,
        gl.CLAMP_TO_EDGE);
    gl.texImage2D(gl.TEXTURE_2D,0,gl.RGBA,width,height,0,gl.RGBA,gl.
        FLOAT, null);
    return id;
}
```

17.6.3 Calculating Matrices for Light

During the half-angle slicing process, we have to determine the coordinates of a given slice vertex in the light buffer. To do so, we calculate the vertex coordinates in light space. This is achieved by creating the light's shadow matrix (S) using its modelview (MV_L) and projection (P_L) matrices. Given the light's world space position (light-PosWS), we can use the `lookAt` function to get the light's modelview matrix (MV_L). The light's projection matrix (P_L) is obtained by using the perspective projection function available in the matrix JavaScript library. For our case, we use the glMatrix

JavaScript library.[*] We also require a bias matrix (B). Since the projection matrix (P_L) does not change, we precompute the combined bias and projection matrix (BP). Whenever the light changes, we update the light's modelview matrix (MV_L). The updated modelview matrix is multiplied with the combined bias and projection matrix (BP) to get the shadow matrix (S) as shown in Listing 17.3. A different shadow matrix is required for each light source.

Listing 17.3 Matrices setup for shadow matrix calculation.

```
mat4.identity(MV_L);
mat4.identity(P_L);
mat4.lookAt(lightPosWS, [0, 0, 0], [0, 1, 0.0001], MV_L);
mat4.perspective(45.0, 1.33, 1, 200.0, P_L);
mat4.identity(T);
mat4.identity(B);
mat4.translate(T,[0.5,0.5,0.5], T);
mat4.scale(T, [0.5, 0.5, 0.5], B);
mat4.multiply(B, P_L, BP);
mat4.multiply(BP, MV_L, S);
```

17.6.4 Clearing the Eye Buffer and Light Buffers

After the FBO and matrix setup, we begin the half-angle slicing process by first binding our FBO to enable rendering to offscreen draw buffer. We then clear the eye buffer to (0,0,0,0) and light buffer to light color using the code shown in Listing 17.4.

Listing 17.4 Code showing how to clear the eye and light buffers.

```
gl.bindFramebuffer(gl.FRAMEBUFFER, fboID);
ext.drawBuffersWEBGL([gl.NONE, ext.COLOR_ATTACHMENT1_WEBGL]);
gl.clearColor(0,0,0,0);
gl.clear(gl.COLOR_BUFFER_BIT);

ext.drawBuffersWEBGL([ext.COLOR_ATTACHMENT0_WEBGL]);
gl.clearColor(lightColor[0],lightColor[1], lightColor[2],
            lightColor[3]);
gl.clear(gl.COLOR_BUFFER_BIT);
```

17.6.5 The Slicing Loop

We run a loop for the desired number of slices. These intermediate slices are obtained by slicing a unit cube with a plane perpendicular to the viewing direction as explained in the previous section. In each iteration, we first bind the shadowShader. This shader is responsible for finding the 2D texture coordinates to look up the light buffer for the given vertex coordinates using the light's shadow matrix (S). This is done to look up the light

---

[*] https://github.com/toji/gl-matrix

contribution for the current slice from the light buffer. The modelview (MV) and projection matrices (P) for the eye are passed as shader uniforms as shown in Listing 17.5. The attribute and uniform locations are cached at initialization.

Listing 17.5 Code showing the initial slicing loop.

```
for (var i = 0; i < TOTAL_SLICES; ++i) {
    gl.useProgram(shadowShader.getProgram());
    gl.uniformMatrix4fv(shadowShader.getUniform("uPMatrix"), false, P);
    gl.uniformMatrix4fv(shadowShader.getUniform("uMVMatrix"),
        false, MV);
    gl.uniformMatrix4fv(shadowShader.getUniform("uSMatrix"), false, S);
```

We then bind the light buffer as texture (assuming that texture unit 2 is the currently active texture unit) as the shadowShader assumes that the light buffer is bound to texture unit 2. We then draw slicing polygons from the point of view of eye as shown in Listing 17.6.

Listing 17.6 Code showing how to bind the light buffer and draw slices from the point of view of eye.

```
gl.bindTexture(gl.TEXTURE_2D, lightBufferID);
DrawSliceFromEyePointOfView(i, bIsViewInverted);
```

Next, we use the normal volumeShader, which simply samples the volume data set into the light buffer by using the light's modelview (MV_L) and projection (P_L) matrices. This lets us obtain the light attenuation through the volume data set as seen from the point of view of light. We then draw slicing polygons and the loop terminates as shown in Listing 17.7.

Listing 17.7 Code showing how to render slice from the point of view of light.

```
gl.useProgram(volumeShader.getProgram());
gl.uniformMatrix4fv(volumeShader.getUniform("uPMatrix"), false, P_L);
gl.uniformMatrix4fv(volumeShader.getUniform("uMVMatrix"), false, MV_L);
DrawSliceFromLightPointOfView(i);
```

We then unbind the FBO and restore our viewport. Then we bind the eye buffer as currently active texture since it contains the final blended output as detailed in Listing 17.8.

Listing 17.8 Code showing how to unbind the FBO and use the eye buffer.

```
gl.bindFramebuffer(gl.FRAMEBUFFER, null);
gl.viewport(0,0,gl.viewportWidth, gl.viewportHeight);
gl.bindTexture(gl.TEXTURE_2D, eyeBufferID);
```

17. Real-Time Volumetric Lighting for WebGL

17.6.6 Drawing Slice from the Point of View of Eye

We now detail the drawing function to render slices from the point of view of eye. In this function, we pass two parameters, the index of the current slice (i) and a Boolean flag that tells if the view direction vector is inverted. We first set the appropriate draw buffer (eyebuffer). When the view direction vector is inverted, the Under blending operator is used, which is (gl.blendFunc(gl.ONE_MINUS_DST_ALPHA, gl.ONE)). This ensures that the slices are always sorted in front-to-back order when rendered from the point of view of eye. This blending operator is used when we have associated colors (i.e., the color values that are premultiplied with the alpha) [Porter and Duff 84]. For this to work, support of a separate alpha buffer is required. We do the alpha premultiplication in the volume fragment shader as shown later in Listing 17.13.

If the view direction vector is not inverted, the Over blending operator is used that is (gl.blendFunc(gl.ONE, gl.ONE_MINUS_SRC_ALPHA)). This ensures that the slices are blended in back-to-front order when rendered from the point of view of eye. Here, we do not use the conventional Over blending equation (gl.blendFunc(gl.SRC_ALPHA, gl.ONE_MINUS_SRC_ALPHA) because the alpha values are premultiplied. Finally, we issue the drawArrays call to draw the appropriate slice depending on the currently given index (i) as shown in Listing 17.9.

Listing 17.9 Definition of the function to draw slices from the point of view of eye.

```
function DrawSliceFromEyePointOfView(i, bIsViewInverted) {
    ext.drawBuffersWEBGL([gl.NONE, ext.COLOR_ATTACHMENT1_WEBGL]);
    gl.viewport(0, 0, gl.viewportWidth, gl.viewportHeight);
    if(bIsViewInverted) {
        gl.blendFunc(gl.ONE_MINUS_DST_ALPHA, gl.ONE);
    } else {
        gl.blendFunc(gl.ONE, gl.ONE_MINUS_SRC_ALPHA);
    }
    gl.drawArrays(gl.TRIANGLES, 12*i, 12);
}
```

17.6.7 Drawing Slice from the Point of View of Light

The drawing function to render slice from the point of view of light works similarly. It first sets the appropriate draw buffer (lightBuffer). For light buffer, the conventional Over blending equation is used as the slices are always blended front to back. Finally, we issue the drawArrays call to draw the *i*th slice as shown in the following code snippet. The whole function is defined as shown in Listing 17.10.

Listing 17.10 Definition of the function to draw slices from the point of view of light.

```
function DrawSliceFromLightPointOfView(i) {
    ext.drawBuffersWEBGL([ext.COLOR_ATTACHMENT0_WEBGL]);
    gl.viewport(0, 0, gl.viewportWidth, gl.viewportHeight);
```

```
        gl.blendFunc(gl.SRC_ALPHA, gl.ONE_MINUS_SRC_ALPHA);
        gl.drawArrays(gl.TRIANGLES, 12*i, 12);
}
```

17.6.8 The Volume Shadow Vertex Shader

We now describe the volume shadow vertex shader. The vertex position attribute (aVertexPosition) is passed from the client application. The three shader matrices: the modelview (uMVMatrix), projection (uPMatrix), and the shadow (uSMatrix) matrix are passed in as shader uniforms. Since the given vertex position is from (–0.5,–0.5,–0.5) to (0.5,0.5,0.5), the 3D texture coordinates are calculated by adding (0.5,0.5,0.5) to the given vertex position attribute. This gives us the 3D texture coordinate (vUVW) for looking up the density value in the volume texture sampler in the fragment shader. The vertex position attribute is also multiplied with the shadow matrix (uSMatrix) to get the texture coordinates (vLightUVW) to sample the light buffer. The two obtained texture coordinates (vUVW and vLightUVW) are stored as varying attributes to interpolate them to the fragment shader stage. The whole vertex shader is detailed in Listing 17.11.

Listing 17.11 The volume shadow vertex shader.

```
attribute vec3 aVertexPosition;//input vertex position
uniform mat4 uMVMatrix; //modelview matrix
uniform mat4 uPMatrix; //projection matrix
uniform mat4 uSMatrix; //shadow matrix
varying vec3 vUVW; //volume texture sampling coordinates
varying vec4 vLightUVW; //shadow texture sampling coordinates
void main(void) {
    vec4 v = vec4(aVertexPosition,1);
    vUVW = (aVertexPosition.xyz + vec3(0.5));
    vLightUVW = uSMatrix * v;
    gl_Position = uPMatrix * (uMVMatrix * v);
}
```

17.6.9 The Volume Shadow Fragment Shader

The volume shadow fragment shader first gets the density value from the volume texture using the interpolated volume texture coordinates (vUVW). The volume data set is stored in a 2D flat texture layout whereby slices are laid out in rows and columns. In order to get the density value from this 2D flat texture, a custom sampling function getVolumeValue is used, which is defined as shown in Listing 17.12 [Congote 11]. This is a workaround provided due to lack of 3D textures support in WebGL 1.0. WebGL 2.0 is expected to support 3D textures, so then this function will be replaced by a hardware-supported 3D texture sampling function.

```
float getVolumeValue(vec3 volpos, float totalSlices, float slicesX,
                     float slicesY)
{
        float s1, s2, dx1, dy1, dx2, dy2;
        vec2 texpos1, texpos2;
        s1 = floor(volpos.z * totalSlices);
        s2 = s1 + 1.0;
        dx1 = fract(s1/slicesX);
        dy1 = floor(s1/slicesY)/slicesY;
        dx2 = fract(s2/slicesX);
        dy2 = floor(s2/slicesY)/slicesY;
        texpos1.x = dx1 + (volpos.x/slicesX);
        texpos1.y = dy1 + (volpos.y/slicesY);
        texpos2.x = dx2 + (volpos.x/slicesX);
        texpos2.y = dy2 + (volpos.y/slicesY);
        return mix(texture2D(volumeTexture, texpos1).x,
                   texture2D(volumeTexture, texpos2).x,
                   (volpos.z * totalSlices) - s1);
}
```

The current light contribution is then obtained from the light buffer by calling the texture2DProj function using the interpolated light texture coordinate (vLightUVW). If there is a transfer function sampler, the obtained density value is used in a dependent texture lookup to get the classified color of the given volume sample. Finally, the light intensity is blended with the color of the sample to get the final fragment color. The complete fragment shader is as in Listing 17.13.

Listing 17.13 The volume shadow fragment shader.

```
precision mediump float;
varying vec3 vUVW;  //interpolated volume texture coordinates
varying vec4 vLightUVW;  //interpolated shadow texture coordinates
uniform sampler2D volumeTexture;  //volume sampler
uniform sampler2D transferFunction;  //transfer function sampler
uniform sampler2D shadowTexture;  //shadow texture sampler (light
                                    buffer)
uniform vec4 color;  //color of light
uniform vec3 config;  //contains total slices, slices on x and y axis

void main(void) {
        float density = getVolumeValue(vUVW, config.x, config.y, config.z);
        float alpha = clamp(density, 0.0, 1.0);
        vec2 value = vec2(density, 0.5);
        vec4 sampleColor = texture2D(transferFunction, value);
        alpha = clamp(sampleColor.a, 0.0, 1.0);
```

```
alpha * = color.a;
vec3 lightIntensity = texture2DProj(shadowTexture, vLightUVW.
    xyw).xyz;
gl_FragColor = sampleColor * vec4(color.xyz * lightIntensity *
    alpha, alpha);
}
```

17.7 Experimental Results and Performance Assessment

The volume lighting technique detailed in this chapter was implemented on a laptop ASUS K56CB with an NVIDIA® GeForce® GT 740M GPU. A custom application was written to convert given raw binary CT/MRI data into a 2D texture. For testing purposes, the volume lighting application and the data sets were hosted on a local web server. We used five data sets (Aorta, CTHead, Engine, Skull, and VisibleMale). Details about the data sets and their dimensions are given in Table 17.1.

For performance evaluation, we compare two different rendering modes: direct volume rendering (DVR) and half-angle slicing. Each of these has two slicing modes based on whether the slicing is implemented on the CPU or on the GPU. For each of these modes, the sampling step size is $1/512 = 0.00195$. The camera/viewer is rotated 360° around the data set so that the view dependency of volume rendering can be reduced. The obtained min and max frame time pair (in milliseconds per frame) are in Table 17.2.

As can be seen, the performance of DVR is the fastest. There are two main reasons for this. First, in DVR rendering mode, there is no light attenuation calculation. Second, the result of DVR is calculated in a single pass. On the other hand, half-angle slicing is half as fast, which is mainly because our implementation requires two passes to get the final output. This output contains light attenuation that is not present in DVR. To further assess

Table 17.1 Data Sets Used in the Performance Assessment

Data Set	Raw Data Dimensions	2D Flat Tile Layout	2D Texture Size (pixels)
Aorta	$256 \times 256 \times 100$	10×10	2560×2560
CTHead	$256 \times 256 \times 256$	16×16	4096×4096
Engine	$256 \times 256 \times 110$	16×16	4096×4096
Skull	$256 \times 256 \times 256$	16×16	4096×4096
VisibleMale	$120 \times 120 \times 100$	16×16	1920×1920

Table 17.2 Comparison of Rendering Performance of Different Data Sets[a]

Data Set	Direct Volume Rendering		Half-Angle Slicing	
	CPU Slicing	GPU Slicing	CPU Slicing	GPU Slicing
Aorta	[43.67, 47.62]	[41.49, 47.62]	[64.10, 92.59]	[60.98, 82.65]
CTHead	[63.29, 74.07]	[61.73, 75.19]	[108.70, 196.08]	[97.09, 119.05]
Engine	[62.5, 69.93]	[54.95, 71.43]	[64.94, 102.04]	[62.89, 79.37]
Skull	[60.61, 74.63]	[59.88, 74.07]	[79.37, 102.04]	[78.13, 88.50]
VisibleMale	[45.87, 50.76]	[41.67, 49.75]	[72.99, 96.15]	[72.46, 95.24]

[a] Each cell contains the min, max frame time pair (in milliseconds per frame).

the impact of slicing on frame time, we carried out slicing both on the CPU and the GPU. As can be seen in Table 17.2, offloading the slicing calculation to the GPU improves performance. Snapshots from our WebGL-based real-time interactive visualization application are shown in Figures 17.4 and 17.6.

Usually in volume rendering, different colors are assigned to classify volume densities. This is carried out by creating a transfer function texture. For our case, we implemented a JavaScript-based dynamic transfer function widget (as shown in Figure 17.5) in another

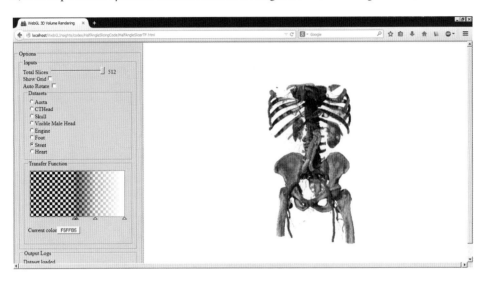

Figure 17.4

Snapshot of our real-time 3D visualization application showing the stent data set.

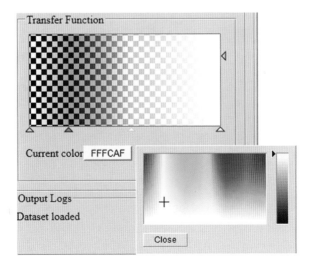

Figure 17.5

The transfer function widget for classifying volumes dynamically.

canvas. This transfer function widget allowed us to add new color keys as desired. Then, dragging the mouse on a color key modified the alpha value for the current color key.

Once the transfer function was modified, the WebGL texture associated to the transfer function widget was also updated. All of this is possible because of a feature in WebGL to create a texture directly from a canvas object. This way we can show or hide different densities in the volume as desired. This method is used to selectively render musculature in Figure 17.6(a, b) and bones in Figure 17.6(c). For the engine block data set in Figure 17.6(d)

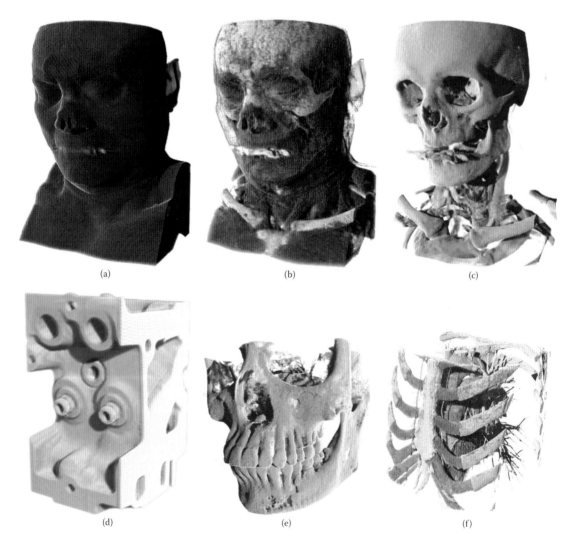

(a) (b) (c)

(d) (e) (f)

Figure 17.6

Rendering results from the volumetric lighting technique detailed in this chapter: (a–c) CTHead data set; (d) engine block data set; (e) skull data set; (f) heart data set. A modified transfer function is used to show musculature in (b) and bones in (c). A similar transfer function is used for the heart data set in (f) to show bones, heart, and blood vessels.

and skull data set in Figure 17.6(e), a simple transfer function was used. The color map similar to a bone's natural color was used for the skull data set. A slightly different transfer function was used for the heart data set in Figure 17.6(f) to highlight vasculature, bones, and soft tissues.

17.8 Conclusion and Future Work

A number of extensions are possible to the technique given here. The first extension is to use the `WEBGL_draw_buffers` extension to render both eye and light buffers in a single pass. In addition, to support platforms that do not support the `WEBGL_draw_buffers` extension, we may use a draw buffer with twice the width so that the light and eye buffers can be accommodated in a single image. The eye and light buffer can then be rendered by adjusting the viewport accordingly. We will still require two passes but the context switch required when color attachments are switched will not be required. We may use ping-pong strategy, using two frame buffer objects to render to one buffer while we read from another buffer, which will ensure that there is no read/write race condition.

We kept the size of our light and eye buffers the same as our back buffer size. We can reduce the offscreen texture size and use the linear filtering support provided by the WebGL hardware to resize the buffers to the size of the back buffer. This can improve performance, especially on smartphones where the memory size is a limit.

17.9 Additional Resources

A number of books have detailed the half-angle slicing technique. For details, we refer interested readers to Ikits [2004] and Engel [2006]. A more modern half-angle slicing implementation (using OpenGL v3.3 core profile) is detailed in Chapter 7 of *OpenGL Development Cookbook* by Packt Publishing [Movania 13b]. A demo implementing this technique in C++ is available from the publisher's website [Movania 13b].

Bibliography

[Congote 11] J. Congote, L. Kabongo, and A. Moreno. "Interactive Visualization of Volumetric Data with WebGL in Real Time." *Proceedings of 2011 Web3D ACM Conference*, pp. 137–146, 2011.

[Engel 06] Klaus Engel, Markus Hadwiger, Joe Kniss, and Christof Rezk-Salama. *Real-Time Volume Graphics*, AK Peters/CRC Press, 2006.

[Exposure 14] Exposure Render. URL: https://code.google.com/p/exposure-render/ (accessed in 2015).

[Ikits 04] Milan Ikits, Joe Kniss, Aaron Lefohn, and Charles Hansen. "Volume Rendering Techniques," Chapter 39 in *GPU Gems*, 2004, URL: http://http.developer.nvidia.com/GPUGems/gpugems_ch39.html (accessed in 2015).

[Kroes 12] T. Kroes, F. H. Post, and C. P. Botha. "Exposure Render: An Interactive Photo-Realistic Volume Rendering Framework." *PLoS ONE* 8(4), 2012.

[Kruger 03] J. Kruger and R. Westermann. "Acceleration Techniques for GPU-Based Volume Rendering." *Proceedings of the 14th IEEE Visualization*, 2003.

[Movania 13a] M. M. Movania and F. Lin. "On-Site Volume Rendering with GPU-Enabled Devices." *Wireless Personal Communications*, 76(4), Springer, pp. 795–812, 2013.

[Movania 13b] M. M. Movania. "Implementing Volumetric Lighting Using the Half-Angle Slicing." Recipe in Chapter 7: *OpenGL Development Cookbook*, Packt Publishing Co. UK, 2013, URL: https://www.packtpub.com/game-development/opengl-development-cookbook (accessed in 2015).

[Porter and Duff 84] Thomas Porter and Tom Duff. "Compositing Digital Images." *Computer Graphics*, 18(3), pp, 253–259, 1984. doi:10.1145/800031.808606.

[Preim and Botha 13] Bernhard Preim and Charl P. Botha. *Visual Computing for Medicine*, 2nd ed.: *Theory, Algorithms, and Applications* (The Morgan Kaufmann Series in Computer Graphics), Morgan Kaufmann Publishers, 2013.

[Stegmaier 05] S. Stegmaier, M. Strengert, T. Klein, and T. Ertl. "A Simple and Flexible Volume Rendering Framework for Graphics-Hardware-Based Raycasting." *Proceedings of Fourth International Workshop on Volume Graphics*, pp. 187–241, 2005.

18

Terrain Geometry—LOD Adapting Concentric Rings

Florian Bösch

18.1 Introduction

Nested concentric rings are a common LOD scheme in geometric clipmapping and other terrain rendering algorithms. We will cover geomorphing, mipmapping and MIP selection, detail mapping, and interpolation methods. For practical reasons (size and scope of this chapter and example code distribution feasibility), we substitute virtual texturing aspects, such as megatextures or clipmapping, etc., with lookups to a small repeatable texture. The texturing function can be exchanged for any more elaborate scheme if so desired.

At the basis of any large-scale terrain-rendering approach lies the observation that things get smaller the further they are from the observer. To exploit this, various level of detail (LOD) schemes are used that have the goal to reduce the amount of information, texels, and vertices needed to render faraway features.

LOD schemes introduce new problems, such as popping and wobbling between detail levels, how to combine several detail levels at a texturing stage, and how to avoid GPU stalling work, for example, for buffer uploads.

The variant of LOD presented here offers the following advantages:

- Minimizes VRAM (video RAM) use (no large set of meshes as in real-time optimally adapting mesh [ROAM])
- Minimizes draw calls (no frustrum culling, one draw call per ring)
- Zero buffer uploads at run-time
- No buffer switching and/or skirt stitching
- Blend between detail levels to minimize visual artifacts such as popping or wobbling

18.2 How to Read This Chapter, Demo, and Source

The chapter, demo, and source are complementary. It will greatly help comprehension when they are read together. Each of the sections is accompanied by a corresponding source. The application is live at http://codeflow.org/webgl/lacr. The source can be found on github https://github.com/pyalot/webgl-lacr; the sections are in the source folder src/sections.

18.3 Rendering the Grid

The grid's vertex buffer is filled with whole numbers. If the grid size is 16, the vertices go from (–8,–8) to (8,8). This is a useful definition for a lot of the math used. To derive a world position for a vertex, a grid position is normalized by dividing it by the grid size. After division, the vertices go from (–0.5,–0.5) to (0.5,0.5). For display, they are multiplied by grid scale, which varies with the world size a grid should cover. A transform function is introduced in Listing 18.1:

Listing 18.1 Transforms a gridspace coordinate to a worldspace coordinate.

```
uniform float gridSize, gridScale;
vec2 transformPosition(vec2 position){
    return (position/gridSize)*gridScale;
}
```

It's useful to think of different coordinate systems in use in a shader as spaces:

- Gridspace: the coordinate system that uses grid coordinates as delivered by the attributes of the grid mesh
- Worldspace: an absolute coordinate system used to transform things for display on screen
- Texturespace: a coordinate system as in texture coordinates

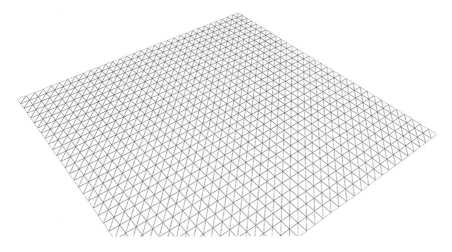

Figure 18.1

Even tessellation.

The first step is to render a regularly subdivided grid. A simple, even tessellation is used, as shown in Figure 18.1.

The grid is drawn by passing the position to the X/Z plane.

```
gl_Position = proj * view * vec4(position.x, 0.0, position.y, 1.0);
```

18.4 Offsetting the Grid

The height field is stored in a png image, which offers the advantage of efficient lossless compression (particularly for empty channels). The high byte is in the red channel and the low byte is in the green channel. To make it possible to filter and mipmap the data, this is converted to a floating-point texture using an off-screen framebuffer before use (Listing 18.2).

Listing 18.2 Height conversion from unsigned short in the off-screen preprocessing step (convertHeights.shader). The scale factor is calculated off-line from the actual height field height.

```
vec4 texel = texture2D(source, coord);
float height = (((texel.r*256.0 + texel.g)/257.0)-0.5)*scaleFactor;
gl_FragColor = vec4(height, 0, 0, 1);
```

The height field is sampled for display at run-time with a linear interpolated texture lookup. The vertex position is offset in the Y direction by the height. The sampling location is calculated by converting the worldspace position to texturespace (using the textureScale parameter) as shown in Listing 18.3.

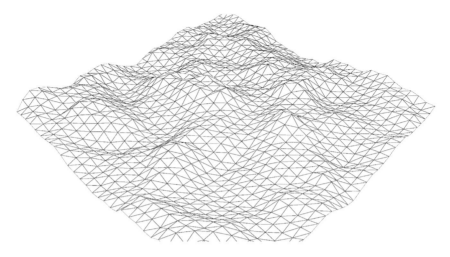

Figure 18.2

Grid offset in Y.

Listing 18.3 Offsetting in the Y direction (Figure 18.2).

```
uniform sampler2D uTerrain;
float getHeight(vec2 position){
    vec2 texcoord = position/textureScale;
    return texture2D(uTerrain, texcoord).x*textureScale;
}

void main(){
    vec2 pos = transformPosition(position);
    float yOffset = getHeight(pos);
    gl_Position = proj * view * vec4(pos.x, yOffset, pos.y, 1);
}
```

18.5 Derivative Maps and Lighting

It is useful to have a definition of normals that allows for high quality per fragment lighting and easy detail normal mapping, is not dependent on the underlying height field resolution, is well behaved for interpolation, and does not require tangent spaces. Partial derivative maps serve that purpose well. A derivative map stores the derivative (central difference) of the height in the S and T direction; it replaces the normal map (Listing 18.4). Derivative maps are discussed among others by Mikkelsen [2010] and Schueler [2013]. By storing the derivative, we can combine several layers of derivative maps easily by addition without having to resort to tangent spaces. The code from Listing 18.2 is modified to also store the derivatives. The derivatives are resolution independent because they are multiplied with the size of the texture (Figure 18.3).

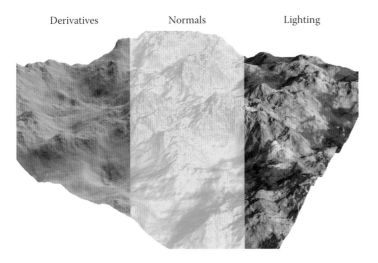

Derivatives   Normals   Lighting

Figure 18.3

Derivatives, normals, and lighting, side by side.

Listing 18.4 Modified height field decoding in the off-screen preprocess to store derivatives as well.

```
float height = getHeight(0.0, 0.0);
float left = getHeight(-1.0, 0.0);
float right = getHeight(1.0, 0.0);
float bottom = getHeight(0.0, -1.0);
float top = getHeight(0.0, 1.0);
float dS = (right-left)*0.5;
float dT = (top-bottom)*0.5;
gl_FragColor = vec4(height, dS*viewport.s, dT*viewport.t, 1);
```

A cross product is used during rendering to compute the normal from the partial derivative map (Listing 18.5).

Listing 18.5 Calculating normal from derivatives, used during rendering.

```
vec3 getNormal(vec2 derivatives){
    vec3 sDirection = vec3(1, derivatives.s, 0);
    vec3 tDirection = vec3(0, derivatives.t, 1);
    return normalize(cross(tDirection, sDirection));
}
```

Lambertian reflectance is used as the lighting term.

18.6 Moving the Grid

To keep the grid centered at the camera it needs to be offset by the camera position (instantaneously). It is also required that the grid be positioned in steps of one cell size in order to avoid popping. The cell size is the size of one cell on the grid. An easy way to achieve this is to transform the camera position into gridspace and shift the grid position by the rounded camera position. `invTransformPosition` is the inverse of `transformPosition` from Section 18.3; it transforms a worldspace position to gridspace (Listing 18.6).

Listing 18.6 Moving the grid position to be centered around the camera. Done by offsetting it by the rounded position of the camera.

```
vec2 cameraPosition = invTransformPosition((invView * vec4(0, 0, 0, 1)).
    xz);
vec2 pos = position + floor(cameraPosition+0.5);
```

18.7 Nesting Grids

To achieve an LOD scheme, successive grids are nested (Figure 18.4). The innermost grid is the square central patch. Rings are rendered around this. Each ring has a hole at its center half its size to make room for the next innermost ring (or the center patch).

The shader is the same as in the previous section. The `gridScale` parameter is used to scale successive rings to twice the size of what they enclose. The draw function maps 1:1 to a `drawArrays` call (Listing 18.7).

Listing 18.7 Drawing rings of increasing size around the central patch.

```
for level in [0...@gridLevels.value]
    scale = @app.gridScale.value * Math.pow(2, level+1)

    @ringState
        .float('gridScale', scale)
        .draw()
```

Each ring, as well as the central patch, is drawn with one draw call.

18.8 Filling the Grid Gaps

Each grid shifts half the distance of its outer neighbor and so gaps appear (Figure 18.5). Due to rounding to the camera position, there's always only one gap, but it can jump left/right or top/bottom.

This could be fixed by adding a skirt and stitching [Asirvatham and Hoppe 2005], to have no cracks with the next bigger LOD level. However, our method fixes this issue by making the grid larger by one row/column. This makes sure that no matter the positioning of the meshes, there is no gap between them because they will overlap.

18. Terrain Geometry—LOD Adapting Concentric Rings

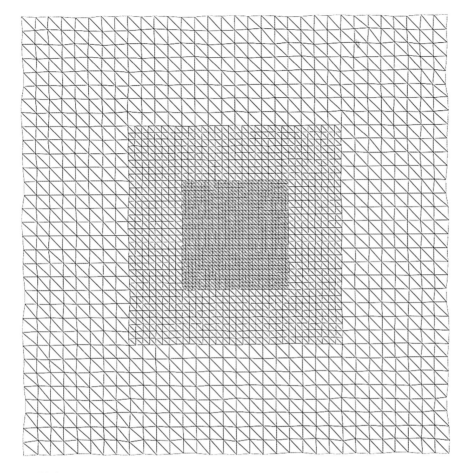

Figure 18.4

Nesting grids.

This will lead to overdraw and z-fighting. But because the terrain is shaded per fragment and identical for each LOD level, z-fighting will not be visible.

Advances in GPU speed (particularly fillrate over upload bandwidth and state changes) made a bit of overdraw an acceptable trade-off to avoid doing more draw calls or updating vertex buffers during run-time.

18.9 Geomorphing

In order to nest grids of different resolution, it's necessary that a grid of finer resolution can mimic a grid of coarser resolution. If a vertex's height is solely determined by the height field at its own position, that corresponds to a morph factor of 0; it is fully morphed/detailed. If the vertex mimics a coarser grid that corresponds to a morph factor of 1, it is fully demorphed/undetailed. A vertex needs to calculate if it falls on the corner

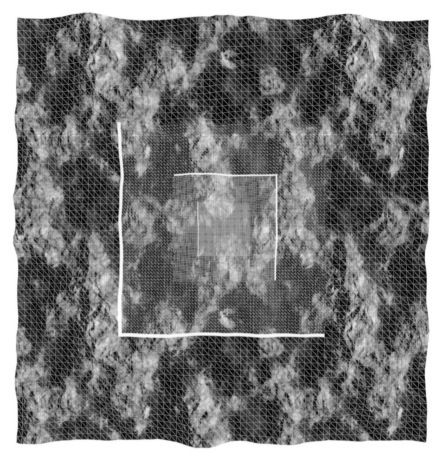

Figure 18.5

Gaps between LOD levels.

of a coarser grid vertex or in between. If it falls on the corner, no change is required. If it falls in between two coarser vertices, its height is averaged between the neighboring vertices depending on the morph factor.

The large violet dot in Figure 18.6 indicates the vertex being displaced. The small orange dots are the neighboring sampling locations on the coarser grid.

18.9.1 Neighboring Function

A neighboring function is introduced to calculate the direction to the next nearest neighbors on the coarser grid. The following method only works on a tessellation where diagonals point the same direction and they go from low X/Z to high X/Z.

In order to locate neighbors, each vertex in gridspace is modulo divided by twice the grid cell size, which is always one in gridspace. The next step is to look up the height at the neighboring vertices and at the vertex being displaced, as in Listing 18.8.

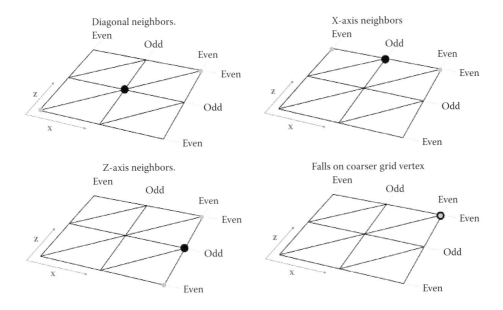

Figure 18.6

Neighboring offsets.

Listing 18.8 Locating neighboring vertices in the coarser grid and looking up their positions. A conditional on the length of `modPos` is used to scale lookup and interpolation if it isn't required.

```
vec2 modPos = mod(position, 2.0);
vec2 ownPosition = vPosition = transformPosition(position);
float ownHeight = getHeight(ownPosition);

if(length(modPos) > 0.5){
    vec2 neighbor1Position = transformPosition(position+modPos);
    vec2 neighbor2Position = transformPosition(position-modPos);

    float neighbor1Height = getHeight(neighbor1Position);
    float neighbor2Height = getHeight(neighbor2Position);

    float neighborHeight = (neighbor1Height+neighbor2Height)/2.0;
    float yOffset = mix(neighborHeight, ownHeight, morphFactor);

    gl_Position = proj * view * vec4(ownPosition.x, yOffset,
        ownPosition.y, 1);
}
else{
    gl_Position = proj * view * vec4(ownPosition.x, ownHeight,
        ownPosition.y, 1);
}
```

18.10 Morph Factor between LOD Levels

Section 18.5 implements geomorphing. To morph between the LOD levels, it is required to vary this morph factor such that it is one (fully detailed) at the inner edge of each ring, and zero at the outer edge of each ring.

The Chebyshev distance can be used for this purpose. Unlike Euclidian distance, it regards the larger axis difference as the distance value (Figure 18.7).

The morph factor is calculated in gridspace. In gridspace, the size of the grid is one-half around the center (0,0). To get a correct morph factor we interpolate between one at the outer edge (distance one-half) to zero at the inner edge (distance one-fourth).

Camera position in gridspace is taken into account to modify the distance metric. This is done so that the grid can shift, but the morph factor advances with the camera. This keeps the morphing smooth and consistently on the camera so as not to introduce popping.

To avoid cracks between LOD morph levels, a bias of one cell size is applied in its calculation. This ensures that the borders of the LOD level match (Listing 18.9) (Figure 18.8).

Listing **18.9** Calculation of the LOD morph factor.

```
vec2 cameraDelta = abs(pos - cameraPosition);
float chebyshevDistance = max(cameraDelta.x, cameraDelta.y);
morphFactor = linstep(
  gridSize/2.0-1.0,
  gridSize/4.0+1.0,
  chebyshevDistance
);
```

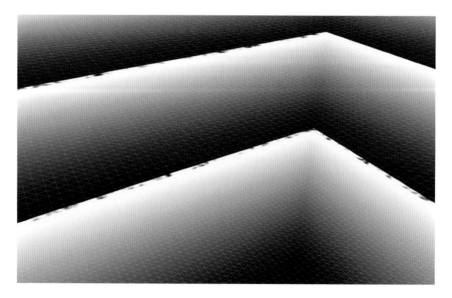

Figure 18.7

The morph factor visualized.

18. Terrain Geometry—LOD Adapting Concentric Rings

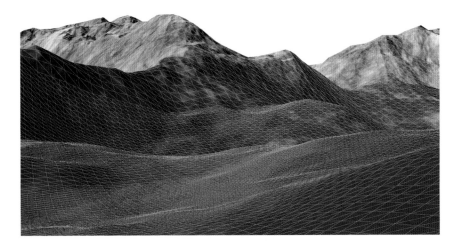

Figure 18.8

LOD levels align with no cracks.

18.11 Texture MIP Level Selection

If we do not choose an appropriate texture MIP level selection in the vertex shader, the terrain may look odd and exhibit more visible morph transitions due to aliasing. At the current setting, the innermost grid matches the resolution of the texture. We will use a MIP factor that matches the grids closely.

Listing **18.10** Calculating the MIP level.

```
float getMiplevel(vec2 position, float texelSize){
  float dist = max(abs(position.x), abs(position.y));

  float cellSize = startGridScale/(gridSize*2.0);

  float correction = log2(cellSize/texelSize);
  float distanceLevel = max(0.0, log2(dist*4.0/startGridScale));

  return distanceLevel+correction;
}
```

To use this function, we pass in the position relative to the world camera position and the size of a texel.

Listing **18.11** Using the miplevel function.

```
float texelSize = textureScale/terrainSize;
float miplevel = getMiplevel(abs(position - camera), texelSize);
```

The uniform `startGridScale` is used to pass in the scale around which the innermost grid size is found. The MIP level is calculated in respect to the `startGridScale`. A correction factor is introduced to account for the size difference between a grid cell and a texel; this avoids under- or oversampling the texture, depending on if a grid cell is smaller or bigger than a texel.

18.12 Magnification

If the mesh is finer than the texture, then linear interpolation will display saddle and facet artifacts. This issue is addressed by replacing the lookup into MIP level zero of the texture with a higher-order interpolation. A good choice for this is the Catmull–Rom spline as it produces plausible height field variation between height posts.

Here is the modified terrain lookup function (`getHeight`) (Listing 18.12):

Listing 18.12 Modified terrain height lookup.

```
uniform sampler2D uTerrain;
float getHeight(vec2 position, vec2 camera){
  float texelSize = textureScale/terrainSize;
  float miplevel = getMiplevel(abs(position - camera), texelSize);
  vec2 texcoord = position/textureScale;

  float mipHeight = texture2DLod(
    uTerrain, texcoord, max(1.0, miplevel)
      ).x*textureScale;

  if(miplevel > = 1.0){
    return mipHeight;
  }
  else{
      float baseHeight = texture2DInterp(
        uTerrain, texcoord, vec2(terrainSize)
          ).x*textureScale;
      return mix(
        baseHeight, mipHeight, max(0.0, miplevel)
        );
  }
}
```

The `texture2DInterp` function implements the higher-order texture interpolation. Figure 18.9 shows the difference between linear and Catmull–Rom interpolation.

18.13 Detail Mapping

Up close to the terrain, the detail isn't very good. This is because the texels of the basis map get fairly large and the detail is insufficient (Figure 18.10).

Detail mapping [Wolfire 2009] is used to address this issue. Three materials are used for this (rock, dirt, and grass), and they are mixed according to a material mix map.

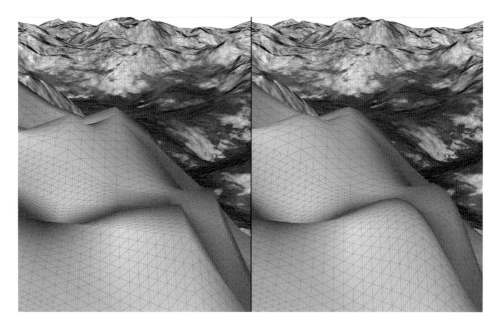

Figure 18.9

Linear interpolation (left) and Catmull–Rom (right).

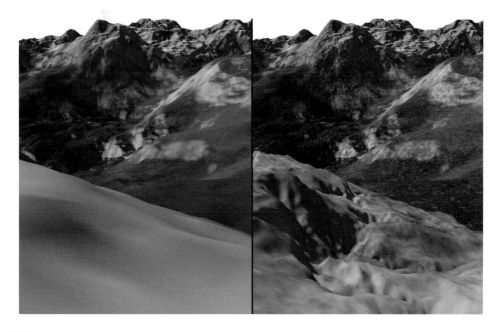

Figure 18.10

The difference between no detail mapping (left) and detail mapping (right).

To look up the material mix, we introduce a function that fetches the mix map texel and makes sure the sum of the result is one (Listing 18.13).

Listing 18.13 Fetching the material mix and normalizing sum to one.

```
uniform sampler2D uMaterialMix;
vec3 getMaterialMix(vec2 position){
  vec2 texcoord = vPosition/textureScale;
  vec3 mixFactors = texture2D(uMaterialMix, position/textureScale).rgb;
  return mixFactors/= (mixFactors.r + mixFactors.g + mixFactors.b);
}
```

18.13.1 Mixing Basis and Detail Albedo

The idea behind albedo mixing is that the basis color is divided by the average color of the detail and then multiplied by the detail color. The getDetailColor function looks up the detail color, and mixing is done according to the material mix (Listing 18.14).

Listing 18.14 Mixing the basis color with the detail color.

```
vec3 rockAlbedo = (albedo/rockAvg)*getDetailColor(uRockColor);
vec3 dirtAlbedo = (albedo/dirtAvg)*getDetailColor(uDirtColor);
vec3 grassAlbedo = (albedo/grassAvg)*getDetailColor(uGrassColor);

vec3 detail = (
    materialMix.x*dirtAlbedo +
    materialMix.y*grassAlbedo +
    materialMix.z*rockAlbedo
);
```

This tints the detail color by the basis color, so as to provide a greater variety of shades.

18.13.2 Adding the Basis and Detail Derivatives

As mentioned in Section 18.6, derivatives are convenient to combine by addition and they are stored independently of resolution. We fetch and mix the derivatives according to the mixmap and add them to the basis derivatives (Listing 18.15).

Listing 18.15 Mixing the basis derivatives with the derivatives.

```
vec2 rockDerivatives = getDetailDerivatives(uRockHeight);
vec2 dirtDerivatives = getDetailDerivatives(uDirtHeight);
vec2 grassDerivatives = getDetailDerivatives(uGrassHeight);

vec2 detailDerivatives = (
    materialMix.r*dirtDerivatives +
    materialMix.g*grassDerivatives +
    materialMix.b*rockDerivatives
);

return derivatives + detailDerivatives*showDetail;
```

18. Terrain Geometry—LOD Adapting Concentric Rings

18.13.3 Adding the Basis and Detail Height

The same MIP level calculation in Section 18.11 is used in the `getDetailHeight` function; the only change is that the texel size passed to `getMiplevel` is the one of the detail texel, not the basis texel. Detail heights are mixed similarly to how derivatives in Section 18.13.2 are mixed. Detail height and basis height are added in the modified `getHeight` function by addition (Listing 18.16).

Listing 18.16 Adding basis and detail height in the modified getHeight function.

```
float detailHeight = getDetailHeight(position, camera);
if(miplevel > = 1.0){
    return detailHeight+mipHeight;
}
else{
    float baseHeight = texture2DInterp(
        uTerrain, texcoord, vec2(terrainSize)
    ).x*textureScale;

    return detailHeight+mix(
        baseHeight, mipHeight, max(0.0, miplevel)
    );
}
```

18.14 Conclusion

We have explored the ways in which it is possible to efficiently render a complex height field without resorting to CPU work and to solve important issues such as how to provide additional detail, minimize LOD-morph artifacts, perform per-fragment lighting and coloring operations, select appropriate miplevels, and structure rendering to avoid state changes and uploads.

Since all the logic to produce a correct rendering is contained in a shader that works instantaneously for any camera position without time delay, this method is suitable for static and dynamic height fields. It can also be extended to more elaborate texturing schemes such as virtual texturing, megatexturing, or clipmapping.

It is also possible to extend to nonsquare grids, like simplex/hex grids, and to offsets in other directions than the vertices.

Optimizations such as index buffers, which would make better use of vertex cache, and frustum culling are also recommended for production settings.

Bibliography

[Asirvatham and Hoppe 2005] *GPU Gems 2*, Chapter 2, "Terrain Rendering Using GPU-Based Geometry Clipmaps." (http://http.developer.nvidia.com/GPUGems2/gpugems2_chapter02.html).

[Kent 2013] "WebGL Terrain Rendering in Trigger Rally." (http://www.gamasutra.com/blogs/JasmineKent/20130904/199521/WebGL_Terrain_Rendering_in_Trigger_Rally__Part_1.php).

[Losasso and Hoppe 2004] "Geometry Clipmaps: Terrain Rendering Using Nested Regular Grids." (http://research.microsoft.com/en-us/um/people/hoppe/geomclipmap.pdf).

[McGuire 2014] "Fast Terrain Rendering with Continuous Detail on a Modern GPU." (http://casual-effects.blogspot.ch/2014/04/fast-terrain-rendering-with-continuous.html).

[McGuire and Sibley 2004] "A Heightfield on an Isometric Grid." (http://graphics.cs.brown.edu/games/IsoHeightfield/mcguiresibley04iso.pdf).

[Mikkelsen 2010] "Bump Mapping Unparametrized Surfaces on the GPU." (https://dl.dropboxusercontent.com/u/55891920/papers/mm_sfgrad_bump.pdf).

[Schueler 2013] "Normal Mapping without Precomputed Tangents." (http://www.thetenthplanet.de/archives/1180).

[Wolfire 2009] "Detail Color Matching." (http://blog.wolfire.com/2009/12/Detail-texture-color-matching).

SECTION VI
Visualization

Some of the most impactful WebGL uses are in areas such as mapping, news, and data visualization. This section looks at prominent data visualization applications.

In Chapter 19, "Data Visualization Techniques with WebGL," Nicolas Belmonte walks through two data visualizations. The first is a visualization of global temperature changes using data from NASA with a focus on the multipass postprocessing rendering stage. The second is real-time color decomposition using the video and camera APIs, Web Workers, and ANGLE_instanced_arrays. In both cases, the algorithms are light on the CPU and offload massively parallel work to the GPU.

Visualizing massive data sets is a common challenge for WebGL applications. Data acquisition has been getting easier and cheaper, which leads to more and more data—everything from OpenStreetMap to CAD to neurological data.

In Chapter 20, "hare3d—Rendering Large Models in the Browser," Christian Stein, Max Limper, Maik Thöner, and Johannes Behr describe their highly adaptive rendering environment (hare3d) with a focus on rendering massive CAD models on any device. Based on factors such as a target frame rate or user movement, hare3d adjusts the run-time behavior of a user-defined graph of pipeline stages on the fly. To do so, objects are quickly binned according to their shader, state, screen-space size, etc., and rendered iteratively. This chapter also includes a discussion of the shape resource container (SRC) format for progressive streaming, balancing size, and fast decoding.

In Chapter 21, "The BrainBrowser Surface Viewer: WebGL-Based Neurological Data Visualization," Tarek Sherif discusses the visualization requirements of neurological research and how WebGL can be used to address them. The architecture of the BrainBrowser Surface Viewer is described, including geometry and per-vertex data load pipelines that utilize Web Workers and three.js BufferGeometry, user interaction, and visualizing massive data sets using on-demand loading.

19

Data Visualization Techniques with WebGL

Nicolas Belmonte

19.1 Introduction

Along with game development and creative coding, data visualization is one of the main use-cases for WebGL. In this chapter, we cover two examples of exploratory data visualizations that couldn't have been accomplished in the Web without WebGL. The first example analyzes weather data taken from NASA showing temperature changes across the globe (Figure 19.1). The second example explores real-time color decomposition from videos and camera input. Both examples use modern browser APIs as part of the data-gathering process, and then use WebGL for the data treatment and visualization process.

These examples show a work flow for data visualization using WebGL. We start by describing a framework to prototype and develop data visualizations for the Web quickly. Then we dive into the data visualization process.

Just like the visualization step is only one part of the data visualization process (there's data gathering, analysis, etc.), WebGL is just one of the APIs we use to create a data visualization. For example, when we talk about the data-gathering process in the web, some might just think about loading a text or JSON file asynchronously, but there are many

Figure 19.1

Temperature anomalies visualization.

other data entry points in the web: binary files, images, video, camera input, geolocation of the user, audio, device orientation, etc.

For the data treatment process, the main goal is to offload as much computation as possible to the GPU. Sometimes this can be done in one pass; sometimes we need more passes. Sometimes it is not practical to achieve this on the GPU, so in that case we can solve many performance issues by using typed arrays in JavaScript. These examples cover both situations.

Finally, for the visualization process we cover APIs of WebGL 1.0 but also of WebGL 2.0. In particular we talk about two extensions: floating-point textures and instanced arrays. We show how these are used to speed up rendering and perform data transformations.

19.2 Framework Setup

The Web has come a long way in producing tools to speed up prototyping and development of Web applications. In particular, Yeoman (http://yeoman.io/) is a great set of tools for developing and deploying static websites. Yeoman consists of three main tools to improve productivity when creating a Web application:

- Yo: scaffolds out a new application, creating the directory structure: index.html and 404.html files, images, videos, shaders, scripts and styles folders, etc. This is all the boilerplate needed to start a new application.
- The build system: used to build, preview, and test our project—a local server, compressing and concatenating JavaScript files, compiling SASS and LESS files into CSS, running image optimizers, etc. The current options are "grunt" and "gulp."

19. Data Visualization Techniques with WebGL

- The package manager: used to handle dependencies so we don't need to manually go to a website, find the download link for the version we're looking for of a lib, and then copy and paste the files into the directory structure. Bower and npm are two popular options for package management.

A typical scaffolded project would look a bit like what is shown in Figure 19.2.

There is an "app" folder that contains source files. The "bower_components" are the JavaScript libraries installed through Bower. The "data" folder contains data in various formats, in this case, TSV (tab-separated values) and JSON files. We have asset folders like "images" and "fonts," a folder containing JavaScript files "scripts" and another containing style (SASS in this case) files called "styles." In the "app" folder, we finally have a "shaders" folder in which we place our custom fragment and vertex shaders. Outside the "app" folder we have the "dist" folder that contains the built site. The "node_modules" include locally installed node modules through npm and probably used by grunt tasks (specified in the "Gruntfile").

From the command line, Grunt lets us do a bunch of interesting things. We can spawn a local server with "grunt serve." This is a development server. It triggers a page reload every time we save a JavaScript file so that we don't have to. It also updates the CSS files if we change any SASS files. The command "grunt build" will create the "dist" folder with the optimized site. Finally, if we wanted to add a task, for example, to convert TSV files into JSON files, then we could create a custom task and call it with "grunt tsv2json."

We can create our own generators so that the scaffolding adapts to our own needs. A special scaffolding by Google called Web Starter Kit* is built on the same principles but is more straightforward to use since it allows for less customizations.

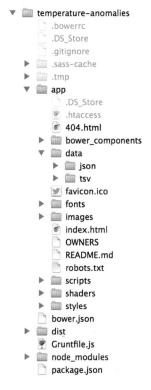

Figure 19.2

Yo directory structure generation.

19.3 Choosing a Data Visualization Framework

19.3.1 PhiloGL

For these examples, we're using PhiloGL,[†] which is a thin layer on top of WebGL that provides boilerplate code to create a WebGL application. This framework is expressive, which means that one concise call to set a texture, for example, will trigger a set of calls in the WebGL API that would be very verbose, but don't abstract the whole WebGL

---

* https://developers.google.com/web/starter-kit/
† http://philogl.org/

API, meaning it doesn't get in our way. The framework has the bare minimum basic concepts of scene, camera, and 3D primitives. This is how an application is created in PhiloGL:

Listing 19.1 Example of application creation with PhiloGL.

```
PhiloGL('canvasId',     {
  program:            {
    from:          'uris',
    path:    './shaders/',
    vs:      'lighting.vs',
    fs:      'lighting.fs'
  },
  camera: {
    position: {
      x: 0, y: 0, z: -50
    }
  },
  textures: {
    src: ['arroway.jpg', 'earth.jpg']
  },
  onError: function(e) {
    console.error("There was an error creating the app.");
  },
  onLoad: function(app) {
    /* Do things here */
  }
});
```

This snippet creates a program from given shader URLs, sets up a camera, asynchronously loads textures, and, if everything goes well, `onLoad` is called. The first argument for `onLoad` is the application object. This object contains interesting properties like the following:

```
app.{gl, program, scene, camera}
```

We can access the `WebGLContext` directly but also use high-level abstractions like a scene or camera. The code in the examples will be mostly low-level code to easily replicate with the bare-bones WebGL API. You can find more simple examples of PhiloGL at http://philogl.org/demos.html#lessons.

19.4 Example 1: Temperature Anomalies

NASA collects year-by-year data about temperature changes around the globe. This information has been collected since 1880 and tracks temperature anomalies (i.e., changes) in different points of the Earth as a 2D heatmap. By loading these images into textures and then mapping them into a 3D histogram, we are able to track the temperature changes around the globe interactively (Figure 19.3). Smooth animations between date ranges enable us to spot the overall differences in temperature across the years. Access this visualization at http://philogb.github.io/page/temperature-anomalies/.

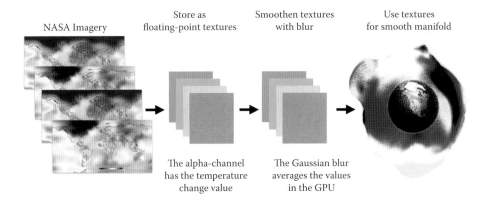

Figure 19.3

A pipeline diagram of the data processing and rendering process. The NASA images that contain the temperature anomaly data are loaded first. These images are stored into floating-point textures where temperature change values are mapped from the red/blue color spectrum to the actual temperature change value, and this value is stored in the alpha channel of the floating-point texture. Next a Gaussian blur is applied to the texture, averaging the temperature points and allowing for a smoother temperature change transition. Finally, these textures are used to shape the smooth manifold wrapping the Earth.

In this example, we cover

- Data loading: The input data are images with a linear color scale indicating temperature anomalies at different positions.
- Data processing: In order to fit a smooth spherical shape, we need to smoothen (i.e., average the data set stored in the floating-point textures with a Gaussian blur).
- Rendering: We'll cover the postprocessing stage.

19.4.1 Loading Data

The NASA imagery is loaded directly. It consists of images showing an equirectangular map projection and a color coding indicating a decrease of 3°C (blue) to an increase of 3°C (red) (Figure 19.4).

The images are loaded asynchronously, and the image loading progress is shown by a progress bar. In PhiloGL, this looks like:

Listing 19.2 Loading multiple images with PhiloGL.

```
var images = new IO.Images({
    src: imageUrls,
    onProgress: function(perc) {
      console.log('loaded ' + perc + '%');
    },
    onComplete: function() {
      /*do stuff with images array here*/
    }
});
```

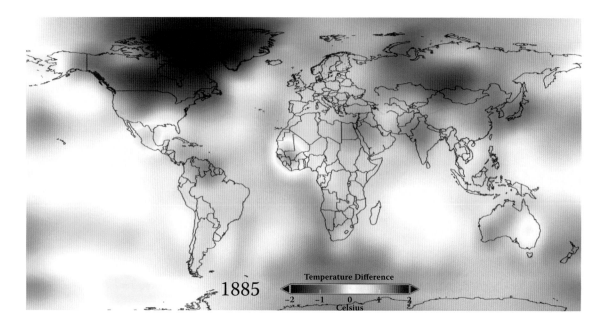

Figure 19.4

Temperature anomalies imagery used in this visualization. (NASA/Goddard Space Flight Center Scientific Visualization Studio. Data provided by Robert B. Schmunk, NASA/GSFC GISS).

IO.Images is a convenience function to load an array of images. The function uses the native Image API, which loads asynchronously. For example:

Listing 19.3 Native image API loading.

```
var image = new Image();
image.src = 'myimage.jpg';
image.onload = function() {
  console.log('Image loaded');
};
```

The next step is to set these images as textures and to process and average the data set to get a smooth spherical manifold.

19.4.2 Processing Data

In order to smoothen the data, we first load the image as an RGBA (red green blue alpha) floating-point texture, calculate the temperature change number in the alpha channel based on the RGB colors of the image, and then smoothen the data by applying a two-pass Gaussian filter on the alpha channel.

19. Data Visualization Techniques with WebGL

Let's start with texture loading. In PhiloGL, this looks very similar to the actual WebGL API except that it is less verbose:

Listing 19.4 Setting up a texture in PhiloGL. All the texture options can be mapped directly to raw WebGL API calls.

```
app.setTexture('img', {
  width: textureWidth,
  height: textureHeight,
  parameters : [{
    name : 'TEXTURE_MAG_FILTER',
    value : 'LINEAR'
  }, {
    name : 'TEXTURE_MIN_FILTER',
    value : 'LINEAR'
  }, {
    name: 'TEXTURE_WRAP_S',
    value: 'CLAMP_TO_EDGE'
  }, {
    name: 'TEXTURE_WRAP_T',
    value: 'CLAMP_TO_EDGE'
  }],
  data: {
    width: textureWidth,
    height: textureHeight,
    value: img
  }
});
```

In this example, we're setting the image `img` as a texture. In PhiloGL, we can provide identifiers (i.e., ids) for the textures using the first argument of `setTexture`. For the texture parameters, we use linear filters to interpolate the image when it is scaled. We also use `CLAMP_TO_EDGE` texture wrap mode so that we can use non-power-of-two images as textures.

The postprocessing passes use a set of framebuffers. In PhiloGL, these are set like the following:

Listing 19.5 Setting up a framebuffer and binding a texture to it. In this case the texture is a floating-point texture.

```
app.setFrameBuffer('framebuffer-id', {
  bindToTexture: {
    data: {
      type: gl.FLOAT,
      width: textureWidth,
      height: textureHeight
    }
  }
});
```

This means that we are attaching a texture to the framebuffer as well. We can store the result of rendering into a framebuffer attachment and then bind the texture for processing.

The first postprocessing pass stores the actual temperature changes in the alpha-channel of the texture. The fragment shader for this pass is pretty simple:

Listing 19.6 Map temperature anomaly changes to the alpha channel.

```
uniform sampler2D sampler1;
varying vec2 vTexCoord1;
//map from [-1, 1] to [0, 2]
float scale(vec3 color) {
  return color.r - color.b + 1.0;
}
//read color from texture and store value in alpha channel
void main() {
  vec3 color = texture2D(sampler1, vTexCoord1).rgb;
  gl_FragColor = vec4(color, scale(color));
}
```

Since the colors gradually change from blue to red, we can create a scale that will map to transition colors (varying from blue to red) to [0, 2] range.

The next step is the multipass smoothing procedure. This technique performs something close to a Gaussian blur in an inexpensive way. The algorithm first performs a blurring on the x-axis, and then the intermediate horizontal smoothing result is smoothed on the y-axis. The algorithm is explained in more detail in Rákos [10] (Figure 19.5).

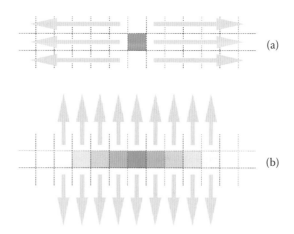

Figure 19.5

(a) Blur the texture horizontally first. (b) Then blur the result vertically. The result is a Gaussian blur.

19. Data Visualization Techniques with WebGL

The fragment shader is:

Listing 19.7 Gaussian blur fragment shader.

```
uniform float width;
uniform float height;
uniform float blurX;
uniform float blurY;
uniform sampler2D sampler1;

varying vec2 vTexCoord1;

void main() {
  vec4 sum = vec4(0.0);
  vec2 dim = vec2(width, height);
  vec2 blurSize = vec2(blurX, blurY)/dim;
  vec2 p = vTexCoord1;

  if (blurX != 0. || blurY != 0.) {
    sum + = texture2D(sampler1, p - 4.0 * blurSize) * 0.05;
    sum + = texture2D(sampler1, p - 3.0 * blurSize) * 0.09;
    sum + = texture2D(sampler1, p - 2.0 * blurSize) * 0.12;
    sum + = texture2D(sampler1, p - 1.0 * blurSize) * 0.15;
    sum + = texture2D(sampler1, p                  ) * 0.16;
    sum + = texture2D(sampler1, p + 1.0 * blurSize) * 0.15;
    sum + = texture2D(sampler1, p + 2.0 * blurSize) * 0.12;
    sum + = texture2D(sampler1, p + 3.0 * blurSize) * 0.09;
    sum + = texture2D(sampler1, p + 4.0 * blurSize) * 0.05;
  } else {
    sum + = texture2D(sampler1, p);
  }

  gl_FragColor = sum;
}
```

A more performant way would be to have two separate programs: one to blur on the x-axis and another one to blur on the y-axis instead of having a `blurSize` variable and a conditional in the fragment shader.

The `blurX` and `blurY` uniforms indicate the blur direction. If either is not `0.0`, then the next step samples pixels in a Gaussian distribution across the selected axis. We achieve the result in three passes: one that copies the values to the alpha value in the texture, and the other two that apply the Gaussian smoothing to the texture. The JavaScript code to call the postprocessing passes is:

Listing 19.8 Postprocessing passes in PhiloGL.

```
Media.Image.postProcess({
  //use original image as sampler
  fromTexture: 'img',
    //store using 'to-float' framebuffer
```

```
    toFrameBuffer: 'to-float',
        //use the 'float-scale' program
    program: 'float-scale',
}).postProcess({
    fromTexture: 'to-float-texture',
    toFrameBuffer: 'blurX',
    program: 'blur',
    uniforms: {
        width: textureWidth,
        height: textureHeight,
        //perform smoothing on x-axis
        blurX: 1,
        blurY: 0
    }
}).postProcess({
    fromTexture: 'blurX-texture',
    toFrameBuffer: 'to-float',
    program: 'blur',
    uniforms: {
        width: textureWidth,
        height: textureHeight,
        blurX: 0,
        //perform smoothing on y-axis
        blurY: 1
    }
});
```

Each pass is executed when `postProcess` is called. The first one sets `img` as the input texture, which references the original image. The result of the pass is stored in the texture bound to the framebuffer `to-float`. The texture bound to a framebuffer with id "id" has `id-texture` as id. The second pass reads from the bound texture, `to-float-texture`, and stores the result in the `blurX` framebuffer. We set `blurX` to `1` to execute the Gaussian blur in the x-axis, and then set `blurY` to `1` for the y-axis pass.

19.4.3 Visual Display

Finally, we need to render the globe and the surrounding mesh. For the neon effect, we reuse the same Gaussian blur shaders as before, and also add a multiplier before blending to the original image by adding the following code to the bottom of the Gaussian blur shader in Listing 19.1:

Listing 19.9 We add a mixing code to the bottom of the Gaussian blur shader so that we can get proper blending.

```
gl_FragColor = sum;

if (multiplier > 0.0) {
    sum * = multiplier;
    gl_FragColor = sum + texture2D(sampler2, p);
}
```

This new shader accepts a second sampler and blends the blurred result of the first sampler to the second sampler to create the neon effect. The `multiplier` variable sets how pronounced the neon effect will be.

The whole procedure renders twice to framebuffers. The first rendering is done in a downsized viewport and only renders the spherical manifold. A Gaussian blur is applied to the manifold. The second rendering renders both the globe and the surface and both renderings are blended together with additive blending. The result is rendered to the screen. Finally, we also render to the screen the globe and mesh with the actual colors taken from the texture. Both images rendered in the screen are blended. See Figure 19.6.

The first step renders a downsized manifold to a framebuffer:

Listing 19.10 We scale the rendering by a "factor" and then render the manifold to a texture.

```
app.setFrameBuffer('grid', true);
gl.lineWidth(2);
gl.clear(gl.COLOR_BUFFER_BIT | gl.DEPTH_BUFFER_BIT);
earth.display = false;
gl.viewport(
    viewportX/factor,
    viewportY/factor,
    viewportWidth/factor,
    viewportHeight/factor
);
scene.renderToTexture('grid');
app.setFrameBuffer('grid', false);
```

This code first binds the framebuffer and then sets some options like the viewport and line width before rendering to the bound texture of that framebuffer with

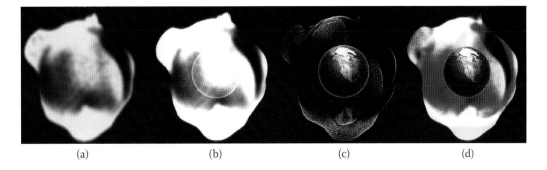

(a)   (b)   (c)   (d)

Figure 19.6

Intermediate rendering passes: (a) blur the manifold and (b) add this to the original image. (c) Then render the color image without any postprocessing. Finally (b) and (c) are rendered to the screen with blending mode on. The end result is (d).

renderToTexture. The next texture rendering is similar but includes the globe and is not downsized:

Listing 19.11 Render the globe and the manifold in actual scale.

```
app.setFrameBuffer('planet', true);
gl.lineWidth(1);
gl.clear(gl.COLOR_BUFFER_BIT | gl.DEPTH_BUFFER_BIT);
earth.display = true;
gl.viewport(
    viewportX,
    viewportY,
    viewportWidth,
    viewportHeight
);
scene.renderToTexture('planet');
app.setFrameBuffer('planet', false);
```

Finally, just as we did before, we create the two-pass filter to get a Gaussian blur and add the two renderings. The whole code for the process is on github.*

19.5 Possible Extensions

An interesting way to extend this project would be to get the latitude and longitude location for the user and center the globe at that location. We could also provide extra information regarding the temperature anomalies in that specific location using the navigator. geolocation API:

Listing 19.12 Using the navigator.geolocation API to infer the user's location.

```
if (navigator && navigator.geolocation) {
  navigator.geolocation
    .getCurrentPosition(function yep(position) {
    console.log(
        position.coords.latitude,
        position.coords.longitude
    );
  }, function nope() {
    console.log('no position available');
  });
}
```

The API is pretty straightforward: getCurrentPosition returns a position object with the latitude and longitude of the user. Given this, we can rotate the globe to the current position. The full code to perform the globe rotation, given a lat/lon pair, is on github.†

---

* https://github.com/philogb/remix/blob/master/temperature-anomalies/app/scripts/globe.js#L302
† https://github.com/philogb/remix/blob/master/temperature-anomalies/app/scripts/globe.js#L559

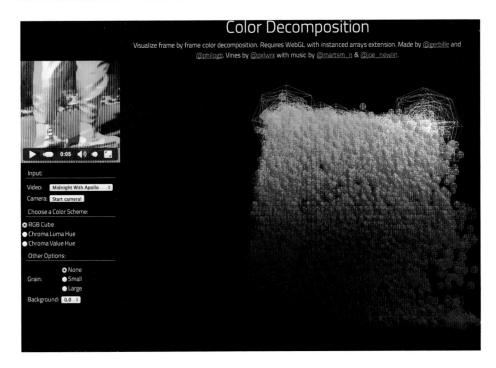

Figure 19.7

Color decomposer UI image.

19.6 Example 2: Color Decomposer

In the previous example we looked at loading a dataset and processing and displaying it with WebGL. This next example takes a dynamic data source and performs real-time updates to the visualization.

Color decomposition on RGB or other color schemes takes a 3D form. In order to show the color decomposition of an image, each dot on a 3D color histogram is assigned a different diameter. The following example[*] renders fine-grain and precise color decomposition schemes of each frame of a video in real time. Color varies on each take. When viewing, change the color scheme or pause the video to have a closer look at the color decomposition. Interact with the visualization by using drag and drop and zooming in/out with the mouse wheel (Figure 19.7).

In this example we cover the following:

- Loading data: The input data are video frames and the user's camera.
- Processing data: In order to create a color histogram we count the number of pixels in each color bucket. Also, we use Web Workers to load the 3D primitives asynchronously.
- Rendering: The interpolation process happens on the GPU and instanced arrays are used to quickly render millions of elements.

---

[*] http://philogb.github.io/page/color-cube/

19.6.1 Loading Data

In this example, data comes in two flavors: a preselected set of videos that we loop, and then the camera input. Not all browsers support the same video formats. The best choice today is to convert videos in both WebM and MP4 formats. A simple way to include both sources in the video tag is:

Listing 19.13 Adding multiple video formats to a video element.

```
<video id = "movie" controls width = "240" loop = "true">
  <source src = "movie/file.webm" type = "video/webm"/>
  <source src = "movie/file.mp4" type = "video/mp4"/>
</video>
```

The video element provides a straightforward API to handle video:

Listing 19.14 Using the video API.

```
var movie = document.querySelector('#movie');
movie.play();
move.pause();//etc.
```

For capturing the camera input, we also use the video element, but the procedure is trickier. We need to capture the media stream and set it as the source for the video element. The standard way to capture the local media stream is:

Listing 19.15 Capturing the user camera, and audio (optionally).

```
navigator.getUserMedia(
  {video: true},
  function success(stream) {
    movie.src = window.URL.createObjectURL(stream);
    movie.play();
  },
  function error() {
    console.warn('An error occurred!');
  }
);
```

This code uses the `getUserMedia` method on the `navigator` object to request the user to provide video information. The first argument is an object with the information being requested. For example, besides requesting video, we could also request audio. The next two arguments are callbacks. The first gets called when everything works correctly and the user accepts to send video information. The `error` function gets called when the user declines using the camera or an error occurs.

In the `success` function, we create a `URL` object out of the media data we're getting so that we can set it as the source of the video element. The code is on github.[*] This is the standard way of doing it but each browser has some special case. Once we are playing our video or capturing information from the camera, the next step is to read each pixel to form the histogram.

19.6.2 Processing Data

The best way to read the pixels of an image (or video/camera in this case) is to use the canvas element. By using the `drawImage` canvas method we can capture the pixels in a typed array. The canvas API allows drawing an image into it and then reading the pixels as an RGBA `Uint8ClampedArray`. The process to do this goes like this:

Listing 19.16 Using the navigator.geolocation API to infer the user's location.

```
var movie;//movie element
var width;//movie width
var height;//movie height

//create a canvas element
var canvas = document.createElement('canvas');
canvas.width = width;
canvas.height = height;
//get context
var ctx = canvas.getContext('2d');
//draw image into canvas
ctx.drawImage(movie, 0, 0, width, height);
ctx.getImageData().data;//contains the Array
```

A different technique that makes use of the GPU for creating histograms can be seen in Scheuermann [2007]. Let's explore here a few optimizations we could do in the JavaScript side for this.

Since this capturing is done on the JavaScript side and not through the GPU, we can reduce the computation time by creating a canvas with a size smaller than the actual frame. If the image needs to be reduced to fit in the canvas, most browsers use bilinear interpolation by default.[†]

Once we have a handle to the pixels, we iterate through each pixel of the array and populate a `Float32Array` with the count of pixels in each color. If we had a full RGB cube, then we would need 256 elements per color, which would give us a `Float32Array(256 * 256 * 256)`. In this case we use `dim` as the dimension of the cube.

Listing 19.17 Creating a histogram data structure from the RGBA pixel color array.

```
var ans = new Float32Array(
    (dim + 1) * (dim + 1) * (dim + 1)
),
    pixels = ctx.getImageData(0, 0, width, height).data,
    i, len;
```

---

[*] https://github.com/philogb/remix/blob/master/color-cube/app/scripts/histogram.js#L270
[†] http://entropymine.com/resamplescope/notes/browsers/

```
for (i = 0, len = pixels.length; i < len; i + = 4) {
  var r = round(pixels[i    ]/255 * dim),
      g = round(pixels[i + 1]/255 * dim),
      b = round(pixels[i + 2]/255 * dim),
      index = r +
          g * (dim + 1) +
          b * (dim + 1) * (dim + 1);

  ans[index] = (ans[index] || 0) + 1;
}
```

First we define the histogram data structure: a long array that will represent the three-dimensional matrix containing each RGB component in the graphic. We call this variable `ans`. Then we get the typed array containing pixels of the video frame. We call this `pixels`. Then we iterate through each RGBA component of the pixels array and convert the [0-255] value of each RGB component in the pixels' array to the [0-dim] domain where `dim` is the dimension of our cube. Finally, we get the proper index of the histogram array by generating it as one would do a lookup in a matrix. Once we get the index, we increase the count of the component by one.

The other type of data processing we need to do includes generating the shapes for the position of the elements in RGB, HSL, and HSV shapes. This is not very costly, but since it's being done client-side, we can offload this processing to a Web Worker so it doesn't interrupt the UI thread and the application is still responsive. We won't go over calculating the 3D positions of an RGB cube and hue maps, but we describe here how to use a Web Worker. The worker code resides in a separate file. The worker API is pretty simple. We send and receive messages in the worker code by listening to `onmessage` and sending messages back with `postMessage`. For example, the worker code looks like this:

Listing 19.18 The worker code creates models in a background thread.

```
onmessage = function() {
  createSphere();
  createRGB();
  createHueMaps();

  postMessage({
    sphereVertices: sphereVertices,
    sphereNormals: sphereNormals,
    colors: colors,
    rgb: rgbPositions,
    hsv: hsvPositions,
    hsl: hslPositions,
    indices: sphereIndices
  });
};
```

When the code receives a message, it creates a sphere and returns `sphereVertices`, `sphereNormals`, and `indices`. This sphere is used as the encoding for the number of pixels in each color. For example, if we'd like to have a $256 \times 256 \times 256$ RGB cube, then we would render $256 \times 256 \times 256$ point primitives. After creating the sphere, we create arrays of positions for the RGB, HSV, and HSL shapes. We send all that information back to the main UI thread by calling `postMessage`. In the main thread we create a Worker object that points to our worker file and then set a listener for getting messages from it and send an initial `postMessage` call:

Listing 19.19 Calling the worker from the UI thread.

```
var worker = new Worker('scripts/histogram-models.js');
worker.onmessage = function(e) {
  //e.data will contain the data
};
worker.postMessage('init');
```

For more on Web Workers, see Chapter 4. Now that we have all the data we need, we combine the histogram information with the color scheme layout information to produce the result.

19.6.3 Visual Display

The main rendering challenge is to render a large number of spheres for each one of the color components in the histogram. There are three ways to do this:

- Loop through each component generating a `drawElements` call. This would be `dim * dim * dim` calls every frame. Unacceptable.
- Generate all the `dim * dim * dim` point primitives in the Web Worker from the beginning. For this we would also need to add the proper offsets, colors, etc., to the spheres. If each sphere has about 500 vertices, then we would have `500 * dim * dim * dim` elements in the screen. Not only would this be slow to compute, it would allocate a ton of memory.
- Use the `ANGLE_instanced_arrays` extension. Generate only one sphere. Then, for the offsets, generate as many points as spheres we want. Instance those vertices and perform only one `drawElementsInstancedANGLE` call that will draw as many spheres as there are positions in the RGB cube at once. This option is both fast and memory efficient.

We will explain the last option. The first step is to set the instanced buffers:

Listing 19.20 Setting the instanced buffers.

```
program.setBuffers({
    'colors': {
    value: new Float32Array(histogram.colors),
    size: 3,
    instanced: 1
  },
```

```
    'rgb': {
      value: new Float32Array(histogram.rgb),
      size: 3,
      instanced: 1
    },
    'hsl': {
      value: new Float32Array(histogram.hsl),
      size: 3,
      instanced: 1
    },
    'hsv': {
      value: new Float32Array(histogram.hsv),
      size: 3,
      instanced: 1
    }
});
```

The buffers we instance are the ones that are not going to be repeated during the rendering phase. With `instanced: 1` we specify that each sphere will use each component of the array only once. `instanced: 2`, for example, would create a pair of spheres with the same color component. Under the hood, we get the extension object first:

```
var ext = gl.getExtension('ANGLE_instanced_arrays');
```

and then right after setting the `vertexAttribPointer` GL call, we call `ext.vertexAttribDivisorANGLE`:

```
gl.vertexAttribPointer(
    loc, size, dataType, false, stride, offset
);
ext.vertexAttribDivisorANGLE(loc, instanced);
```

The sphere buffers are not instanced:

Listing 19.21 Sphere buffers are not set to be instanced, since they are the ones that are going to be repeated.

```
program.setBuffer('sphereVertices', {
  value: new Float32Array(histogram.sphereVertices),
  size: 3
});
program.setBuffer('sphereNormals', {
  value: new Float32Array(histogram.sphereNormals),
  size: 3
});
program.setBuffer('indices', {
  bufferType: gl.ELEMENT_ARRAY_BUFFER,
  drawType: gl.STATIC_DRAW,
  value: new Uint16Array(histogram.indices),
  size: 1
});
```

Finally, on the rendering side, instead of calling `drawElements` we call:

```
ext.drawElementsInstancedANGLE(
    gl.LINE_LOOP,
    histogram.indices.length,
    gl.UNSIGNED_SHORT,
    0,
    (dim + 1) *
    (dim + 1) *
    (dim + 1)//# of spheres to render
);
```

The last argument of the draw call specifies the number of spheres to be rendered.

19.7 Resources

The working code for the projects is described at Garcia Belmonte [2014]. There are many great libraries to dive into WebGL and data visualization, D3[*] by Mike Bostock is one of them. D3 [Bostock 2011] uses SVG but has many convenience functions to use with WebGL; three.js[†] is a high-level layer on top of the WebGL API and it's been used by Google and other organizations. The library glfx.js[‡] provides an array of interesting post-processing techniques worth exploring.

Bibliography

[Garcia Belmonte 2014] Nicolas Garcia Belmonte. "WebGL Insights Examples." https://github.com/philogb/page/#README
[Rákos 10] Daniel Rákos. "Efficient Gaussian Blur with Linear Sampling." http://rastergrid.com/blog/2010/09/efficient-gaussian-blur-with-linear-sampling/
[Scheuermann 2007] Thorsten Scheuermann and Justin Hensley. "Efficient Histogram Generation Using Scattering on GPUs."

---

[*] http://d3js.org/
[†] http://threejs.org/
[‡] https://github.com/evanw/glfx.js

20

hare3d—Rendering Large Models in the Browser

Christian Stein, Max Limper, Maik Thöner, and Johannes Behr

20.1 Introduction

Across the majority of business areas, Web technology seems to have become the predominant solution for creating powerful, platform-independent applications. On almost any kind of client device and software platform, a Web browser is able to provide standardized access to shared resources, stored in the cloud or local intranet. Lately, browsers even became able to utilize the power of modern GPUs, using WebGL. As a consequence, it is not surprising that many visualization and graphics applications are being ported to the Web environment. Even high-performance visualization of complex and highly irregular 3D models, such as CAD and BIM data, has become feasible inside the browser.

While many WebGL-based frameworks, such as three.js or X3DOM, have been introduced over the last years, providing scalable rendering solutions for all kinds of models and devices has not been their primary focus. To deal with data sets of arbitrary size on any potential client device, a flexible client–server environment becomes necessary. An example for such an environment is *WebVIS/instant3DHub* [Jung et al. 12]. Providing services for large model visualization, such as 3D data transcoding, server-side rendering, server-side culling, spatial indexing and search, the architecture of WebVIS/instant3DHub is tailored toward scalable solutions that match a wide range of constraints. Depending on a particular setting, the fitting combination of client- and server-side services is chosen. One constraint might, for instance, be security, allowing to stream 3D data only through trusted networks and preferring streams of remotely rendered images in other contexts.

This chapter focuses on our client-side visualization component, entitled *hare3d* (highly adaptive rendering environment). The main task of hare3d is the browser-based rendering of potentially massive 3D data sets, on any device, using WebGL. While the general architecture is not limited to a specific use case, we discuss our optimizations for rendering large sets of CAD data on the client-side with WebGL.

20.2 System Overview

While WebGL provides a common rendering API on a diversity of platforms, the capabilities of the devices vary widely. Therefore, a scalable solution has to be able to adjust to the client device's capabilities by choosing the most fitting internal setup depending on application-specific goals such as rendering performance or minimal visual error.

20.2.1 Pipeline Design

hare3d ensures a high degree of scalability and flexibility by supporting individual rendering pipelines, based on the assembly of stages. For technically adept users it is possible to freely create their pipeline configuration themselves. However, hare3d also provides a number of configurations for typical application scenarios, such as CAD and augmented reality (AR) (Figure 20.1). Internally, the pipeline is set up to best match the device's capabilities.

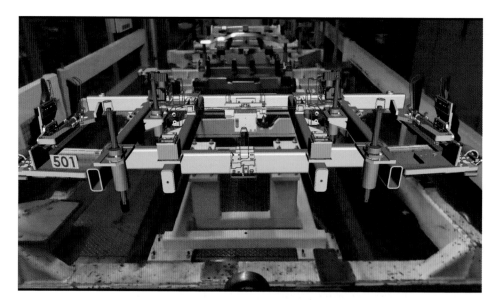

Figure 20.1

An example AR review application that combines several pipeline stages to render parts of the scene with a postprocessing effect.

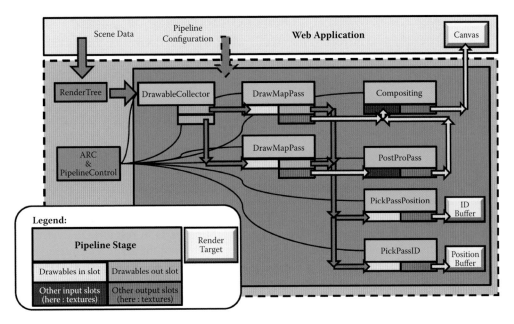

Figure 20.2

Example pipeline and integration of hare3d into a web application.

Figure 20.2 depicts a possible pipeline setup for such CAD/AR applications. In this example, a selection of the model's parts is first rendered to a separate offscreen buffer. The result is then postprocessed, rendering only the parts' edges and outlines, so they can be compared with their real-world equivalents, displayed as a background video or image. Furthermore, object IDs and positions are written to textures, allowing for picking and scene measurements. After the optional pipeline creation, the Web application communicates with hare3d primarily through modification of the scene represented by the *RenderTree*. This structure will be explained in Section 20.3.

hare3d manages all internal data in the form of resources. Resources act as lightweight wrappers that add simple states and versioning to their data, for instance, to the buffer behind a TypedArray view. The pipeline stages are the nodes in a graph; the edges are the connections of a stage's output slot to another stage's input slot. By default all stages hold slots for sets of *Drawables*, the essential items running through the pipelines. Yet, slots are generic and can be used for arbitrary resources. Drawables are instantiated renderable objects, containing all drawing-related information. They are extracted from the scene in the initial *DrawableCollector* stage (Section 20.4). In the depicted AR pipeline, the Drawables are split up to run through separate paths, rendering to different buffers, before being joined again for the picking passes. After the postprocessing has taken place, the resulting image textures are finally composited.

On target devices supporting the WEBGL_draw_buffers[*] extension, rendering the scene and picking data can be performed in a single pass, reducing render costs.

---

[*] http://www.khronos.org/registry/webgl/extensions/WEBGL_draw_buffers/

20.2.2 Runtime Optimization and Adaptive Render Control

hare3d aims at increasing the overall performance by decreasing the workload per frame. By using time-stamped event flags (e.g., *ViewChanged* or *GeometryChanged*) of the so-called *PipelineControl*, hare3d adjusts the runtime behavior of its pipelines on the fly. The different stages are only run as required, when commanded by the PipelineControl, or whenever hare3d of a slot has changed or has not been processed entirely in the context of iterative rendering (Section 20.5.3). Compared to boolean flags, the use of time stamps offers much more flexibility. Each stage can decide when to react to specific events and, for example, delay its execution for several frames. Thus, approaches like employing hysteresis thresholding for the triggering of specific render passes can easily be realized.

By default, this behavior is highly coupled with the integrated interaction modes, like *walk-through* or *inspection*. For example, during a user's navigation, there is no need to update the data in the picking buffer(s), which happens only once after a continuous navigation has finished. Likewise, highlighting parts in response to a user's hovering interaction needs to result in only a partial redraw of the scene. Of course, these example optimizations are not always applicable, yet very beneficial in our CAD visualization scenarios, with camera movement usually being the only motion within the scene.

For the majority of large CAD models, the fine tessellation of the original free-form surfaces leads to a high amount of geometry data. The car model shown later in Figure 20.7a, for instance, consists of 86 million triangles. Conversely, our target applications' requirements on lighting and shading are relatively straightforward. These constraints, in contrast to those expected for games, for instance, result in the rendering performance being limited by either the number of vertices or the number of draw calls. Consequently, reducing the amount of rendered vertices or issued draw calls is the most important optimization in this case, especially for devices with comparably low-powered GPUs, such as handhelds or tablets (see Chapter 8). If an interactive visualization of the complete models is simply not possible, we have no alternative but to reduce the GPU's workload in order to reach a defined target frame rate.

Starting with adjustments of the culling techniques or respective thresholds, this is ultimately accomplished by reducing the quality of the rendering, starting with the removal of insignificant details using POP buffers* [Limper et al. 13] up to the entire discard of those parts of the scene that appear to be small on the screen or are expected to be least important. Therefore, the *Drawables* are evaluated based on view-dependent criteria as well as user-defined attributes. For example, the priority for the binning of the Drawables results from both their screen-space footprint as well as a user-assigned importance factor.

The Adaptive Render Control [Stein et al. 13] (ARC) dynamically controls rendering speed during run-time. As a linear closed-loop control system, it tries to minimize the error between the system's output, being the current frame rate, and a given target frame rate. To accomplish this task, it automatically adjusts a variety of parameters, like the thresholds for culling techniques or iterative rendering, while having the strength of these adjustments proportionally dependent on the size of the error. Therefore, each parameter has to be defined by a validity range, a weighting, a preferable step size, and a direction indicating how the parameter needs to be modified to positively influence the rendering performance (Figure 20.3).

---

* http://x3dom.org/pop

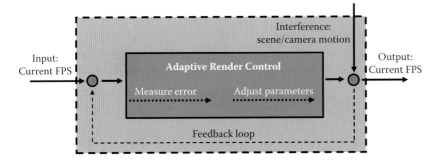

Figure 20.3

The adaptive render control's feedback loop.

20.3 RenderTree Structure and Spatial Processing

Considering its internal scene representation, hare3d takes a different approach than other scene-graph-based WebGL frameworks like X3DOM or three.js. As depicted by Tavenrath and Kubisch [2013], the traditional approach of scene graph traversal can easily result in the CPU being the bottleneck of the whole rendering pipeline, given a deep hierarchy and/or a huge number of objects. In contrast, to be output driven, hare3d's performance has to be independent of the size of its input data.

Therefore, an optionally existing application graph such as X3D is mapped to the RenderTree structure and "rolled out" by cloning nodes with multiple parents, to create unique paths to each node. This allows caching all context-sensitive information, like matrix transformations or the current world coordinates of the bounding volume. Changes to the scene structure will flag the affected information as dirty, while the data will only be updated on demand (i.e., the next time the traversal reaches the respective nodes). The RenderTree is optimized for fast traversal during which the nodes' lists of Drawables are collected and serve as input for the directly connected pipeline stages (cf. DrawableCollector in Figure 20.1). This traversal includes different culling operations, like hierarchical view frustum and small feature culling, which are not only used to exclude parts of the scene from rendering, but also augment the drawables with frame-dependent information, like a screen-space footprint approximation based on the world space bounding volume.

Figure 20.4 details hare3d's internal structure. Each Drawable holds a number of *Forms*, for example, to distinguish different levels of detail. The *Forms* combine the (mesh) data of a *GeometryDescriptor* with the shading information of an *AppearanceDescriptor*. This structuring allows for a Drawable to instantly switch between representations, while supporting various ways of data sharing within a scene.

The descriptors provide their data as resources, for example, vertex or index buffers of a GeometryDescriptor. The actual data is downloaded on demand—in this case, once the GeometryDescriptor is used for rendering for the first time. This obviously introduces a delay before the data is available for drawing, defined by the time needed for downloading the data from the server and uploading it to the GPU. Section 20.4 details our approach to deal with this issue.

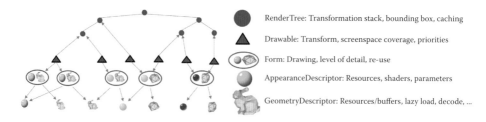

RenderTree: Transformation stack, bounding box, caching

Drawable: Transform, screenspace coverage, priorities

Form: Drawing, level of detail, re-use

AppearanceDescriptor: Resources, shaders, parameters

GeometryDescriptor: Resources/buffers, lazy load, decode, ...

Figure 20.4

Data sharing between hare3d's internal representations.

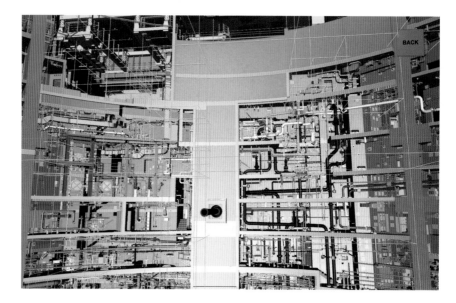

Figure 20.5

Bounding box visualization for the nodes of an octree.

Typically, the input of large CAD models is given in a semantic structure (e.g., as classical scene graph or scene tree), which also serves as interface for the application developer to interact with its content. As such input data does not necessarily provide any spatial relationships, and hierarchical traversal might not derive the maximal benefit from culling operations.

Therefore, hare3d chooses a hybrid approach, where different subtree configurations can be mixed in a single application, by allowing spatial data structure nodes to be inserted at any point in the hierarchy. The collected Drawables of such encapsulated subtrees are inserted into adaptive spatial data structures. hare3d supports octrees (Figure 20.5) and bounding interval hierarchies so far.

While both data structures adapt well to the geometric density of the scene, the octree allows for simpler incremental updates.

Subsequently, the optimized structure is traversed, while the original semantic structure stays untouched. By default we utilize spatial data structure root nodes for subtrees with static CAD data. As these subtrees frequently contain a large number of (very) small objects, they generally profit quite heavily from this optimization.

We wrapped this functionality in a cross-compiled *asm.js* library (Chapter 5) to gain additional performance, compared to hand-written JavaScript, for the computationally expensive culling operations. This resulted in a significant speedup, halving the time needed for constructing and hierarchically traversing an octree [Stein et al. 14].

Additionally, the spatial structuring allows us to efficiently employ hierarchical occlusion culling, a technique particularly well suited for large-scale models. In our paper, we evaluated different traversal algorithms, like CHC++ [Mattausch 08], and showed that great benefits are achievable from combining them with comparably small spatial data structures.

We approximate the occlusion properties of geometries that have not been downloaded yet by rendering their bounding boxes as placeholders. As a first, at a very rough level of detail, these *proxy shapes* are fed into the rendering system. As soon as a proxy shape becomes visible, it triggers the download of the actual geometry data. Consequently, occluded geometry will not be considered as visible and thus not downloaded at all.

We implemented occlusion culling based on the *hardware occlusion queries (HOQ)* using *Firefox's* experimental *WebGL2RenderingContext*, available in the *Nightly* builds since summer 2013[*] and very recently in standard *Firefox* browsers. As our hierarchical culling (frustum culling as well as HOQs) is based on the bounding volumes of the nodes of the spatial data structure, the use of proxy shapes does not lead to any deterioration of its granularity.

20.4 Delivery Format and Transmission

One of the key aspects of existing 3D rendering frameworks for the Web is the use of optimized transmission formats, which attempt to minimize both the time spent downloading 3D model data from the network as well as the time used for transferring that data to the GPU. Besides the successful binary formats used by the X3DOM framework, the GL transmission format *glTF*, as proposed by the Khronos Group, is certainly the most important technology in this context.[†] When designing hare3d, we first considered glTF as candidate technology, but came to the conclusion that it does not fulfill all of our requirements. This led to the development of our *Shape Resource Container* (SRC) format. An extensive overview of the SRC format, along with a proposal on how it could be integrated into declarative 3D languages, can be found in Limper et al. [2014]. Minimizing the number of downloads, as well as progressive, interleaved transmission of different kinds of mesh data, was the most important demand for our target applications.

Minimizing the number of downloads and having mesh data stored in binary buffers became possible through the use of a self-contained file format consisting of a structured header and a binary file body. Compared to glTF (which only allows base64 encoding to embed mesh data into the structured description), the use of binary chunks saves memory as well as decode time, since they can directly be pushed to the GPU. Using a well-defined addressing scheme for mesh data inside an SRC file, we are able to pack multiple meshes into a single container. Inside the scene description, meshes can be addressed using a hash delimiter, for instance *mySRCfile.src#myMesh1*. This is an important property for

---

[*] https://wiki.mozilla.org/Platform/GFX/WebGL2
[†] http://gltf.gl/

the overall performance of our application, since it allows us to balance the number of HTTP requests against the granularity of the progressive scene download.

Besides reducing the number of downloads, the actual amount of data that has to be transmitted must be reduced by applying compression methods. We currently rely on simple quantization of vertex attributes to reduce the size of encoded mesh data, as well as to reduce the memory requirements on the GPU. In addition, we use GZIP compression, supported by HTTP and every standard browser. To avoid any explicit decoding step in JavaScript on the client side, we always use byte-aligned formats that can be directly pushed to the GPU. In addition, for optimal rendering performance on all platforms that support WebGL, attribute components must be aligned to 4 byte boundaries, which might introduce additional padding bytes [Vanik and Russell 11]. A typical configuration uses 16 bits of precision for each position component (x, y, z), and 8 bits of precision for each normal component. Since vertex positions are specified as local object coordinates, while the full-precision transformation of each object is still available through the RenderTree, the loss of precision due to local coordinate quantization is neglectable.

Another important aspect, in terms of user experience, is progressive transmission of individual pieces of 3D mesh data. While many methods have been proposed within the past two decades, only a few recent approaches have considered the important trade-off between download time and decode time. Important works in this area are, for instance, the method of Lavoué et al. [2013], and the POP buffer method. We have chosen the latter for our current implementation, but we have designed our SRC format in such a way that various kinds of progressive transmissions are possible. To do so, SRC introduces the concept of *BufferChunks*. In addition to the low-level concept of a *buffer*, as known from glTF, SRC allows one to split a buffer into several chunks. Other approaches, like *Blast*, are using per-chunk headers to make chunks self-contained, which has some advantages for parallel decoding [Sutter et al. 14]. Nevertheless, we have designed the chunks to simply represent parts of a binary buffer, which can directly be pushed to the GPU. Using this simple concept, splitting all relevant buffers into chunks and interleaving them allows us to progressively stream mesh data to the client's GPU. For instance, by concatenating buffer chunks that correspond to index and vertex data inside the binary file body, we are able to render a coarse representation of the model, even if neither index data nor vertex data have been fully transmitted.

The question remains as to when a mesh should actually be downloaded. Generally, we are assuming large data sets while aiming for near-zero startup times, so always downloading everything is simply not possible. A first improvement is to download only the visible objects, given that their bounding box is already known at startup. We ensure this by including this information inside our structured scene description, along with the URL, which addresses the mesh inside its respective SRC file. Until the geometry becomes available, a proxy shape is rendered, using a special color or texture to indicate the status of the geometry download and upload to the GPU, as shown in Figure 20.6.

Currently, we use a yellow color for the proxy shape to indicate that the download is in progress. If the download fails, we draw a red wireframe box (instead of a filled one), to show the rest of the scene despite the errors. Additionally, by using the color green, we illustrate that data have been successfully downloaded already, but the GPU upload has been deferred. This is primarily the case during interaction, but it can generally occur as soon as our GPU upload manager decides to keep within a certain CPU time budget per frame (Chapter 14). In a similar fashion as a download manager, this GPU upload

Different proxy objects, indicating the loading state for each model part.

manager controls the data stream between main memory and GPU memory to guarantee that rendering always happens at an interactive rate and user interactions are never blocked. This proxy rendering behavior can be configured differently or disabled entirely, depending on the user's preference. Alternatively, global indicators, like a spinning wheel with a percentage number, or a progress bar, could be used.

20.5 Fast and Dynamic Rendering

Our target applications often require rendering large scenes where individual parts may be dynamically updated. This rules out time-consuming preprocessing of the scene data.

When these changes occur, the updated parts must be re-encoded on the server from their original authoring format (e.g., a CAD format). As such, this process should be kept as straightforward as possible, in order to quickly deliver the updated parts to the web application.

20.5.1 Priority-Based Rendering

With the mentioned constraints—fast updates and delivery of single CAD parts—in mind, the question is how to deliver the performance required for a consistent and fluid user interaction within the Web application. A lot of common, large-model visualization methods require time-consuming preprocessing steps [Kasik et al. 07; Brüderlin et al. 07], so they are not fit for our target applications. Thus, we chose a different approach to ensure high frame rates, especially during user interaction. We will refer to our dynamic method as *priority-based rendering*.

The basic idea is simple: If there are too many visible parts that should be rendered within the next frame, given a limited budget of frame time, we simply don't draw all of them. Instead, we render only the most important objects, which will give enough clues about what the user is currently interested in. Of course, the crucial question at this point is: What makes a visible object really important? While we don't have the perfect answer yet, we have identified some important criteria for our use cases:

- Screen-space footprint. If available through the hardware, this would be the exact amount of pixels that were generated by an object's Drawable(s) during the last frame. Currently, this is just an approximation that is calculated using it's bounding box (Figures 20.7a and 20.7b).

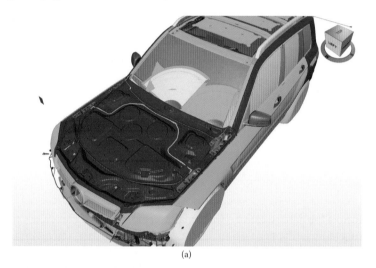

(a)

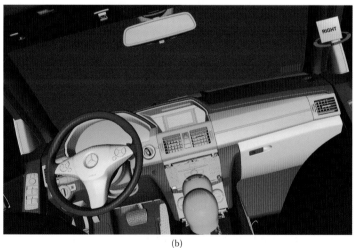

(b)

Figure 20.7

Priority visualized for each drawable, based on the screen-space footprint of the bounding box, colored using a cool–warm schema (blue indicates low priority, red indicates high priority). (a) Exterior view of an SUV CAD model; (b) cockpit view. Data courtesy of Daimler AG.

- Selection. If the user selects an object and then navigates, the selected object's Drawables should stay visible.
- Screen-space to render cost ratio. If there is a large object that is very cheap to render—for instance, because it consists of just a few polygons or has a very simple material—it is particularly interesting as a cheap visual orientation during interaction.
- Updating costs. This metric captures whether data are fully prepared for rendering (e.g., if the data have been downloaded from the network or if the data are present in GPU memory).
- Prominent position. The object is in an important position—for example, close to the current view's center point or closer to the camera than other objects. If the focus of the user is known (for example, through the use of eye tracking), distance from the current focal plane could also be taken into account.

Currently, we mainly rely on the first two criteria, but we plan to explore how an overall formula could look to integrate as much information as possible.

20.5.2 Draw Map Data Structure

As soon as the priority of an object has been computed, all Drawables that have not been discarded (due to our culling operations) need to be sorted by their priority before rendering.

The ARC can be used to decide when to terminate the rendering process. Therefore, we are experimenting with different criteria like a maximum number of Drawables or a priority threshold. In addition to priority-based sorting, common rendering optimization techniques, like sorting by shader and GPU states, have to be applied [Hillaire 12].

Since the performance of complex CPU-side operations based on JavaScript is much worse than native code, we have to be cautious at this point: In the worst case, we might spend more time on prioritizing and sorting objects than we would need to simply render them all. This would cause not only a slower application, but also a worse visual result (as fewer objects are rendered). Therefore, we have employed the concept of a Draw Map that is used for both fast sorting and fast sorted rendering.

In an initial experiment, we found that JavaScript's standard functions, like the array sorting function `sort` or the function `indexOf`, have a surprisingly slow execution speed. The main reason for this is that JavaScript needs to take care of the many special features of the language—for example, the possibility of having only parts of an array defined. This problem was also the reason for the development of fast.js,* a library that provides JavaScript-based reimplementations of JavaScript's standard functions, but with a few constraints on data. Still, even with a faster sort implementation, performance was not sufficient, even for only a single sorting criterion. Therefore, we decided to implement binning with direct insertion into a map, a much faster strategy than explicit sorting. The downside of this approach is the trade-off of sorting granularity for performance; yet, we found it to be an acceptable solution. For the priorities, using around 10 different bins gave satisfying results. The amount of shader configurations of our target applications is limited to a manageable amount anyway, and state changes are

---

* https://github.com/codemix/fast.js/tree/master

clustered in such a way that the most expensive changes, such as, for example, texture changes, are always taken care of.

The Draw Map is currently organized as follows:

- There are 10 bins for different object priorities.
- For each priority bin, there is a small number of shader bins.
- For each shader bin, there is a small number of state bins.
- For each state bin, the last element of a linked list of Drawables is stored.

Since Drawables directly act as linked list elements, each storing a pointer to its successor, the numer of elements per list is not limited, and lists do not consume additional memory except for the single additional pointer attribute per Drawable.

At the beginning of each frame, all Drawables are cleared from the Draw Map by setting the linked lists' head pointer to null. This step is accelerated by caching a list of references to all these pointers, so that clearing takes only a negligible amount of time. We have found this method, as well as the usage of linked lists instead of arrays in general, to be especially beneficial because it does not require any expensive frame-wise allocation of memory or any copy operations. Before rendering, we compute the priority of each Drawable that was not culled in previous stages and insert it into the Draw Map. Given that the respective bins already exist, this can be done in O(1). Each time we create a new shader program or encounter a new state configuration, we call an explicit registration function to set up the corresponding entries inside the Draw Map.

As can be seen in Table 20.1, these efforts more than pay off, as soon as a medium or large number of Drawables has to be processed. For the experiment shown in the table, 12 random shader IDs and 2,500 random GPU state-set IDs have been assigned to the Drawables. Even though the version using the JavaScript standard sort only sorts Drawables by their (integer) shader ID, ignoring priority and state set, it is already significantly outperformed by the Draw Map in all major browsers. The Draw Map, sorting by priority and shader ID, as well as state-set ID, is even able to process one million Drawables within acceptable time. This is mainly due to the fact that no expensive computations,

Table 20.1 Our Draw Map's Sort Performance, Compared to JS Array-Based Sort by Shader ID

Browser	No. Drawables	JS Standard Sort Time (Shader Only)	Draw Map Sort Time (Priority, Shader, and State)
Chrome 39	1M	144 (158)	1 (16)
Chrome 39	2K	32 (41)	0 (6)
IE 11	1M	195 (233)	3 (5)
IE 11	2K	39 (51)	3 (4)
Firefox 33	1M	220 (1911)	3 (5)
Firefox 33	2K	54 (357)	3 (5)

Note: All numbers are given in milliseconds; timings in parentheses were taken during the first run. Timings were taken using an i3-2120 CPU @ 3.30 GHz.

no memory allocations, and no copy operations are involved. We have noticed that the first run of either method (shown in parentheses in Table 20.1) is, in general, expected to consume more time than subsequent runs. This is due to run-time optimizations of the JavaScript engine, which is why we have decided to exclude timings from the first run from the average numbers shown in Table 20.1.

An interesting side effect of the Draw Map's insertion sorting mechanism is that we can also employ a fixed priority threshold (possibly on-the-fly adjusted by the ARC) to discard Drawables upon insertion, comparable to a relatively small feature threshold. Nevertheless, for a relative ranking of priorities we rely on the binning provided by the Draw Map. During rendering, we traverse the map and draw its entries, from the highest priority to the lowest. If we exceed a certain budget of frame time or a maximum number of entries to draw, the traversal of the Draw Map is simply terminated. Traversing a simple list would require explicit checks for shader changes or state changes after each draw. In contrast, the Draw Map inherently provides this knowledge; hence such checks become obsolete.

20.5.3 Iterative Rendering

In order to keep the application always interactive but nevertheless draw all parts of a very complex scene, an iterative rendering approach can be used. Similarly to progressive rendering, a term commonly associated with ray-based rendering techniques [Ou et al. 12], a complete picture of the scene is rendered across multiple frames, thus splitting up the workload.

However, in contrast to such methods, we explicitly omit a large part of our scene from the initial drawing. The image in the framebuffer is then accumulated over a series of frames, each processing only a fraction of the Drawables. This fraction has to be set to match the devices' capabilities or can adaptively be adjusted by the ARC.

Figure 20.8 shows several stages of the iterative rendering of a large model over multiple frames. The advantage of this approach is that the accumulation process can be stopped after any "subframe" to ensure a high frame rate. For static, large CAD models, generally no motion occurs among the objects of the scene, so that the frame is going to be completed iteratively. Only when the view changes due to camera motion, is the render target cleared and the generation process restarted. As a result, during a continuous interaction, only a certain fraction of the scene is visible.

Of course, this technique has its shortcomings as well. For example, noncamera animations would result in a frequent restart of the frame accumulation process, even in cases where no progress might actually be visible as the moving parts are not important or even occluded.

As a possible solution we have to preserve the accumulated frame buffer of the static objects in a separate render target, re-render only the dynamic objects upon change, and composite the resulting frame buffers in a postprocessing step. Correct handling of transparent objects would introduce yet another render target to such a compositing process.

However, this improvement can be expanded to have minor changes resulting in merely a re-rendering of the affected parts—for instance, as reaction to a material change of an opaque object due to highlighting.

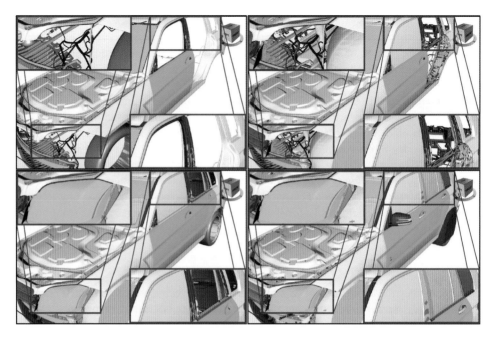

Figure 20.8

Several stages of the accumulation process for the iterative rendering of a Mercedes-Benz C 204 (from top left to bottom right): 20%, 40%, 80%, and 100%.

20.6 Roadmap

As our CAD target applications are inherently bound by the number of vertices or by the number of draw calls, the Draw Map helps us to gain performance by temporarily reducing both. In contrast, another common, loss-free solution is to reduce the number of draw calls by batching multiple draws together. So far, we are relying on server-side batching during SRC conversion. To still be able to modify partwise properties, such as visibility, highlighting, or coloring, on a sub-buffer level, we use per-vertex IDs to index into textures, which store the respective information. We are also working on a client-side solution that batches the loaded geometry on the fly, with respect to its importance, to spatial placement and to the integrated memory management. However, this is still work in progress and out of scope for this chapter. We are also looking forward to getting an additional performance boost from using WebGL 2's instancing where appropriate.

To reduce storage costs of geometry data, as well as upload time, we want to investigate alternative formats for vertex and normal data. At this point in time we are using the straightforward normal encoding in Cartesian space, but consider using more efficient 2D normal parameterizations [Cigolle et al. 14; Meyer et al. 10]. Connectivity compression, using strips or fans, is also an important topic that has to be investigated.

So far, we do not have any common API at hand, implemented in all major browsers, that enables progressive downloads of binary data; however, there have been proposals, such as *moz-chunked-arraybuffer*; a text-based workaround, with an additional copy operation, is also possible.

A key part of our system, which still demands additional research, is the assignment of object priorities for rendering. Our current estimation of each object's screen-space footprint can be improved by using better algorithms for this approximation, or by employing advanced GPU-based techniques.

Although our shaders are currently relatively simple, we found it important to make them conformant to a wide variety of mobile platforms, involving great efforts (Chapter 8). For instance, in the case of batched rendering, storing part-related information in textures and fetching this information inside a vertex shader will fail on WebGL-capable platforms that do not provide the optional feature of vertex texture access. Furthermore, the use of dynamic shader components, like specific numbers of clip planes or lights, raises the need for a smart system to dynamically create and composite shaders. A recent concept that could be employed in this context is *shade.js* [Sons et al. 14], but we are still investigating candidate technology.

Acknowledgments

We would like to thank Daimler AG for the C204 automotive model, as well as EDF for their highly complex power facility models. Images are courtesy of Daimler AG and the PLM project of EDF–DIN.

Bibliography

[Brüderlin et al. 07] Beat Brüderlin, Mathias Heyer, and Sebastian Pfützner. "Interviews3D: A Platform for Interactive Handling of Massive Data Sets." *IEEE Computer Graphics and Applications*, Vol. 27, 48–59, 2007.

[Cigolle et al. 14] Quirin Meyer, Jochen Süßmuth, Gerd Sußner, Marc Stamminger, and Günther Greiner. "A Survey of Efficient Representations for Independent Unit Vectors." *Journal of Computer Graphics Techniques* (JCGT), 3 (2), 2014.

[Hillaire 12] Sébastien Hillaire. "Improving Performance by Reducing Calls to the Driver," in Patrick Cozzi and Christophe Riccio, ed. *OpenGL Insights,* CRC Press, Boca Raton, FL, 353–363, 2012.

[Jung et al. 12] Yvonne Jung, Johannes Behr, Timm Drevensek and Sebastian Wagner. "Declarative 3D approaches for distributed web-based scientific visualization services." *Proceedings Dec3D for the Web Architecture* (at WWW2012), Lyon, France, 2012.

[Kasik et al. 07] Dave Kasik. "Visibility-Guided Rendering to Accelerate 3D Graphics Hardware Performance." *ACM SIGGRAPH 2007 courses*, San Diego, CA, USA, 2007.

[Lavoué et al. 13] Guillaume Lavoué, Laurent Chevalier, and Florent Dupont. "Streaming Compressed 3D Data on the Web Using JavaScript and WebGL." *Proceedings Web3D*, San Sebastian, Spain, 19–27, 2013.

[Limper et al. 13] Max Limper, Yvonne Jung, Johannes Behr, and Marc Alexa. "The POP Buffer: Rapid Progressive Clustering by Geometry Quantization." *Proceedings Pacific Graphics*, Singapore, 197–206, 2013.

[Limper et al. 14] Max Limper, Maik Thöner, Johannes Behr, and Dieter W. Fellner. "SRC—A Streamable Format for Generalized Web-Based 3D Data Transmission." *Proceedings Web 3D*, Vancouver, BC, Canada, 35–43, 2014.

[Mattausch 08] Oliver Mattausch, Jiri Bittner, and Michael Wimmer. "CHC++: coherent hierarchical culling revisited." *Proceedings Eurographics*, Crete, Greece, 221–230, 2008.

[Meyer et al. 10] Quirin Meyer, Jochen Süßmuth, Gerd Sußner, Marc Stamminger, and Günther Greiner. "On Floating-Point Normal Vectors." *Proceedings EGSR*, Saarbrücken, Germany, 1405–1409, 2010.

[Ou et al. 13] Jiawei Ou, Ondrej Karlík, Jaroslav Krivánek, and Fabio Pellacini. "Evaluating Progressive Rendering Algorithms in Appearance Design Tasks." *IEEE Computer Graphics and Applications* 33 (6), 58–68, 2013.

[Sons et al. 14] Kristian Sons, Felix Klein, Jan Sutter, and Philipp Slusallek. "shade.js: Adaptive Material Descriptions." *Proceedings Pacific Graphics*, Seoul, Korea, 51–60, 2014.

[Stein et al. 13] Christian Stein, Max Limper, and Arjan Kuijper. "Spatial Data Structures for Efficient Visualization of Massive 3D Models on the Web." Masters thesis 2013, TU Darmstadt.

[Stein et al. 14] Christian Stein, Max Limper, and Arjan Kuijper. "Spatial Data Structures to Accelerate the Visibility Determination for Large Model Visualization on the Web. " *Proceedings Web 3D*, Vancouver, BC, Canada, 53–61, 2014.

[Sutter et al. 14] Jan Sutter, Kristian Sons, and Philipp Slusallek. "Blast: A Binary Large Structured Transmission Format for the Web." *Proceedings Web3D*, Vancouver, BC, Canada, 45–52, 2014.

[Tavares 11] Gregg Tavares. "WebGL Techniques and Performance." San Francisco, CA, USA, Google I/O 2011.

[Tavenrath and Kubisch 2013] Markus Tavenrath and Christoph Kubisch. "Advanced Scenegraph Rendering Pipeline." GPU Technology Conference, 2013.

[Vanik and Russell 11] Ben Vanik and Ken Russell. "Debugging and Optimizing WebGL Applications." New Game Conference, San Francisco, CA, USA, 2011 SF.

21

The BrainBrowser Surface Viewer
WebGL-Based Neurological Data Visualization

Tarek Sherif

21.1 Introduction

BrainBrowser* is a lightweight, open-source, high-performance JavaScript visualization library built to facilitate visualization of the large, distributed data sets that have become common in neuroimaging research. BrainBrowser leverages WebGL, Web Workers, and other HTML5 technologies to allow visualization with any modern web browser. While BrainBrowser consists of two visualization tools, the Surface Viewer and the Volume Viewer, this chapter focuses on the WebGL-based Surface Viewer, which is used to visualize 3D surface (Figure 21.1) or tractography (Figure 21.2) geometry extracted from volumetric neuroimaging data.

---

* https://brainbrowser.cbrain.mcgill.ca/

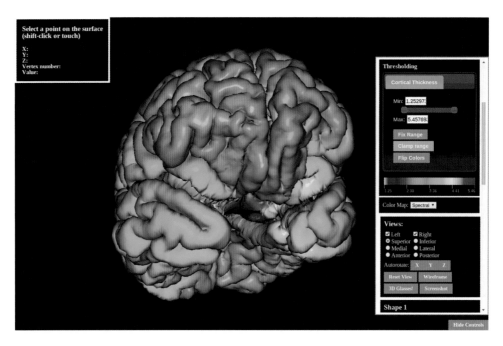

Figure 21.1

Surface data with colors representing cortical thickness.

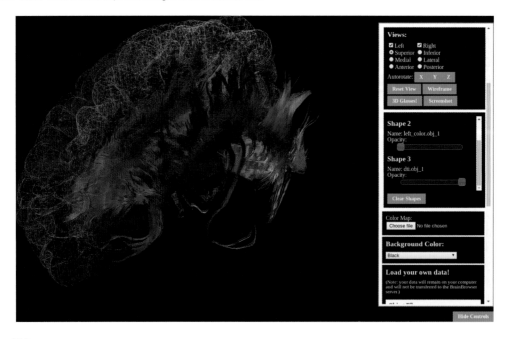

Figure 21.2

Diffusion tensor imaging tractography data.

In recent years, neuroimaging research has seen itself inundated by large, distributed data sets that have, in many cases, made traditional visualization tools, requiring local installation and local data, impractical, if not completely unusable. The amount of data available has necessitated a shift in how scientists approach their research: Guiding hypotheses are often articulated after analyzing the mass of available data, and data sharing has become a necessity for scientific discovery.

In this new research environment, web-based visualization tools present significant benefits as they make transitioning from acquisition to exploration more seamless. Data need not be transferred completely to the machine on which they are visualized. The fact that these technologies are now a part of the web platform has also made it possible to consider using visualization tools for novel research purposes:

- **Publication:** With scientific articles now being published online, publishers are looking for ways to disseminate data sets and present them in more meaningful ways. It is now possible to replace the static visual media of traditional print publications, such as static figures or charts, with more dynamic, interactive 2D or 3D visualizations, such as in this online version of an article on Broca's area: http://onpub.cbs.mpg.de/role-brocas-area-language.html.
- **Distribution:** Online visualization tools can also be used to share data in a more accessible manner. Researchers can use web-based tools to explore remote data sets directly. The Surface Viewer was used, for example, to share the MACACC data set [Lerch et al. 06] discussed in Section 21.4.1.

The BrainBrowser Surface Viewer was built to address these new challenges and opportunities in neuroimaging research by leveraging several web technologies that have either recently been introduced or have recently become viable for the development of sophisticated applications: a more robust JavaScript language and more performant JavaScript engines, parallel processing using Web Workers, and access to the GPU through WebGL.

21.2 Background

21.2.1 Data

Data for a neurological study are typically acquired from some form of scanner, usually a magnetic resonance imaging (MRI) scanner due to the low risk the procedure poses to subjects. The data are typically in the form of a three- or four-dimensional array of scalar values: three spatial dimensions with a fourth dimension, if present, representing a variable such as time or diffusion direction. Each element in the array represents a voxel in the volume at a given point in time, and the value at each voxel represents the intensity of the magnetic resonance signal measured at that point. The significance of the signal depends on the scanning protocol being used. A T1-weighted image, for example, tends to differentiate between white and gray matter, while a blood-oxygen-level-dependent (BOLD) contrast image is sensitive to the relative levels of oxygenated and deoxygenated blood in an area, which is an indirect indicator of neuronal activity.

Once data are acquired, they are usually processed in various ways to extract features of interest. Especially pertinent to this discussion are tools such as the CIVET processing pipeline [Ad-Dab'bagh et al. 06] or Freesurfer [Fischl 12] that extract geometry from volumetric data. The geometry might be a mesh representing a surface, such as the cortex, or, in what are referred to as tractography data, it might be a set of line strips representing neuronal tracts through the brain. These tools often also extract other information from the volume that can be visualized along with the geometry. CIVET, for example, which extracts the cortical surface as part of its processing, also calculates the thickness of the cortex at all points along the surface. This information is encoded as a list of scalar values, each associated with a vertex in the surface mesh.

Visualization of the geometry itself is fairly straightforward, as it simply entails rendering the mesh or line strips to the screen. Additional per-vertex data, such as the cortical thickness measures described before, however, require a supplementary step to make them visually meaningful: They are mapped to colors. A color map, in this context, is simply a list of colors into which the scalar values can be interpolated. Controlling how this interpolation is done is a key part of exploring the data, and researchers often manipulate various parameters in order to focus on information of interest. Different color maps are used to accentuate different areas of contrast, thresholds can be set on the values to be colorized, and various data maps can be blended to combine multiple layers of information.

21.2.2 History

The BrainBrowser Surface Viewer began development in April 2010 as part of the CBRAIN project [Sherif et al. 14] at McGill University. It was originally built as a stand-alone application using Google's now-defunct O3D plug-in. When Google chose to discontinue development of O3D in May 2010, the Surface Viewer incorporated the WebGL implementation of the O3D API. As the latter project's development began to wind down, however, the decision was made in April 2013 to port the code base to the three.js library.

While the port to three.js was generally painless, there were some operations that took a significant hit in performance. Particularly problematic were per-vertex color updates. Manipulations of the color mapping function being used to visualize per-vertex data are a critical part of the exploration process for neuroimaging researchers. Researchers must be able to modify parameters of the mapping function and immediately see the results. The problem with three.js was that its default handling of geometry involves creating JavaScript objects to represent each vertex and, in mesh data, to represent each face. This means, first of all, that an enormous number of JavaScript objects must be allocated to represent a given model, but even more importantly, that updates to vertex colors have to traverse a tree-like structure of faces and vertices, severely hampering performance. This situation was eventually remedied when three.js introduced BufferGeometry, which represents geometry as typed arrays of vertices, colors, normals, and indices. In November 2013, when the BufferGeometry mechanism became sufficiently mature, the Surface Viewer was restructured to take advantage of it.

It eventually became clear that the Surface Viewer had a potential utility outside the CBRAIN project, and that its implementation as a stand-alone application was no longer ideal. The core functionality was extracted and refactored into a reusable library in

December 2013 and, in March 2014, the code base was packaged and released as an open-source project.*

As WebGL became more ubiquitous, in particular with its recently being enabled in iOS, some of the Surface Viewer's users began expressing interest in visualizing their data on mobile devices. The primary use case was for clinical studies, which generally involve questionnaire data, often acquired using tablets, in addition to image data acquired from a scanner. Researchers felt that it could greatly streamline their work flow if they were able to view image data on the same devices they were using to fill in questionnaires. Throughout the latter half of 2014, the Surface Viewer was made more usable on mobile devices (Figure 21.3), primarily through optimizations to critical code paths, by limiting unnecessary usage of the GPU, and by ensuring that all functionality was accessible to touch controls.

21.3 Architecture

At its core, the BrainBrowser Surface Viewer is a JavaScript library exposing an API that is simple enough to begin visualizing a data set in 5–10 lines of code (Listing 21.1), while at the same time being robust enough for the creation of full-featured applications.

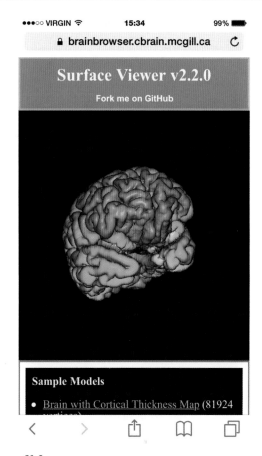

Figure 21.3

The BrainBrowser Surface Viewer running on an iPhone 6.

Listing 21.1 Starting a visualization with the BrainBrowser Surface Viewer.

```
BrainBrowser.SurfaceViewer.start("visualization-div",
  function(viewer) {
  viewer.render();
  viewer.loadColorMapFromURL("color-map.txt");
  viewer.loadModelFromURL("brain.obj" {
    complete: function() {
      viewer.loadIntensityDataFromURL("cortical-thickness.txt");
    }
  });
});
```

Behind the simple function calls of the API are several layers interacting to optimize the flow of data and allow users to interact with them in complex and meaningful ways.

---

* https://github.com/aces/brainbrowser

21.3.1 Data Flow

Data coming into the Surface Viewer proceed through several steps to prepare them for display. Heavier data parsing steps are handled by Web Workers to avoid blocking the main UI and rendering thread. Whenever possible, data are passed to and from Web Workers as transferable objects[*] to avoid the overhead associated with structured cloning[†] (Listing 21.2). For more details on transferable objects, refer to Chapter 4.

> **Listing 21.2** Passing parsed geometry data as transferable objects from a Web Worker to the main thread.
>
> ```
> var result = {};
> var transfer = [];
>
> result.vertices = new Float32Array(data.vertices);
> transfer.push(result.vertices.buffer);
>
> if (data.normals) {
>   result.normals = new Float32Array(data.normals);
>   transfer.push(result.normals.buffer);
> }
>
> //Etc...
>
> self.postMessage(result, transfer);
> ```

The flow of geometry data into the Surface Viewer (Figure 21.4) proceeds through the following steps:

1. Geometry data are loaded asynchronously over the network using AJAX, or from the local file system using the FileReader API. Geometry data can be described as several independent objects, depending on the data format,

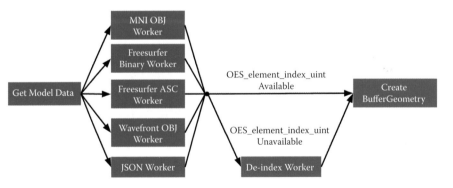

Figure 21.4

Surface Viewer work flow for loading geometry.

---

[*] http://www.w3.org/html/wg/drafts/html/master/infrastructure.html#transferable-objects
[†] https://developer.mozilla.org/en-US/docs/Web/Guide/API/DOM/The_structured_clone_algorithm

and these are parsed into what are referred to as separate "shapes" in the Surface Viewer object model. Different shapes can be manipulated either individually or collectively (step 4), and if the data are not de-indexed (step 3), all shapes in a model will share the same vertex buffer. Data can be in one of several supported binary or text formats commonly used to describe neurological geometry: MNI OBJ, Freesurfer binary, and Freesurfer ASC. The Surface Viewer also supports models described in Wavefront OBJ and a custom JSON-based format we developed to facilitate exporting geometry out of external applications and into the Surface Viewer.

2. Geometry data are sent to a Web Worker for parsing. Each supported file format in the Surface Viewer is associated with a Web Worker script that can be spawned to convert the format's geometry description into the geometry object model that the Surface Viewer uses internally (Listing 21.3). This architecture, in which one Web Worker is responsible for each data format, exposes a plug-in framework that can be used to easily add support for other data formats. Adding support for a new format requires only the creation of a Web Worker script that can convert the new format into the Surface Viewer object model.

3. The next step in the data flow will depend on the capabilities of the client browser and machine. One of the weaknesses of the core WebGL 1.0 specification is that indices describing how to build a mesh from individual vertices are limited to 16 bits in size. This puts a limit of 65,536 vertices on indexed models that can be rendered with a single draw call. The Surface Viewer is meant to handle data sets that fall well outside this limit (the tractography visualization shown in Figure 21.2, for example, contains 560,674 vertices). Fortunately, there is a widely supported WebGL extension, `OES_element_index_uint`*, that enables support for 32-bit indices. If this extension is available, the Surface Viewer will simply pass the parsed data on to the next step. As a fallback option, if `OES_element_index_uint` is not available on the client machine, the Surface Viewer will send the indexed geometry data to a second Web Worker that "de-indexes" them, essentially "unrolling" the indices to create unindexed geometries that can be drawn using `gl.drawArrays()`, rather than `gl.drawElements()`. While not optimal, this fallback approach was chosen for the simplicity of both its implementation and of mapping back to the original geometry description. Furthermore, the `OES_element_index_uint` extension will be promoted to core in WebGL 2.0, so this branching logic for handling indices will eventually become unnecessary.

4. Once geometry data are prepared, they are used to create a three.js BufferGeometry object to prepare them for display. The Surface Viewer supports displaying models made out of triangles, used for most surfaces, or lines, used mainly for tractography data. If normals or vertex colors are provided in the model description, they are also passed to the BufferGeometry. Otherwise, the vertex colors are set to gray, and the normals are inferred from the geometry using available three.js utility methods. If several independent shapes are described in the input data, each

---

* https://www.khronos.org/registry/webgl/extensions/OES_element_index_uint/

shape produces a separate BufferGeometry object. This allows each shape to be manipulated individually (to apply different opacity levels to different parts, for example) or collectively (to rotate the model as a whole, for example).

Listing 21.3 The Surface Viewer object model.

```
{
  type: ("line" | "polygon"),
  name: "...",
  vertices: [x1, y1, z1, x2, y2,...],
  normals: [nx1, ny1, nz1, nx2, ny2,...],
  colors: [r1, g1, b1, r2, g2,...],
  shapes: [
    {
      name: "...",
      indices: [i1, i2, i3, i4, i5,...]
    },
    {
      name: "...",
      indices: [j1, j2, j3, j4, j5...]
    }
  ]
}
```

If additional per-vertex scalar data, such as the cortical thickness data mentioned in Section 21.2.1, are to be mapped to a loaded model, four additional steps are taken (Figure 21.5):

1. A color map file is loaded. Color maps are simply text files with a list of up to a few hundred RGB (red, green, blue) values listed one per line. Parsing these files is a relatively inexpensive operation, so it is done directly in the main thread, rather than in a Web Worker.

2. The per-vertex data are loaded and parsed. Parsing is done similarly to the geometry data, with separate Web Workers being implemented for each supported data format. As with the geometry, this creates a plug-in framework for

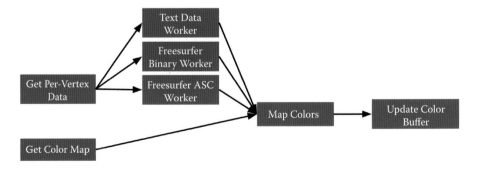

Figure 21.5

Surface Viewer work flow for colorizing a surface based on per-vertex scalar data.

data format support, making it straightforward to add support for new formats. Currently, plaintext, Freesurfer binary, and Freesurfer ASC per-vertex scalar data are supported.

3. The color map is used to assign a color to each vertex based on the per-vertex scalar value. Currently, color mapping is performed on the CPU, and while this has proved to be an adequate solution thus far, we would likely see significant performance gains by moving it to the GPU.

4. The mapped colors are then used as per-vertex colors for rendering the geometry.

21.3.2 Interaction

Once a model is loaded and colorized, it can be manipulated in several ways. The Surface Viewer implements mouse and touch controls for basic rotation, panning, and zooming of the model. More complex interactions are available through the API (Listing 21.4):

- Access to mouse and touch coordinates relative to viewing canvas
- Wireframe rendering providing a clearer view of a model's geometry
- Setting opacity on different shapes to reveal internal elements
- Manipulating per-vertex data in various ways such as setting input thresholds for the color mapping function, blending multiple data sets, and programmatically correcting the data set itself

Listing 21.4 Examples of interactions with loaded data.

```
viewer.mouse.x;
viewer.mouse.y;

viewer.touches[0].x;
viewer.touches[0].y;

viewer.setWireframe(true);

viewer.setTransparency(0.5, {shape_name: "left"});

viewer.setIntensityRange(1.5, 3.5);
```

One of the more significant interactions the Surface Viewer allows with loaded data, however, is its vertex-picking mechanism: `viewer.pick(x, y)`. This method takes as arguments x and y coordinates relative to the viewing canvas (defaulting to the current mouse position) and returns information about the vertex rendered closest to the selected point. It uses the raycasting functionality available in three.js to project a ray from the selected point into the scene and retrieve information about the triangles with which the ray intersects (Listing 21.5).

Listing 21.5 Raycasting code in the Surface Viewer's picking mechanism.

```
var raycaster = new THREE.Raycaster();
var vector = new THREE.Vector3(x, y, camera.near);
var intersects;
```

```
vector.unproject(camera);
raycaster.set(camera.position, vector.sub(camera.position).
  normalize());
intersects = raycaster.intersectObject(model, true);
```

The `intersects` array at this point contains a list of triangles with which the ray intersected, sorted by depth, as well as the actual points of intersection with those triangles in world coordinates. Since neuroimaging data are generally described per vertex, rather than per primitive, the Surface Viewer selects the nearest vertex to the point of intersection in the nearest intersected triangle and returns the following information about it:

- The index of the vertex
- The x, y, and z coordinates, in model space, of the selected vertex
- The specific shape in the model, as a three.js object, that contains the selected vertex

With this information, it is possible to implement much more complex interactions with a rendered model based on the specific vertex with which a user is interacting. The MACACC data set viewer discussed in Section 21.4.1, for example, uses picking to dynamically load correlational data to colorize a cortical surface based on the selected vertex.

21.3.3 Event Model

A key to building applications on top of the BrainBrowser Surface Viewer is the communication that occurs between the calling application and the visualization. Since a great deal of the data flow occurs asynchronously, the Surface Viewer implements a straightforward event model to communicate key events in its life cycle to the calling application. Examples of life cycle events triggered by the Surface Viewer include:

- **displaymodel:** a new model has been rendered.
- **loadintensitydata:** new per-vertex data have been loaded.
- **loadcolormap:** a new color map has been loaded.
- **changeintensityrange:** new thresholds have been set for the color mapping function.
- **updatecolors:** per-vertex colors have been updated.
- **draw:** the scene has been redrawn.

The application can attach listeners to these events and use them to update the interface or perform other required tasks (Listing 21.6).

Listing 21.6 Creating an event handler.

```
viewer.addEventListener("updatecolors", function() {
  document.getElementById("info").innerHTML = "Colors updated";
});
```

The event model can also be used to create and trigger custom events (Listing 21.7).

Listing 21.7 Listening for and triggering a custom event.

```
viewer.addEventListener("mycustomevent", function() {
    document.getElementById("info").innerHTML = "Custom event
        triggered";
});

viewer.triggerEvent("mycustomevent");
```

21.3.4 Rendering

As we made the move toward mobile, it became necessary to rethink the Surface Viewer's usage of device resources. Earlier in its development, the Surface Viewer would render the scene once per frame in a straightforward animation loop. This meant that if nothing had changed in the scene, the Surface Viewer would be sending useless draw commands to the GPU. Some phones on which we tested would become physically hot while the Surface Viewer was running. We remedied this by implementing a flag to explicitly indicate to the Surface Viewer that the scene should be redrawn. Any user interactions or API calls that might affect the scene will set this flag automatically, and the calling application can set it manually to force a redraw.

21.4 The Surface Viewer in Production

The BrainBrowser Surface Viewer has already been deployed for web-based visualization on a variety of platforms. Here, we discuss three cases that we believe highlight the Surface Viewer's strengths.

21.4.1 The MACACC Data Set Viewer

The MACACC data set is a database of structural correlations across 81,924 points on the cortex for three morphological variables: thickness, area, and volume. Essentially, each map in the database encodes information about the correlations between a given vertex and all other vertices on the surface for a given set of parameters. The total number of permutations encoded in the data set is 81,924 vertices × 9 blurring kernels × 3 morphological variables × 3 statistical indices for a total of over 6.3 million data maps requiring over 1 TB of storage space.

The MACACC data set presents an ideal use case for the Surface Viewer in that its size makes it extremely inconvenient to distribute or to visualize using traditional, locally installed tools. These types of tools would generally necessitate the transfer of the entire data set to the machine on which it is to be visualized. The BrainBrowser Surface Viewer, on the other hand, makes it trivial to remotely explore this data set in an intuitive manner. The MACACC Data Set Viewer[*] (Figure 21.6) uses the Surface Viewer's picking mechanism to select a vertex based on where the user is clicking on the screen. A network request

---

[*] https://brainbrowser.cbrain.mcgill.ca/macacc-viewer

Figure 21.6

The MACACC Data Set Viewer.

is made for the correlation map associated with the selected vertex, and the map is used to colorize the model. This dynamic fetching means that only the portion of the 1 TB data set associated with the selected vertex, approximately 455 kB, is transferred over the network, thus minimizing bandwidth usage and increasing accessibility.

21.4.2 CBRAIN

CBRAIN is a distributed, collaborative, web-based, grid computing platform developed at McGill University. CBRAIN has been in active production since 2009 and currently has over 250 users from 53 cities in 21 countries. CBRAIN is a complex system comprising many interconnected components, but there are four components in particular that are relevant to the discussion here:

- Users interact with CBRAIN through a web interface that can be accessed through any modern web browser.
- Data are connected to CBRAIN through data providers, which are essentially storage devices that might be at any controlled, secure, network-accessible location.
- CBRAIN is connected to several high-performance computing centers (HPCs) at several locations around the world.
- Installed on the HPCs are data processing tools that researchers use to analyze or process their data.

A typical work flow for a CBRAIN user might include the following steps:

1. Register data with CBRAIN. This can be done either by creating a new data provider or by uploading data to one that already exists.
2. Visualize data as a preparatory quality control step before submitting them to an HPC.
3. Submit a job to an HPC through the web interface. This typically involves selecting the processing tool one wishes to use and the HPC on which to run the job.
4. Once the job is complete, the results are saved back to the user's account.
5. Visualize results as a means of ensuring their quality or otherwise interpreting them.

Early in CBRAIN's development, visualizing data for the second and final steps was found to be problematic. It would often require exporting the data from the platform to visualize them locally. The Surface Viewer was an ideal solution in this situation as it could be integrated directly into the web-based user interface of the platform without requiring a plug-in. The Surface Viewer is now commonly used to visualize surface files produced by running the CIVET processing pipeline through CBRAIN.

21.4.3 LORIS

LORIS [Das et al. 11] is a web-based database system providing infrastructure to automate the flow of clinical data for complex multisite neuroimaging studies. Initially developed to manage data for the NIH MRI Study of Normal Brain Development (NIHPD) [Evans 06], LORIS has since been used in numerous decentralized, large-scale international projects related to the study of child brain development, autism, and Alzheimer's disease.

As an example of the type of data LORIS handles, the NIHPD project used LORIS to house and distribute 3 TB of native and processed data for over 2,000 MRI acquisitions. Medical doctors visualized these data throughout the project to assess the quality of incoming scans. As with CBRAIN, visualization using traditional tools was problematic for users of LORIS, and the Surface Viewer is now used to visualize any compatible data that are stored in the system.

21.5 Conclusion

The advent of big data in neurological research and the new methodological approaches it necessitates have led to an exploration of ways to use newly available web technologies, such as HTML5 and WebGL, to facilitate visualization of the large, distributed data sets that have become the norm in the field. Conversely, these newly available technologies have opened doors to using research data in ways that were not previously possible.

We presented the BrainBrowser Surface Viewer as a lightweight, flexible means to easily exploit modern web technologies for a variety of uses in the domain of neuroimaging research. Its simplicity makes it possible to embed an interactive visualization into a web page with a few lines of JavaScript code, while its robustness can be exploited to create full-featured visualization applications. By exploiting the dynamic connectivity of the web platform, we were able to share the massive MACACC data set with other researchers in a highly accessible and intuitive manner. And by relying only on standards-based web technologies, we were able to easily integrate the Surface Viewer into existing web platforms, such as CBRAIN and LORIS.

Over the course of BrainBrowser's history, we have seen 3D graphics on the web evolve considerably. The early transition from O3D to WebGL, at a time when browser support for WebGL was still limited and unreliable, meant sacrificing the stability of a controlled plug-in platform. It was a necessary move, however, even before O3D was discontinued, as many of our users lacked the administrative privileges necessary to install plug-ins on their machines. Looking back now, the current ubiquity of WebGL has made the transition a clear win for us in terms of simply giving our users access to the tools we have created.

Moving forward, one of our key goals will be to add support for new file formats using the plug-in framework described in Section 21.3.1. There are an enormous number of data formats in use in neuroimaging research, and support for new formats is one of the more common feature requests we receive. There are some areas of the Surface Viewer that could be further optimized for performance. One example of such an improvement would be to move color-mapping operations to the GPU.

In our discussion of the BrainBrowser Surface Viewer, we have attempted to show that the potential of modern web technologies is not simply in allowing us to take native applications online. By exploiting the connectivity, ubiquity, and power of the web platform, we can now build applications that are qualitatively different from anything that was previously available.

21.6 Resources

- The BrainBrowser website: https://brainbrowser.cbrain.mcgill.ca/
- Source code on GitHub: https://github.com/aces/brainbrowser
- API documentation: https://brainbrowser.cbrain.mcgill.ca/docs

Acknowledgments

Development of BrainBrowser has been funded by CANARIE, Canada's Advanced Research and Innovation Network,* and McGill University. We would like to thank Nicolas Kassis, who designed and developed the original O3D implementation of the BrainBrowser Surface Viewer. Thanks are also due to our colleagues on the CBRAIN team, Pierre Rioux, Natacha Beck, Marc-Étienne Rousseau, Reza Adalat, and our principal investigator, Alan Evans, for their support of BrainBrowser's development. Finally, we would like to thank all members of BrainBrowser's open-source development community for their contributions.†

Bibliography

[Ad-Dab'bagh et al. 06] Y. Ad-Dab'bagh, O. Lyttelton, J. S. Muehlboeck, C. Lepage, D. Einarson, K. Mok, et al. "The CIVET Image-Processing Environment: A Fully Automated Comprehensive Pipeline for Anatomical Neuroimaging Research," in *Proceedings of the 12th Annual Meeting of the Organization for Human Brain Mapping*, S45, 2006.

---

* http://www.canarie.ca
† https://github.com/aces/brainbrowser/graphs/contributors

[Das et al. 11] S. Das, A. P. Zijdenbos, J. Harlap, D. Vins, and A. C. Evans. "LORIS: A Web-Based Data Management System for Multicenter Studies." *Frontiers in Neuroinformatics*, 5, 2011.

[Evans 06] A. C. Evans., "The NIH MRI Study of Normal Brain Development." *Neuroimage*, 30 (1), 184–202, 2006.

[Fischl 12] B. Fischl. "FreeSurfer," *Neuroimage*, 62 (2), 774–781, 2012.

[Lerch et al. 06] J. P. Lerch, K. Worsley, W. P. Shaw, D. K. Greenstein, R. K. Lenroot, J. Giedd, et al. "Mapping Anatomical Correlations across Cerebral Cortex (MACACC) Using Cortical Thickness from MRI." *Neuroimage*, 31 (3), 993–1003, 2006.

[Sherif et al. 14] T. Sherif, P. Rioux, M.-E. Rousseau, N. Kassis, N. Beck, R. Adalat, et al. "CBRAIN: A Web-Based, Distributed Computing Platform for Collaborative Neuroimaging Research." *Frontiers in Neuroinformatics*, 8, 2014.

SECTION VII
Interaction

WebGL enables the next natural media type for the web. There were text, images, and video. With WebGL, now there is interactive 3D. This final section covers the often overlooked and sometimes subtle area of interaction.

In Chapter 22, "Usability of WebGL Applications," Jacek Jankowski breaks down usability principles for fitting 3D into the traditional web interface of hypertext, covering both prominent and subtle interactions ranging from reading text to navigation, selection, and manipulation of 3D objects.

In Chapter 23, "Designing Cameras for WebGL Applications," Diego Cantor-Rivera and Kamyar Abhari discuss elements and principles of camera design. The chapter begins with a review of the fundamental model-view and projection matrices as the starting point for the definition of the camera matrix, followed by a survey of JavaScript libraries available for implementing camera matrix operations. The chapter delves into basic navigation strategies, including tracking, orbiting, and birds-eye cameras, followed by an analysis of cameras in popular WebGL engines, highlighting two contrasting design strategies: object-oriented encapsulation and the use of transparent, public attributes. Afterward, landmark-based navigation is introduced as a mechanism to facilitate interaction with the scene, providing effective cinematic transitions. The chapter concludes with an analysis of future directions in camera development for WebGL applications.

22

Usability of WebGL Applications

Jacek Jankowski

22.1 Introduction

Computer graphics has reached the point where 3D models can be rendered, often in real time on commodity desktop and mobile devices, at a fidelity that is almost indistinguishable from the real thing. Thanks to WebGL it is now possible to use 3D models rather than 2D images to represent various objects on the web.

Nevertheless, 3D is still not as smoothly integrated in the everyday web experience as images or video. We argue that one of the reasons for this situation is the fact that the current R&D effort involved in the 3D web is focused mostly on 3D graphics and 3D graphics alone, largely ignoring usability. However, what good is a realistic 3D environment if an average web user cannot easily interact with it?

22.1.1 What Is Usability?

A growing number of researchers and developers are considering usability as a major focus of interactive system development. ISO defines usability as *the extent to which a product can be used by specified users to achieve specified goals with effectiveness, efficiency, and satisfaction in a specified context of use.* It is the result of several, sometimes conflicting, characteristics:

- Easy to learn: Learnability determines how easy is it for users to accomplish basic tasks the first time they encounter the design.
- Efficient to use: Efficiency is a measure of how quickly users can perform tasks once they have learned the design.
- Easy to remember: Memorability is a measure of how easily users can re-establish proficiency when they return to the design after a period of not using it.
- Error tolerant: Error tolerance is determined by how well errors are prevented and how easily users can recover from the errors when they occur.
- Subjectively pleasing: Satisfaction is related to how pleasant it is to use the design.

Since characteristics of usability can conflict with one another, the goal of the design should determine the priority with which each characteristic is applied. For example, a tourist information system that assumes no prior training would probably have learnability as a primary usability goal. On the other hand, a computer game would have to be fun to use or, in other words, subjectively pleasing, in order to succeed; in this case the user does not want to accomplish any goals beyond being entertained.

22.1.2 What Is 3D Web Site/Environment?

A virtual environment (VE) can be defined as a three-dimensional world made up of geometry, colors, textures, and lighting. It contains purely visual information. Traditional hypertext applications, on the other hand, consist of symbolic information (text) and navigational means (hyperlinks). 3D web sites/environments can be defined as HTML applications employing 3D graphics deployed on the web and therefore accessible using web browsers. Figure 22.1 presents most common user interfaces for accessing 3D content on the web.

From a human–computer interaction (HCI) perspective, viewing and interacting with such 3D web sites might be complicated: Users need to simultaneously deal with hypertext and three-dimensional graphics—two very different media that should both be appreciated

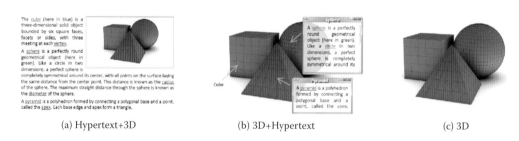

(a) Hypertext+3D   (b) 3D+Hypertext   (c) 3D

Figure 22.1

Most common user interfaces for accessing 3D content on the Web: *Hypertext+3D*, where a 3D scene is embedded in hypertext; *3D+Hypertext*, where hypertextual annotations are immersed in 3D; and *3D*, which contains purely visual information.

equally. Users need to be able to browse the text—look through the general information and search for more specific information. They also need to navigate freely through a three-dimensional space and examine and manipulate virtual 3D objects, to gain a better understanding of the data. An interface for the exploration of this kind of information spaces must provide adequate controls for the user to support all these tasks.

In this chapter, we discuss usability issues involved in all these tasks. More specifically, based on our previous work, Nielsen's and Krug's work on web usability [Nielsen 00; Krug 05], as well as observation of interfaces of popular 3D games, several works dedicated to game design (e.g., Rouse [05]), and design guidelines for virtual environments [Shneiderman 03], we summarize what we consider to be the main 3D web usability principles.

22.2 Use 3D with Care

Implementation of WebGL in all major web browsers provides us with an opportunity to build experiences that would otherwise be impossible—for example, sites that allow users to enter digital re-creations of places that are too dangerous to visit physically, historical sites that no longer exist, interesting places that are simply inaccessible for an average Internet user, experiences that allow for architecture walkthroughs, simple product configuration, augmented reality experiences, and, of course, engaging games. WebGL opens up countless other interesting application scenarios in various areas.

However, while 3D interfaces have a role in certain applications, most web applications (in the current era of the flat screens) do not innately require 3D. For serious applications, consider using 3D graphics for visualization of objects that need to be understood in their multidimensional form.

Moreover, remember to avoid emulating the physical world when it is not really necessary (e.g., by building a virtual shopping mall, a library, or a museum). As Nielsen pointed out, the goal of web design is to be better than reality. Asking users to "walk around the mall" is really putting your interface in the way of their goal, which, in this case, is shopping.

22.3 Make Text Readable

In WebGL applications, 3D representation often needs to be accompanied by text to effectively and comprehensively convey the information. How should we integrate (hyper)text with 3D graphics? Figure 22.1 shows two major techniques: (a) *hypertext+3D*, where a 3D scene is embedded in hypertext and (b) *3D+hypertext*, where hypertextual annotations are immersed in 3D. Which technique should we use and when? The study reported by Sonnet et al. [05] suggests that setting (a) seems to be applicable for extensive texts, because a user can explore a scene without any occlusions from the text, while the setting presented in Figure 22.1(b) works well for short annotations.

Annotation techniques can be divided into two categories:

- Object–space techniques, where annotations are embedded into a 3D scene (e.g., by placing information onto object surfaces or using billboards)
- Screen–space (viewport–space) techniques, where annotations are placed on a 2D plane that overlays a 3D scene, thus ensuring better legibility of text

Figure 22.2

Plain, billboard, anti-interference, and shadow text drawing styles with positive and negative image polarity studied in Jankowski et al. [10].

In one of the experiments evaluating depth and association cues between objects and their labels in information-rich VEs, Polys et al. [07] showed that screen–space interfaces outperformed object–space layouts, where text projected onto faces of objects in a 3D scene undergo pitch, yaw, and roll transformations.

As for typography, designers should choose one that communicates. Our study of text readability in 3D environments [Jankowski et al. 10] showed that the negative presentation, where light characters appear on a dark background, outperformed the positive presentation, where dark characters appear on a light background, and that the billboard drawing styles supported the best performance; subjective comments also showed a preference for the billboard style. See Figure 22.2.

If possible, enable standard operations for text manipulation like copy-and-paste *(Ctrl+C/Ctrl+V)*, increase or decrease of the size of web pages to improve readability *(Ctrl+/Ctrl–)*, and searching text *(Ctrl+F)*. With regard to searching, if the browser locates a given keyword or phrase in an annotation of a 3D object, the virtual camera should show that object; in other words, users should be able to navigate a virtual scene by means of querying its content.

22.4 Simple Navigation

Interactive 3D environments usually represent more space than can be viewed from a single point. Users have to be able to get around within the environment in order to obtain different views of the scene. However, developing a simple and effective technique for navigation for interactive 3D environments is difficult. First of all, viewpoint control involves six degrees of freedom (6DOF): three dimensions for positional placement (translation) and three for angular placement (rotation). The problem is the number of parameters to be controlled by the user; the major limitation inherent in using a mouse is that at least one change of state is required to cover all directions of translation and rotation. Another problem is the type of viewpoint control required by a given task that can be as different as simply exploring large-scale 3D environments to high-precision inspection of some 3D objects.

When designing a navigation interface, try to accommodate as broad an audience as possible by offering multiple ways to control the viewpoint. For scenes with selectable objects, consider reserving a single left mouse button click while the cursor is over the object for targeted movement navigation [Mackinlay et al. 90; Tan et al. 01]. The click on the object of interest should smoothly animate the camera from its current position to the selected object. This preserves the primary interpretation of clicking in the web browser window as following a hyperlink. The technique is easy to use (users simply have to select the point to which they want to fly), fast (it operates within controlled completion times), and cognitively friendly (enhances the perception of the virtual environment); it can also be easily integrated with other techniques. On the other hand, it has a major drawback: The target is always a selectable object.

When choosing a navigation metaphor, we should consider emulating real-world behaviors and taking into account information about the scene and the task at hand. For example, geographic VEs often employ a walking metaphor of camera motion, where user positions are restricted to the 2D plane of the terrain; the examine metaphor is often used to view different sides of objects and is suitable for tasks where the user's goal is to view an object as though he or she were holding it. If the user's goal can be reliably determined, the mode switching between the navigation techniques should be automated.

One problem with the unconstrained movement techniques is that the user may move to unknown and unwanted locations, look at things from awkward angles, or miss seeing important locations/features. As a result, one cannot ensure that the user receives the intended message. Empowering the author to bring some structure to the interaction experience can make VEs more suitable for the *new-to-3D* users [Galyean 95]. Such guided or constrained 3D navigation limits the user's freedom while traveling through a virtual world. It constrains the audience's movement to interesting and compelling places and thus avoids the classic problem of users getting "lost in cyberspace." See Figure 22.3. The simplest way to support this technique is to provide users with the play/pause/stop

Figure 22.3

The experience in the interactive film *3 Dreams of Black* is a good example of guided/constrained 3D navigation.

or previous/next buttons that can be used to navigate between the viewpoints that are defined for a 3D scene.

Three-dimensional viewing applications often provide a viewpoint menu in order to allow easy navigation for users so that they can navigate from one point of interest to another simply by selecting a menu item. The viewpoint menu turned out to be an important navigation tool in one of our recent studies [Jankowski and Decker 12]. We also learned that teleportation to 3D viewpoints should be possible through selecting a destination in an overview (in our case a minimap). Such viewpoints are usually static, and while very useful, they often do not show the "3D-ness" of virtual objects; as Andy van Dam mentioned, "If it ain't moving, it ain't 3D." Our initial evaluation of static versus animated views in 3D web user interfaces [Jankowski 12] indicates that users prefer navigating in 3D using a menu with animated viewpoints rather than with static ones.

22.5 Support Wayfinding

Wayfinding, difficult in the real world, is similarly difficult in virtual worlds. It is related to how people build up an understanding (mental model) of a virtual environment and is significantly affected by technological constraints, among which are small fields of view and the lack of vestibular information. This problem, also known as a problem of users getting *lost in space*, may manifest itself, especially in large virtual worlds, in a number of ways:

- Users may wander without direction when attempting to find a place for the first time.
- They may then have difficulty relocating visited places.
- They are often unable to grasp the overall topological structure of the space.

Efficient wayfinding is based on the navigator's ability to conceptualize the space. This type of knowledge is based on *survey knowledge* (i.e., knowledge about object locations, interobject distances, and spatial relations) and *procedural knowledge* (i.e., the sequence of actions required to follow a particular route). Based on the role of spatial knowledge in wayfinding tasks, we should be aware of the following methodologies that aid navigation:

- *Use of maps and orientation widgets*: Maps have proved to be an invaluable tool for acquiring and maintaining orientation and position in a real environment and this is also the case in a 3D virtual environment. Minimaps, an example of overview-plus-detail interface, are now very popular interface components in computer games. These miniature maps, typically placed in a corner of a user interface (see Figure 22.5 later in chapter), should display terrain, important locations and objects, and dynamically update the current position of the user with respect to the surrounding environment.
- *Use of landmarks*: Studies show that wayfinding strategies and behaviors are strongly influenced by the environmental cues. Characteristic and easily

recognizable landmarks in a virtual environment, exactly like in the real world, support navigation.

- *Use of bookmarks and personal marks*: Web browsers let users add bookmarks as they work. The users should be able to define their own 3D bookmarks with locations to which they can easily teleport when necessary. Users could also be provided with visual markers or "virtual breadcrumbs," which they could drop as a means of assisting navigation. Refer to Chapter 23 for more information about this subject.

Web browsers' back and forward buttons allow users to return to previously visited web pages. They make the users more confident, since they know they can always backtrack. Similarly to the standard behavior of saving the usage of local hyperlinks into navigation history of web browsers, recording a history of the user's motion and allowing him or her immediate and convenient travel back (and forth) along a trace of the previously traveled path should also be implemented using, for example, HTML5 History/State APIs or History.js. Moreover, make it easy to *go home*, so users can start over.

22.6 Selection and Manipulation

Selection is a process of identifying an object, a set of objects, or parts of objects that are targets for subsequent action. The most common way to select objects in 3D space is positioning a mouse cursor over a given object and clicking a mouse button. Techniques to compute the selected object are based most often on ray casting or back-buffer selection (see Chapter 20 for more on back-buffer selection). To show whether something is clickable, remember to change the cursor from an arrow to a pointing hand. An even better approach is to additionally draw the clickable objects in the scene with visible outlines. Applications that are designed to allow for interaction with many objects in one operation should include some variant of rectangle selection and shift-clicking to support group selection. Indirect selection of objects from a collection of objects (e.g., using a ListBox GUI component), the technique used in CAD, and 3D editing software should be considered for applications where selection of objects is difficult.

In some 3D web applications there might be a need for manipulation of objects—a need to virtually reach into the VE and specify object's position, orientation, and scale. A simple and efficient solution for explicit and direct manipulation is to provide users with manipulators also referred to as 3D gizmos (see Figure 22.4). Manipulators have been adopted by most CAD and 3D editing software applications and can be defined as visible graphic representations of an operation on, or state of, an object that is displayed together with that object. This state or operation can be controlled by clicking and dragging on the graphic elements (handles) of the manipulator. Manipulation can be simplified by exploiting knowledge about the real world (e.g., gravity) and natural behavior of objects (e.g., solidity). If possible, provide snap-dragging support. Lastly, manipulation can also be augmented by automatically finding an effective placement for the virtual camera.

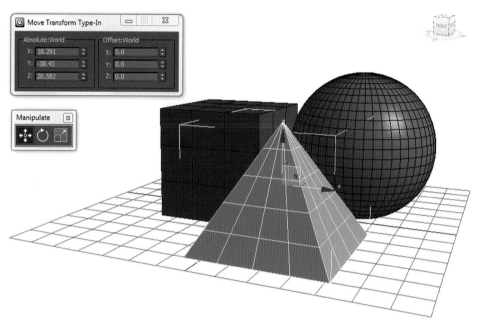

Figure 22.4

Manipulation interface implemented in 3ds Max.

22.7 System Control

System control refers to communication between a user and a system that is not part of the virtual environment. It refers to a task in which a command is applied to change either the state of the system or the mode of interaction. Most game interfaces are designed in a way in which application control interface components are grouped in a dashboard and placed in screen space on a 2D plane called HUD (head-up display) that is displayed side by side with a 3D scene or overlays a 3D scene.

When designing the dashboard, make sure the controls are clear; their physical appearance should tell how they work. Help should be embedded throughout the interface to allow users to click on any UI component for an explanation. A good application control technique should be easy to learn for novice users and efficient to use for experts; it has to also provide the means for the novice users to gradually learn new ways of using the interface. Therefore, for the first interaction with the application, provide only the necessary minimum of UI components; as the users gain more experience, their interfaces can be gradually enhanced and more functionality added. Figure 22.5. shows an example of an article from *Copernicus* [Jankowski and Decker 12] displayed both in the novice-oriented hypertext mode, where the dashboard limits interaction with 3D to a guided tour, and in the 3D mode designed for users with some 3D interaction experience, where the dashboard gives much more freedom with regard to 3D navigation.

Keyboard shortcuts should be provided to allow experienced users to bypass the dashboard. Lastly, if the dashboard is not needed at all times, it should be faded out;

Figure 22.5

An example of one virtual environment displayed in both hypertext and 3D mode of the dual-mode user interface of *Copernicus*. (Jankowski and Decker, 2012, *Proceedings of WWW '12*.)

the users should also be able to hide the dashboard at any time to create more space in a 3D viewing window.

To learn more about interaction techniques for 3D virtual environments, please refer to our recent review of the state of the art [Jankowski and Hachet 13].

22.8 Download and Response Time

Download time is one of the most important issues in web usability. During the early days of the web, slow (in that time usually dial-up) Internet access resulted in users having to wait for long periods of time, sometimes even minutes to download a simple page with few images. While the users did not like to wait, it was acceptable in those times because the web was a new and exciting technology. Nowadays, the web is a tool, a routine, and a slow download time often results in users leaving the site [Nielsen 00]. Nielsen suggests to keep in mind three time limits, which are determined by human perceptual abilities:

- 0.1 second is an ideal response time. The user does not sense any interruption.
- 1 second is about the limit for the user's flow of thought to stay uninterrupted, even though the user will notice the delay.
- 10 seconds is an unacceptable response time. The user experience is interrupted.

Nah's review [Nah 04] of the literature on computer response time and users' waiting time for download of web pages, and her own assessment of web users' tolerable waiting time shows that web users expect a response in about 2 seconds for simple information retrieval tasks. Her study also suggested that the upper bound for web users' tolerable waiting time is 15 seconds when the system does not provide any indication or feedback concerning the download.

The problem of the long data downloading time strongly affects WebGL applications, which very often are content heavy. Solutions to this problem include using compression for faster data transmission and the support of progressive content loading so that the user does not see a blank screen or a frozen scene but sees some progress going on. Several such solutions have been recently proposed, including Google's webgl-loader* [Chun 12], Khronos's glTF,† and the method of Lavoué et al. [13] (see Figure 22.6). Also see Chapter 20.

In addition to the fast download time, predictability is another important issue for response time usability [Myers 85; Nah 04]. Continuous feedback, for example, in the form of an animated progress bar, should be provided in situations where users are expected to face a long waiting time for downloading a page. Such feedback reassures the user that the web browser has not crashed; it indicates approximately how long the user can be expected to wait, thus allowing the user to do other activities; and it finally provides something for the user to look at. Research has shown that if the users are informed to expect the download of the site to be slow, they will be more tolerant of delay. Lastly, before the users decide to invest in a long 3D content download, it is

---

* Simple, fast, and compact mesh compression for WebGL: https://code.google.com/p/webgl-loader/
† The run-time asset format for WebGL, OpenGL ES, and OpenGL: http://gltf.gl/

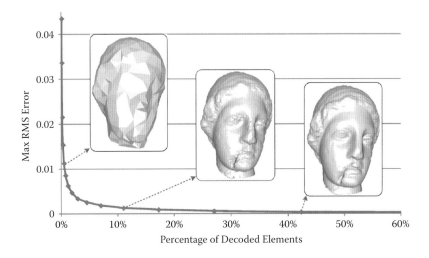

Figure 22.6

LOD quality in the technique proposed in Lavoué et al. [13]; the illustration shows geometric error associated with the different levels of detail according to the percentage of decoded vertices. (Image courtesy of Guillaume Lavoué.)

necessary for them to understand what they will be getting. It is important to provide the users with previews (e.g., previously prepared renderings of the incoming content).

22.9 Usability Evaluation

Websites should undergo evaluation before releasing them to the public. If possible, testing should take place periodically at different points of the design and development cycle, to ensure that the project is staying on track and to minimize the probability and the cost of fixing possible problems. Evaluation of a WebGL application can be done in a number of ways, from simply observing a user interacting with the application through usability inspections to user testing.

Usability inspections are usually performed in the form of cognitive walkthrough, which is an approach to evaluating user interfaces based on stepping through common tasks that a user would perform and evaluating the interfaces' ability to support each step, and heuristic evaluation, which is a method in which usability experts separately evaluate a user interface design by verifying its compliance with recognized usability principles: *the heuristics*. Nielsen's heuristics [Nielsen 94] are the most widely used usability principles for user interface design. Comprehensive sets of heuristics that can help to carry out usability inspections of WebGL applications, originally proposed for evaluation of virtual reality (VR) applications and video games, are reported in Sutcliffe and Gault [04], Sweetser and Wyeth [05], and Pinnele et al. [08].

The most insightful usability evaluation method is user testing, as no other method provides more information than seeing users actually trying to accomplish the tasks the application was designed to support and not succeeding. Generally, user studies involve measuring how well test subjects perform when interacting with the user interface in terms of efficiency, precision/recall, and subjective satisfaction. Such studies can be made in the form of formative, summative, or comparative evaluation (i.e., a statistical comparison of two or more configurations of user interface designs or interaction techniques. In Jankowski and Hachet [13] we describe some of the user study techniques that can be used to evaluate WebGL applications.

22.10 Input and Output

The input devices that are currently most often used for interaction with 3D web applications are a mouse and a touch screen. When designing a user interface for a WebGL application, exactly like for normal web applications, it is important to consider how the UI will work using both input techniques. Remember to emphasize the strengths (e.g., direct engagement, multitouch for a touch screen) and alleviate the weaknesses (e.g., inaccurate pointing, obstructed view of the screen for a touch screen) of the available input devices.

The big advantage of these input devices is that most users are very familiar with using them and novice users learn their usage in few minutes. The main disadvantage is that they are two-dimensional input devices and therefore the interaction metaphor must build a relationship between 2D input space and 3D virtual space.

Luckily, in recent years, we could witness the development of many new input devices more tailored for 3D applications. For example, Mouse 2.0 [Villar et al. 09] introduces multitouch capabilities to the standard computer mouse, giving it four degrees of freedom. The advances in computer-vision-based human motion capture and analysis allow for novel interaction techniques based on tracking, pose estimation, and understanding of human actions and behavior. One of the results of these works was the development of very sophisticated and affordable motion-sensing input devices, such as Microsoft's Kinect and Leap Motion, capable of body motion sensing that truly set a new standard for motion control and opened new exciting possibilities for more natural 3D interaction. Their nature makes them perfect input devices for games. On the other hand you need to remember that their occasionally inaccurate tracking, lack of standardized gestures, and tiring physical motion make the devices difficult to use for productivity tasks.

Similarly with the output devices, we need to consider how our WebGL application will work on different device types with different screen sizes with different pixel densities, ranging from small phones to large TV screens, so that our application is available to as many users as possible. While being compatible with different device types is very important, each screen size offers different possibilities and challenges for user interaction. Therefore, in order to truly satisfy and impress our users, our application must optimize the user experience for each screen configuration.

With regard to interaction with 3D interactive graphics, compared to traditional 2D displays, 3D displays (e.g., stereoscopic monitors and head-mounted displays) have been shown to enhance performance on a variety of depth-related tasks including navigation as well as selection and manipulation of objects. From a cognitive point of view, stereoscopy also proved to improve the spatial understanding of 3D information and helped to conceptualize the space [Ware and Franck 96].

With the emergence of affordable virtual reality headsets from Oculus and other companies, many predict that these gadgets might someday become our display devices of choice, fundamentally changing our everyday lives. Some developers from Mozilla and Google believe that the WebGL-enabled web might be a perfect VR platform and therefore they have begun adding a WebVR API to the Firefox and Chrome browsers,* making it possible for web developers to experiment with VR. From the usability perspective, while many of the principles presented in this chapter still hold, it is not entirely clear what the best practices for bringing VR to web content will be. VR in itself requires new ways of thinking about space, immersion, usability, and experience.† Adding the web to the equation brings the problem to a whole new level. Making WebVR experience engaging and comfortable requires years of exciting research and experimentation.

Bibliography

[Bowman et al. 04] Doug A. Bowman, Ernst Kruijff, Joseph J. LaViola, and Ivan Poupyrev. *3D User Interfaces: Theory and Practice.* Addison–Wesley, Boston, 2004.

[Chun 12] Won Chun. "WebGL Models: End-to-End," in *OpenGL Insights*, P. Cozzi and C. Riccio, ed. CRC Press, Boca Raton, FL, 2012.

[Galyean 95] Tinsley A. Galyean. "Guided Navigation of Virtual Environments." *Proceedings of I3D '95*, 1995.

[Jankowski 11] Jacek Jankowski. "A Taskonomy of 3D Web Use." *Proceedings of Web3D '11*, 2011.

[Jankowski 12] Jacek Jankowski. "Evaluation of Static vs. Animated Views in 3D Web User Interfaces." *Proceedings of Web3D '12*, 2012.

[Jankowski and Decker 12] Jacek Jankowski and Stefan Decker. "A Dual-Mode User Interface for Accessing 3D Content on the World Wide Web." *Proceedings of WWW '12*, 2012.

[Jankowski and Hachet 13] Jacek Jankowski and Martin Hachet. "A Survey of Interaction Techniques for Interactive 3D Environments." *Proceedings of Eurographics '13*, 2013.

[Jankowski et al. 10] Jacek Jankowski, Krystian Samp, Izabela Irzynska, Marek Jozwowicz, and Stefan Decker. "Integrating Text with Video and 3D Graphics: The Effects of Text Drawing Styles on Text Readability." *Proceedings of the 28th International Conference on Human Factors in Computing Systems (CHI '10)*, Atlanta, USA, 2010.

[Krug 05] Steve Krug. "Don't Make Me Think: A Common Sense Approach to the Web," New Riders Publishing, 2005.

[Lavoué et al. 13] Guillaume Lavoué, Laurent Chevalier, and Florent Dupont. "Streaming Compressed 3D Data on the Web Using JavaScript and WebGL." *Proceedings of the 18th International Conference on 3D Web Technology (Web3D '13)*, San Sebastian, Spain, 2013.

[Mackinlay et al. 90] Jock D. Mackinlay, Stuart K. Card, and George G. Robertson. "Rapid Controlled Movement through a Virtual 3D Workspace." *Proceedings of the 17th Conference on Computer Graphics and Interactive Techniques (SIGGRAPH '90)*, Dallas, USA, 1990.

---

* First steps for VR on the web: http://blog.bitops.com/blog/2014/06/26/first-steps-for-vr-on-the-web/; Bringing VR to Chrome: http://blog.tojicode.com/2014/07/bringing-vr-to-chrome.html

† Oculus best-practices guide: http://developer.oculus.com/best-practices

[Myers 85] Brad A. Myers. "The Importance of Percent-Done Progress Indicators for Computer–Human Interfaces." *Proceedings of the 3rd International Conference on Human Factors in Computing Systems (CHI '85)*, San Francisco, USA, 1985.

[Nah 04] Fiona Fui-Hoon Nah. "A Study on Tolerable Waiting Time: How Long Are Web Users Willing to Wait?" *Behavior and Information Technology*, 23 (3), 153–163, 2004.

[Nielsen 94] Jakob Nielsen. *Usability Engineering.* Academic Press, New York, 1994.

[Nielsen 00] Jacob Nielsen. *Designing Web Usability: The Practice of Simplicity*, 2000.

[Pinnele et al. 08] David Pinelle, Nelson Wong, and Tadeusz Stach. "Heuristic Evaluation for Games: Usability Principles for Video Game Design." *Proceedings of the 26th International Conference on Human Factors in Computing Systems (CHI '08)*, Florence, Italy, 2008.

[Polys, et al. 05] Nicholas F. Polys, Seonho Kimand, Doug A. Bowman. "Effects of Information Layout, Screen Size, and Field of View on User Performance in Information-rich Virtual Environments." *Proceedings of the ACM Symposium on Virtual Reality Software and Technology (VRST '05)*, Monterey, USA, 2005.

[Rouse 05] Richard Rouse. "Game Design—Theory and Practice," Wordware Publishing, 2005.

[Shneiderman 03] Ben Shneiderman. "Why Not Make Interfaces Better Than 3D Reality?" *IEEE Computer Graphics and Applications*, 23 (6), 12–15, 2003.

[Sonnet et al. 05] Henry Sonnet, Sheelagh Carpendale, and Thomas Strothotte. "Integration of 3D Data and Text: The Effects of Text Positioning, Connectivity, and Visual Hints on Comprehension." *Proceedings of INTERACT '05*, Rome, Italy, 2005.

[Sutcliffe and Gault 04] Alistair Sutcliffe and Brian Gault. "Heuristic Evaluation of Virtual Reality Applications." *Interacting with Computers*, 16 (4), 831–849, 2004.

[Sweetser and Wyeth 05] Penelope Sweetser and Peta Wyeth. "GameFlow: A Model for Evaluating Player Enjoyment in Games." *Computers in Entertainment*, 3 (3), 2005.

[Tan et al. 01] Desney S. Tan, George G. Robertson, and Mary Czerwinski. "Exploring 3D Navigation: Combining Speed-Coupled Flying with Orbiting." *Proceedings of the 19th International Conference on Human Factors in Computing Systems (CHI '01)*, Seattle, USA, 2001.

[Villar et al. 09] Nicolas Villar, Shahram Izadi, Dan Rosenfeld, Hrvoje Benko, John Helmes, Jonathan Westhues, Steve Hodges, Eyal Ofek, Alex Butler, Xiang Cao, and Billy Chen. "Mouse 2.0: Multitouch Meets the Mouse." *Proceedings of the 22nd Symposium on User Interface Software and Technology (UIST '09)*, Victoria, Canada, 2009.

[Ware and Franck 96] Colin Ware and Glenn Franck. "Evaluating Stereo and Motion Cues for Visualizing Information Nets in Three Dimensions." *ACM Transactions on Graphics,* 15 (2), 121–140, 1996.

23

Designing Cameras for WebGL Applications

Diego Cantor-Rivera and Kamyar Abhari

23.1 Introduction

An important consideration for any WebGL application is the implementation of the camera. Cameras must perform a series of matrix transformations that *move the world*, translating and rotating the geometry to obtain a vantage point. However, WebGL does not provide a construct to represent cameras, and each developer needs to implement his or her own. This is challenging, in particular, for a JavaScript developer with no experience in computer graphics: The math behind the operation of cameras is not developer friendly and it is a source of confusion and code bugs. Additionally, the JavaScript language does not provide native operations on matrices; therefore, extra coding or relying on a matrix library is required. Although these problems are addressed by several publicly available, high-level WebGL libraries, each library implements cameras in a particular way, which might not necessarily address all the required behaviors in the project at hand.

In this context, we present a bottom-up approach to designing cameras in JavaScript for WebGL applications. This chapter begins by discussing the basic matrix definitions relevant to computer graphics. Then, we delve into implementation details: We discuss linear algebra libraries that allow writing camera operations in JavaScript. Next, we explore basic design guidelines including examples taken from current WebGL libraries. Because JavaScript is a language where encapsulation is not strictly enforced, this allows for designs where camera attributes can be modified freely by code foreign to the camera entity (transparent cameras). In this part, we explore the trade-offs of such designs and what they mean for the JavaScript developer, and we compare it to the case where encapsulation is enforced (responsible cameras).

Next, we present the design and implementation of landmark-based camera navigation. This concept enables the creation of flyovers and/or user-triggered cinematic sequences by creating smooth paths for the camera to follow that can be implemented using any of the existing WebGL libraries or our own camera classes.

We conclude with a summary of cameras available in current WebGL libraries and an overview of active research that will influence the design and role of cameras in future WebGL applications.

23.2 Transforming Scene Geometry

To create a *vantage point*, rotation and translation of the geometry is performed using the ubiquitous model-view transform (Q), which corresponds to the composition of two transformations: the *model transform* (M_m) that converts vertices from *object space* to *world space* coordinates, and the *view transform* (M_v), which converts from world space to *eye space* coordinates. The model-view transformation is applied to *every vertex v* in the scene:

$$v' = Q * v$$

$$v' = (M_v * M_m) * v \qquad (23.1)$$

The model-view transform belongs to a special kind of operations called *affine transforms*. This group of transforms is composed of a linear operation, such as rotation and scaling, followed by a translation.

23.2.1 Matrix Representation of the Model-View Transform

The model-view transform can be represented as a 4×4 matrix in *homogeneous coordinates*. The upper-left 3×3 matrix (R) describes a rotation where the three column vectors x′, y′, and z′ define the axes of the new coordinate system caused by the rotation. The translation vector (T) is stored in the last column and defines the amount of shifting between the world origin and the model-view origin in the direction of the new coordinate system (Figure 23.1).

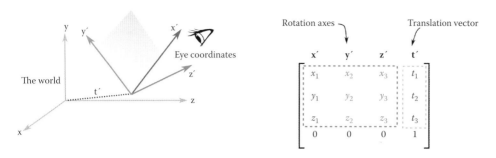

Figure 23.1

Structure of the model-view matrix. This transformation produces eye coordinates.

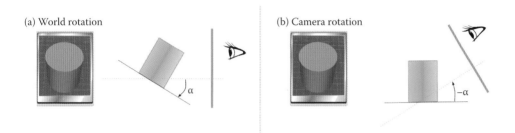

Figure 23.2

Rotating the world (a) is the same as rotating the camera (b) in the *opposite direction*.

23.3 Building the Camera Transform

Since a camera will operate on the scene in a similar way to the model-view transform, by applying rotations and translations, we can correctly assume that it will also be an affine transform. Let's compare what we know of the model-view transform with what we expect from the camera transform.

23.3.1 Rotational Component

The model-view transform rotates the world so that objects can be seen on the screen. In Figure 23.2a, a cylinder stands in the center of the virtual world. The world has been rotated by an angle α on the x-axis. This allows us to see the top face of the cylinder. A virtual camera looking at the object rotates in the same axis by $-\alpha$ to obtain the same scene on screen (Figure 23.2b).

23.3.2 Translational Component

With the model-view transform, it is possible to visualize objects not currently on the screen using translations, which is necessary in most applications as the world does not generally fit in a single viewport. In such a scenario, we need to apply a translation that contains the coordinates of the object, effectively centering the object in the viewport so it can be seen (Figure 23.3a). The translation implemented with a virtual camera translation

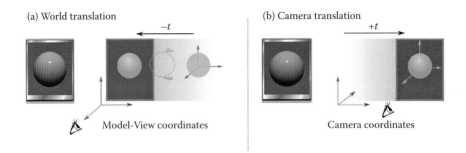

(a) World translation

$-t$

Model-View coordinates

(b) Camera translation

$+t$

Camera coordinates

Figure 23.3

Translating the world (a) is the same as translating the camera (b) in the *opposite direction*.

works differently; the world remains static while the camera position changes, moving toward the object (Figure 23.3b).

23.3.3 Camera Matrix

Mathematically, the complementary behavior of the model-view (Q) and camera (C) transforms is defined by a relationship of *matrix inverses*:

$$Q = TR$$

$$Q^{-1} = (TR)^{-1}$$

$$= R^{-1}T^{-1}$$

$$M = R_t^{-1}T^{-1}$$

$$= \underbrace{\left[\begin{array}{c|c} R^t & 0 \\ \hline 0 & 1 \end{array}\right]}_{R^t} \underbrace{\left[\begin{array}{c|c} I & -T \\ \hline 0 & 1 \end{array}\right]}_{T^{-1}}$$

$$Q^{-1} = C = \left[\begin{array}{c|c} R^t & -R^t T \\ \hline 0 & 1 \end{array}\right]$$

(23.2)

Given that rotation matrices are orthogonal, their inverse is their respective transpose matrix. Therefore, we can go back and forth between the model-view and camera rotations by transposing the respective rotational component. In contrast, the translational component of the camera matrix does not simply correspond to the inverse of the model-view translation. The translation indicates displacement in the coordinate system imposed by the rotation (rotation occurs first). Therefore, it is necessary to premultiply –T (the inverse translation) by R^t to obtain the correct translation in the camera coordinate system (Equation 23.2).

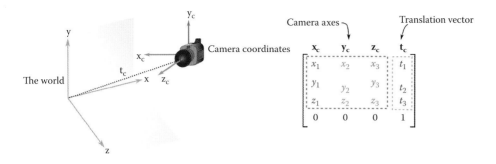

Figure 23.4

Structure of the camera matrix.

Tip: The translation component determines increments *in the direction of the rotated axis*.

Since the camera matrix is the inverse of the model-view matrix, this makes it also *an affine transform*. Therefore, its structure is similar to the model-view matrix with the upper left 3×3 matrix describing the axes of the camera coordinate system: x_c, y_c, and z_c and the last column containing the translation vector t_c (Figure 23.4).

Tip: The camera coordinate system may contain a scaling factor, in which case the column vectors will not have unit length. However, the rotation is still well defined and the coordinate system remains valid (orthogonal). In the absence of scaling, the coordinate system is orthonormal (the dot product of any two elements of the basis is one).

23.3.4 Simulating Camera Properties: The Projection Transform

The projection transform is performed after obtaining a vantage point. It determines the *frustum*, which is the region of the space that will be visible on screen (i.e., the field of view). The geometry of the frustum determines the type of projection: *orthogonal* or *perspective*.

Depending on the metaphor employed, the projection transform may or may not be part of a camera entity in a WebGL application, for example, some implementations can choose instead to model the type of frustum as a property of the scene, view, rendering entities, etc. A rationale for this division is that usually the type of projection transform does not change during the life cycle of a WebGL app, so it can be allocated to an entity separate from the camera. The opposite argument is that a frustum can be seen as being determined by *intrinsic properties* of a camera (e.g., focal length, principal point, and axis skew), suggesting that the corresponding projection transform should belong to the camera entity. In the rest of the chapter we assume that cameras in WebGL applications may contain two matrices: the camera and projection transform.

Tip: A demo of the relationship between the projection transform and the camera's intrinsic properties can be found at http://ksimek.github.io/perspective_camera_toy.html.

Tip: A demo visualizing the effect that frustum properties have on a scene can be found at http://www.realtimerendering.com/udacity/transforms.html.

23.4 Cameras in WebGL Applications

Having a camera matrix allows us to move through the world and look around objects. A moving camera metaphor is more intuitive than shifting and rotating the world in the opposite direction. Nonetheless, geometric transformations, such as motion of 3D objects in the virtual world (e.g., actors, characters, etc.) are *point of view independent*. These transformations need to be multiplied with the model-view matrix. Generally speaking, for every frame, we would:

a. Update the camera matrix based on user interaction events (e.g., mouse, keyboard, gestures, etc.)
b. Obtain the inverse: the model-view transform
c. Use the *model-view* transform to update geometry and the *projection* transform to determine the frustum

This last step is usually performed in the *vertex shader* by passing both matrices as `Mat4` uniforms.

Tip: It is common practice to premultiply the model-view matrix by any *local* transformations (e.g., movement/rotation of individual objects) before passing it along to the vertex shader. This way the matrix multiplication occurs just once per moving object instead of occurring once *per vertex* inside the vertex shader. This optimization is more significant if our application contains very complex geometrical models.

23.4.1 Working with Matrices in JavaScript

Before getting into any implementation efforts, we need to decide *how to* operate with matrices in JavaScript given that matrices *are not native* to the language, and most of the camera operations boil down to operations with 3×3 or 4×4 matrices.

Do we write our own matrix code or do we use a library?

Let's take a look at some of the currently available JavaScript matrix libraries. Any of these could be used to write the code to manipulate a camera in a WebGL application:

- **gl-matrix**[*] is a very complete library with operations such as `lookAt` and `perspectiveFromFieldOfView` to simplify the implementation process
- **mjs**[†] is a good contender providing a variety of methods to operate on the rotational and translational components of 4×4 matrices (e.g., `M4x4.inverseTo3x3`).
- **Sylvester**[‡] is semantically expressive with aliases for operating on matrices (`$M`) and vectors (`$V`). Matrices can be multiplied using the alias for multiplication `.x` (e.g. `var M = M1.x(M2);`), but it lacks basic operations such as translation and scaling. Also, it is not optimized for performance, so it is not ideal for WebGL apps.
- **ewgl-matrices**[§] is a work in progress that is being designed as a response to the slower Sylvester.

---

[*] https://github.com/toji/gl-matrix
[†] https://code.google.com/p/webgl-mjs/
[‡] http://sylvester.jcoglan.com/
[§] https://code.google.com/p/ewgl-matrices/

- **Closure**[*] is a set of JavaScript tools for web application development created by Google. Closure provides efficient and well documented classes to operate on matrices and vectors under the namespace `goog.vec`.
- **Numerics**[†] is intended to be a scientific library providing different kinds of matrix factorizations (SVD, LU, etc.), ordinary differential equations solvers, unconstrained optimization algorithms, etc. Though not designed to be WebGL-aware, it claims to outperform Closure.[‡]

On the other hand, some WebGL libraries, such as three.js, TDL,[§] and Babylon.js (Chapter 9) provide their own matrix and vector classes.

Tip: A JavaScript matrix libraries benchmark can be found at https://github.com/stepheneb/webgl-matrix-benchmarks.

From the benchmark it is clear that TDL, Closure, and gl-matrix are the top contenders in terms of performance as tested on a MacBook Pro (OS X 10.9.5, 2.4 GHz Intel Core i7, 8 GB 1333 MHz DDR3). TDL is a low-level library that focuses more on speed of rendering than usability. If you are just starting to develop your first WebGL application, we recommend considering gl-matrix or Closure.

23.4.2 Camera Types Based on Navigation Strategies

After determining the library, it is necessary to decide the type of navigation. This decision has an enormous impact on the usability of the final product and it is influenced by the specific tasks that the user must complete and the metaphor, such as moving the head, looking forward, walking, flying, etc., that the user is expected to follow when exploring the virtual environment (see Chapter 22).

Table 23.1 contains a list of the most common camera types based on navigation strategies. This list is by no means comprehensive, but rather it provides a general idea of camera mechanics and when each type should be selected.

Unless your application requires a specialized navigation metaphor, orbiting and exploring cameras cover most standard navigation cases: walking and looking at a point (with the tracking camera) and rotating and zooming into an object (with the orbiting camera). Also, consider that it is possible to have more than one type of camera/navigation strategy in the same application. For instance, we may want to switch from the world view (orbiting) to the first-person character view (tracking) depending on the task at hand.

23.5 Designing Cameras for WebGL Apps

Once we have identified the matrix library and the navigation strategy, we can start the design process. Some of the questions that we will be answering in this section are

How do we define a camera entity?
Are camera attributes going to be public or do we make them private?

---

[*] http://docs.closure-library.googlecode.com/git/namespace_goog_vec_Matrix4.html
[†] http://www.numericjs.com/index.php
[‡] http://www.numericjs.com/index.php.html
[§] https://github.com/greggman/tdl/blob/master/tdl/math.js

Table 23.1 Camera Types Based on Navigation Strategies

Camera type	Description
Exploring/tracking 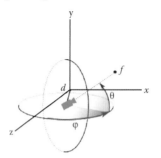	This is used to explore a 3D world. The camera rotates around its position, changing the focal point with every rotation. The camera moves freely around the world and its position is described by world coordinates. This type of camera is used for seeing what a 3D character sees (first-person view) or for tracking 3D characters from a distance (third-person view).
Orbiting (sphere) 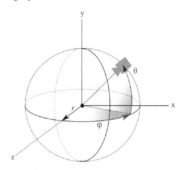	This is used for 3D character design, and in a wide range of applications such as medical imaging. The camera rotates around its focal point, changing its position with every rotation. The camera follows a spherical orbit and its position can be described in terms of its azimuth and elevation.
Orbiting (cylinder) 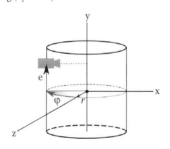	In this type of orbiting camera, the elevation angle is replaced by an altitude measuring the distance from the central plate. This type of coordinate system has been used in projects such as Zygote Body (formerly the Google Body project).
Bird's-eye 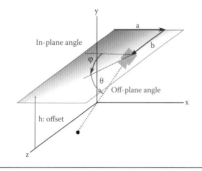	This is used in 3D world design, geographic visualization, real-time strategy (RTS), and role-playing games (RPGs). The camera coordinates are generally defined on an aerial plane parallel to the ground and the focal point changes with the camera position. Rotations are still possible but commonly restricted to looking toward the ground. This category includes top-down and 2.5D cameras.

23. Designing Cameras for WebGL Applications

What kinds of operations does a camera entity provide?
How can we assert a consistent camera state at all times?

In this section, we present the two main approaches to camera design: In the first approach, we explore the traditional object-oriented design where the camera entity is responsible for maintaining a consistent state at all times using private attributes and accessor methods (responsible cameras); in the second, we present a more JavaScript-like alternative where a camera is simply a collection of attributes that can change arbitrarily at any point in the code (transparent cameras). These two design choices are not mutually exclusive, and most of the current WebGL libraries use cameras that combine both responsible and transparent features.

23.5.1 Responsible Cameras

Let's start with a simple idea: a camera entity containing two private attributes, `camera-Transform` and `projectionTransform`, and two public operations, `getModel-ViewTransform()` and `getProjectionTransform()` (Figure 23.5a). In this design the camera entity *assumes the responsibility* of updating the matrices in reaction to events in the application such as user input or events generated by application logic. Therefore, it must provide functions to do so. This basic idea drives us to add public operations to manipulate the internal matrices (Figure 23.5b).

23.5.1.1 Pros and Cons

Responsible cameras follow the object-oriented programming principle of encapsulation: Attributes and operations affecting those attributes must be part of the same entity. Moreover, the camera attributes are private and can only be changed by public operations provided by the camera entity. As the complexity of the application grows, new behaviors can be implemented by creating new functions, and if the behavior represents a significant change in the way the camera operates, then a new type of camera can be created by inheritance (Figure 23.6). In all cases, validation code is written on each of the public operations that update the camera and projection matrices to ensure that the camera state is always consistent. For example, if any operation has the potential to create an inconsistent state, the validation code can veto this operation or reset the camera to a known

(a) Basic Camera	(b) Adding modifier methods

Camera
-cameraTransform: Matrix4×4 -projectionTransform: Matrix4×4
+getModelViewTransform(): Matrix4×4 -getProjectionTransform(): Matrix4×4

Camera
-cameraTransform: Matrix4×4 -projectiontransform: Matrix4×4
+getModelViewTransform(): Matrix4×4 -getProjectionTransform(): Matrix4×4 +rotateX(angle: float): void +rotateY(angle: float): void +rotateZ(angle: float): void +translate(position:Vector3): void

Figure 23.5

Initial camera entity: (a) the camera is responsible for maintaining the camera and projection matrices; (b) with modifier functions.

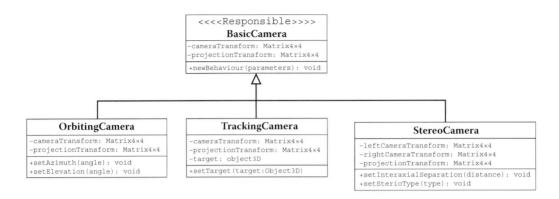

Figure 23.6

Extending the behavior of responsible cameras through new operations or through inheritance.

valid state. Since the state of the camera can only be updated by using the provided camera operations, any code that wants to update the camera needs to use these functions triggering the respective validations every time. Thanks to this design, the rendering loop is shielded from possible, transitional, inconsistent camera states that would occur if the camera attributes were transparent and could be changed at any point in the code.

The principal disadvantage of responsible cameras is that state querying requires extra work; every time the application needs to determine the position, pose of the camera, etc., mathematical operations must be applied on the camera and perspective matrices to extract such information (e.g., calculating Euler angles from the 3×3 rotation matrix; obtaining the near and far planes from the perspective matrix, etc.). This may have a negative impact on more advanced behaviors such as flyovers and animations. However, this can be mitigated by using a cache mechanism (discussed later in this section). Another disadvantage is that the code of the camera entity grows *fast*. Every operation that affects the matrices must check for state consistency and, when any problem is detected, extra code needs to be written to veto the change, return the camera to a valid state, or throw an exception.

Tip:: Although it is possible to implement strict private attributes in JavaScript using closures,[*] they come with some memory overhead.[†] As an alternative, it is common practice to identify attributes assumed private by preceding them with an underscore (_) in the code. This also serves as a warning for other developers: "Use at your own risk."

23.5.1.2 Implementing a State Cache

A responsible camera design can be optimized with caching. The information that is computed from matrices (position, angles, field of view, etc.) can be saved in a set of attributes to facilitate immediate retrieval.

To support caching, any operation that updates the camera or perspective matrices also *needs to be aware of the cached attributes*; for example, a camera `rotate` operation can

---

[*] http://javascript.crockford.com/private.html
[†] https://curiosity-driven.org/private-properties-in-javascript

set the cached attributes `azimuth` and `elevation` in addition to updating the camera matrix (Listing 23.1).

Listing 23.1 Pseudocode of a responsible camera operation with state-caching.*

```
vxlCamera.prototype.rotate = function(azimuth,elevation){

    //_getAngle validates the angle passed as parameter
    this._relElevation = this._getAngle(elevation);
    this._relAzimuth = this._getAngle(azimuth);

    var valid =...//if these are valid angles continue
                //otherwise veto/ignore the change
    if (valid){
        this._elevation + = this._relElevation;
        this._azimuth + = this._relAzimuth;
        this._computeMatrix();//uses the new elevation and
                //azimuth to update the rotation matrix
    }
    return this;
}
```

* https://github.com/d13g0/voxelent/blob/master/source/nucleo/camera/Camera.js

---

Notice that the cached attributes can be set by more than one operation. For example, the `azimuth` and `elevation` can be recalculated after setting a new focal point (Listing 23.2). This is necessary since a new focal point effectively induces a change in the camera orientation/rotation and therefore changes the azimuth and elevation of the camera.

Listing 23.2 Calculating camera attributes after the camera matrix has changed.

```
vxlCamera.prototype.setFocalPoint = function(focalPoint){
    ...//operations to set the new focal point
    v =...//calculates vector from camera to new focal point
    var x = v[0], y = v[1], z = v[2];
    var r = vec3.length(v);
    this._elevation = Math.asin(y/r);//new elevation
    this._azimuth = Math.atan2(-x,-z);//new azimuth
};
```

23.5.1.3 Pros and Cons

As seen, responsible cameras with state-caching attributes allow immediate querying for both the *application* (determining information about camera location, pose, etc.) and the *rendering engine* (obtaining camera and projection matrices). Time-consuming matrix calculations occur only as the result of user interaction or application logic with the advantage of not having to repeat them when querying the camera state.

An obvious downside of this design is the need for maintaining matrices synchronized with state-defining attributes. As shown by the examples in Listings 23.1. and 23.2, the

---

relationship between cached attributes and camera operations is not always one to one. Therefore, it is necessary to perform a careful analysis of possible interactions between the cached attributes and the camera operations (more precisely, which operations update which attributes) in order to maintain data integrity.

23.5.2 Transparent Cameras

A more JavaScript-like approach is to define the camera entity simply as a collection of public attributes. Let's call this design *transparent*. A camera entity holds a set of attributes that defines its location and orientation—and possibly more attributes, depending on the navigation strategy. These attributes can be accessed/changed freely at any point in the code. Then prior to rendering (i.e., before transferring data to shaders and issuing a draw call), the model-view and perspective matrices are constructed using the current camera attributes.

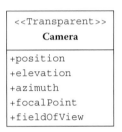

Figure 23.7

Transparent camera: its state is represented by a set of public attributes.

Let's start the design process with a camera entity and a set of public attributes that describe its *state*. This set includes attributes such as the position and orientation of the camera, its focal point, field of view, etc. (Figure 23.7).

In this scenario any changes in the camera state are made by directly changing the value of the camera's public attributes:

```
camera.position = [5,6,7];//sets the camera position
camera.rotation.X = 45;//rotates 45 degrees around the
    x-axis
```

An external entity, such as a renderer object, reads these attributes and creates a matrix representation using methods such as `getModelViewTransform(camera)` and `getProjectionTransform(camera)`. Then, the matrices are passed along to the ESSL (GLSL ES) shaders as needed. As the development process continues and different camera requirements emerge, the renderer can delegate the construction of matrices to an interpreter, that would implement a *strategy* pattern as shown in Figure 23.8.

23.5.2.1 Pros and Cons

The main advantage of using transparent cameras is that they provide *immediate queries* given that the camera state is retrieved directly from the attributes without any additional operation/calculation. Similarly, any desired behavior is simply stated by updating the respective camera attribute. Emerging requirements are implemented by adding new attributes on the camera. In fact, attributes describing different camera types can coexist in the same entity since the camera interpretation is delegated to an external entity. Another advantage of transparent cameras is that they can be easily represented as *nodes* in WebGL applications that are built on the concept of *scene graphs*.

Their main disadvantage is the lack of a state-checking procedure, which can lead to inconsistent behaviors during the lifetime of the application. For example, an operation such as changing the `distance` to the focal point does not automatically update the camera `position` attribute.

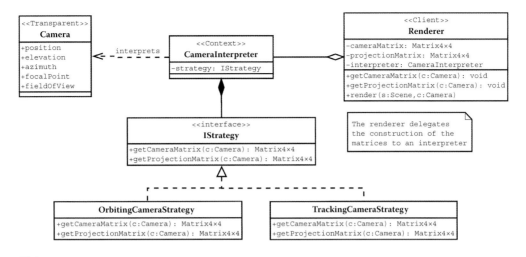

Figure 23.8

Strategy pattern for working with transparent cameras.

23.5.3 Design Continuum

Most WebGL libraries provide camera implementations that share characteristics of responsible and transparent designs. For instance, three.js[*] allows modifying camera attributes directly:

```
camera.rotation.y = 90 * Math.PI/180
```

and provides state-checking using event listeners:

```
var rotation = new THREE.Euler();
var onRotationChange = function(){
  quaternion.setFromEuler(rotation,false);
}
rotation.onChange(onRotationChange);
```

In contrast, WebGL frameworks such as SceneJS[†] and OSG.JS[‡] treat cameras as *scene nodes* containing state-defining attributes to be operated on externally (Listing 23.3).

Listing 23.3 Implementing a camera node in SceneJS.[a]

```
var scene = SceneJS.createScene({
  nodes:[{
    type:"cameras/orbit",
    yaw:40,
    pitch:-20,
```

---

[a] http://scenejs.org/examples/cameras_orbit.html

---

[*] http://threejs.org/
[†] http://scenejs.org
[‡] http://osgjs.org/

```
zoom:10,
zoomSensitivity:1.0,
eye:{x:0, y:0, z:10},
look:{x:0, y:0, z:0},
nodes:[...
```

Also, it is possible to relinquish any responsibility to update the matrices and, instead, provide the developer with functions to do so when required. This approach is followed by libraries such as Turbulenz* (Chapter 10).

On the other side of the spectrum, libraries such as Babylon.JS† (Chapter 9) use responsible cameras with the expected in-place state-checking behavior (Listing 23.4).

Listing 23.4 State-checking behavior in Babylon.JS cameras.* The code shown is written using TypeScript.

```
public _update():void {
    var needToMove = this._decideIfNeedsToMove();
    var needToRotate = Math.abs(this.cameraRotation.x) > 0 || Math.
        abs(this.cameraRotation.y) > 0;
    if (needToMove) {//Move
    this._updatePosition();}...
```

* https://github.com/BabylonJS/Babylon.js/blob/master/Babylon/Cameras/babylon.targetCamera.ts

23.5.3.1 Which Design Is Right for My Project?

Deciding which approach to follow depends on the level of interactivity and the complexity of the navigation that we expect in our application. For simple navigation metaphors (walk, look at), transparent cameras can suffice. Responsible cameras are preferred when the navigation becomes more elaborate and different points in our code try to update the state of the camera in response to simultaneous application events or user interactions. For instance, in a video game where we are hit by our opponent we might want to simulate the blow/fall sequence. This behavior can be added naturally to the set of camera behaviors defined in a responsible design.

23.6 Camera Landmarks: Saving and Retrieving Camera States

Defining camera state attributes enables creating a history of the previous locations where the camera has been, as well as to plan future locations. Each camera state can be recorded into a *landmark*. These are some scenarios where landmarks are useful:

- Going back to a specific savepoint after the hero dies in a game
- Studying anatomical features in medical applications

---

* http://docs.turbulenz.com/jslibrary_api/camera_api.html
† http://www.babylonjs.com/

- Recording vantage points in an architectural application
- Creating cinematic effects where the camera smoothly transitions from one point to the other in the scene

A landmark can be created at any point in the execution of the application by copying the current camera state including position, orientation of the camera, focal point, etc. Subsequently, the camera can be set to any recorded landmark by copying the contents of the landmark directly into the camera state-defining attributes. This causes the scene to be reset to the previous location/orientation defined by the landmark.

Landmarks can be naturally implemented through the use of a *memento* pattern (Figure 23.9)

The following elements constitute this design:

- `Camera`:
  - Produces a `Landmark` every time that `createLandmark` is invoked. Internally, it saves all the state-defining attributes into a `CState` container that then is relayed to the newly created `Landmark`. The parameter name allows saving the landmark in a dictionary held by the `LandmarkManager` entity.
  - The `gotoLandmark` method allows retrieving a `Landmark` from the `LandmarkManager` entity by name.
- `Landmark`:
  - Holds all the state-defining attributes of a camera. Here these attributes are represented by a `CState` entity.
  - A `Landmark` is *immutable*. This means that once created, it cannot be updated or altered.
- `LandmarkManager`:
  - Contains a dictionary where each `Landmark` instance is associated with a string key. This enables us to retrieve landmarks by name.

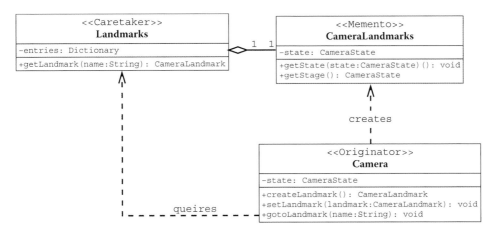

Figure 23.9

Saving and retrieving camera states by defining landmarks.

23.7 Landmark-Based Navigation

Once a landmark is retrieved, the subsequent rendering call automatically reflects the newly set camera state. It is possible, though, to create visually appealing cinematic transitions by not setting the camera state *immediately* but rather *incrementally*. To achieve this, we can use any form of interpolation: linear, cosine, cubic, etc. The basic idea is to interpolate the camera between its current state and the state stored in the landmark. The interpolated state is accessible to the rendering cycle so that the view can be updated as the camera approaches the landmark.

The pseudocode of our implementation* is shown in Listing 23.4. We start by retrieving the camera state stored in the landmark (e.g., camera position). Then, an iteration function that renders the incremental updates is defined and invoked recursively until the camera reaches the desired landmark (i.e., the desired state).

Listing 23.4 Pseudocode for landmark-based animations.

```
//1. getting information from the landmark [lmark]
var camera =...
var finalPosition = lmark.position


//2. Setting up animation
var steps =...;
var currStep = 0;


//3. Define incremental update
function iteration(){
        var interPos = vec3.create();
        if(currStep++ ! = steps){

                //interpolate
                percent = (1-Math.cos(currStep/steps*Math.PI))/2;

                //calculates an interpolated position vector
                interPos = vec3.lerp(interPos, camera.getPosition(),
                    finalPosition, percent);
                camera.setPosition(interPos);
                camera.refresh();//triggers a rendering update

                var distance =...//calculate current distance to
                    landmark
                if (distance >0.01){
                        //If not there yet go onto the next iteration
                        window.setTimeout(iteration,...);
                }
                else{
                        //Set the camera to the final state (landmark)
```

---

* https://github.com/d13g0/voxelent/blob/master/source/nucleo/camera/Camera.js

---

```
                        camera.setPosition(finalPosition);
                        camera.refresh();//triggers a rendering update
            }
        }
};
//3. Execute animation by invoking the first iteration
window.setTimeout(iteration,...);
```

This pseudocode only shows how the camera position is processed. However, the same idea is applicable to any other state-defining attributes, such as the focal point and the roll angle.

In our example, the `vec3.lerp` method offered by the **gl-matrix** library is used for interpolation. The amount of interpolation is set using the variable `percentage`. Notice that instead of setting a linearly increasing percentage every step, we have used a *cosine interpolation* to calculate it every time. This simple hack creates visually appealing, smooth transitions with acceleration/deceleration effects as the camera leaves the current state and approaches the target state. Because the cosine transformation creates many small, indistinguishable increments near the target destination, we quickly adjust the camera to its final destination when it is very close (0,01 in the example), to save processing time.

Tip: The *WebGL Insights* GitHub repository[*] contains the `demo-camera-landmarks.html` (Figure 23.10), which shows landmark-based navigation in action, with the smooth transitions obtained by cosine interpolation. This demo uses the cameras defined by our own WebGL library[†].

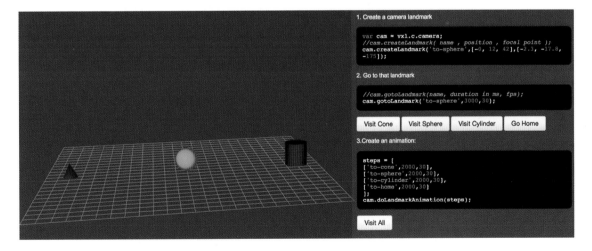

Figure 23.10

Landmark-based navigation in action. This demo is included in the *WebGL insights GitHub repository*.

[*] https://github.com/WebGLInsights/WebGLInsights-1
[†] http://voxelent.com

23.8 Existing Implementations

The tools, the navigation strategies, and the design alternatives are not unique when it comes to implementing cameras in WebGL applications. Each WebGL library provides a different subset of alternatives. Table 23.2 contains a nonexhaustive list of current implementations.

In general most WebGL libraries provide cameras with at least two types of navigation metaphors: orbiting around a static object or exploring the world using a first-person perspective. Libraries such as Babylon.js have put a lot of effort into developing specialized cameras for games that can be controlled with a joystick and, more recently, with the Oculus head-mounted display.* In these cases, the camera maps intuitively one-to-one to the specific gaming device rather than trying to fit a camera-control strategy to the kind of interaction provided by the traditional keyboard and mouse.

23.9 Future Directions

What kinds of cameras can we expect in future WebGL applications?

There are clear challenges to bringing current advancements in virtual camera research to the Web. Although augmented reality applications using WebGL and HTML5 canvas 2D[†] exist, the web still needs image- and video-processing capabilities similar to those currently offered by scientific libraries such as OpenCV,[‡] VTK,[§] and Nvidia CUDA[¶]. In the near future we will see some of these computing capabilities on the web through WebCL.[**] With these capabilities, better and more powerful WebGL applications will be developed.

As processing power increases, more sophisticated camera behaviors will be translated from research into WebGL applications in four areas: *automated path planning*, *assisted navigation*, *affective control*, and *virtual reality*.

With *automatic path planning*, spatial constraints are introduced. Such constraints help avoid object–camera collisions, providing a better user experience. One of the scenarios where it has proven to be useful is in the navigation of immersive virtual museums [Drucker 94]. In this case, a connectivity graph based on accessibility (door location) is generated and traveling paths are calculated from one doorway to the next or from a doorway to a point within the room. Then, the minimal cost path is calculated to re-create the best museum tour. In general, there are many different types of constraints that can be addressed by automatic path planning. Constraint programming extends to other desirable cinematographic properties such as location and orientation, camera-to-object distance, and object-to-objects distance [Christie 03].

Instead of restricting navigation to predetermined paths, some research groups have instead focused on the idea of *assisted navigation*. In this scenario, the camera moves

---

* https://github.com/BabylonJS/Babylon.js/wiki/Using-Oculus-Rift-with-Babylon.js
† http://www.w3.org/TR/2dcontext/
‡ http://opencv.org/
§ http://www.vtk.org/
¶ http://www.nvidia.ca/object/cuda_home_new.html
** https://www.khronos.org/webcl/

Table 23.2 Current Camera Implementations in WebGL Libraries

Library	Description	Camera Classes
three.js	This library supports both orthographic and perspective cameras, and a class (`combinedCamera`) to switch easily between them. For rendering *dynamic reflections* in real time, `cubeCamera` might be employed, which needs to be added to the scene as the second camera positioned at the object with a reflective surface. Furthermore, `CameraHelper` can be employed to visualize the frustum. This class, along with other helper classes such as `DirectionalLightHelper`, can be used to comprehend a scene.	• `OrthographicCamera` • `PerspectiveCamera` • `CombinedCamera` • `cubeCamera` • `CameraHelper`
scene.js	Similar to three.js, this library provides both orthographic and perspective views. Additionally, orbiting and bird's-eye cameras are supported.	• `Orthographic` • `Perspective` • `Orbiting` • `pickFlyOrbit`
babylon.js	This library provides an extensive collection of cameras, including `ArcRotateCamera` (orbiting) and `FollowCamera`. The latter is specifically designed to follow any actor as it moves. This library also provides some device-dependent cameras including `DeviceOrientationCamera`, which responds to a gesture; `VirtualJoysticksCamera`, which reacts to orientation events such as tilting mobile devices; `OculusCamera`, which is a dual-view camera specifically designed for events generated by the Oculus Rift; and `webVRCamera`, which extends the `OculusCamera` and is designed to work with browsers compatible with WebVR API.[a,b]	• `ArcRotateCamera` • `FollowCamera` • `TouchCamera` • `DeviceOrientationCamera` • `VirtualJoysticksCamera` • `DeviceOrientationCamera` • `OculusCamera` • `webVRCamera`
voxelent	This library provides three types of cameras: `tracking`, `orbiting`, and `exploring`. Tracking cameras are first-person cameras that can follow actors using `vxlCamera.follow(actor)` using three different modes: `rotational` (camera does not move but rotates to focus on actor), `translational` (following from a distance), and `cinematic` (camera rotation is determined by the proximity to the actor). Voxelent implements landmark-based animation. Landmarks can be concatenated to produce a fly-through effect. Cameras use bounding boxes to focus on an actor with `vxlCamera.closeUp()` or on the whole scene with `vxlCamera.longShot()`	• `vxlCamera` • `vxlCameraManager` • `vxlLandmark`
C3DL	This library provides basic camera classes such as orbiting and exploring cameras.	• `FreeCamera` • `OrbitCamera`

[a] http://blog.bitops.com/blog/2014/06/26/first-steps-for-vr-on-the-web/
[b] http://tyrovr.com/2014/06/29/three-vr-renderer-tutorial.html

freely, performing automatic small adjustments to avoid collisions or undesired camera angles and redirecting the focus toward the object of interest. This is accomplished by *potential field gradients* [Wojciechowski 13]. These gradients adjust the camera position/angle based on proximity and intensity of the spatial gradient. Such techniques can replace costly conventional collision calculations. Assisted navigation can be useful to train physicians in endoscopic procedures (e.g., colonoscopy and bronchoscopy) within virtual environments [Wood 02].

Another interesting area of research is *affective camera control*; this explores the effect of altering the camera viewpoints on psychophysiology of users, which ultimately can be used to create adaptive camera control in games in order to increase the emotional impact [Yannakakis 10].

In a world where camera-capable mobile devices are standard, the device's accelerometer and video feed are being integrated into mobile WebGL applications to produce *augmented reality* scenes [Barbadillo 13; Feng 13]. In contrast, dedicated hardware solutions, such as the Oculus Rift,[*] for virtual reality environments are being successfully integrated in WebGL applications as a special type of camera. An example of this is provided by the `OculusCamera`[†] in the Babylon.js library.

Improvements in hardware and new, better ways to interact with the GPU, like WebCL, will bridge the gap between research and reality, taking us to a bright new future for 3D applications.

23.10 Resources

Three different demos are provided in the *WebGL Insights* GitHub repository[‡]:

- demo-camera-landmarks.html: shows the concept of landmark-based animation
- demo-camera-management.html: managing multiple cameras, types of cameras
- demo-camera-follow.html: shows different types of object tracking

These demos are also available online at voxelent.com.

Bibliography

[Barbadillo 13] Javier Barbadillo and Jairo R. Sánchez. "A Web3D Authoring Tool for Augmented Reality Mobile Applications." *Proceedings of the 18th International Conference on 3D Web Technology*. ACM, 206, 2013.

[Christie 03] Marc Christie and Eric Languénou. "A Constraint-Based Approach to Camera Path Planning." *Smart Graphics*. Springer Berlin Heidelberg, 2003.

[Drucker 94] Steven M. Drucker and David Zeltzer. "Intelligent Camera Control in a Virtual Environment." *Graphics Interface*. Canadian Information Processing Society, 1994.

---

[*] http://www.oculus.com/
[†] https://github.com/BabylonJS/Babylon.js/wiki/Using-Oculus-Rift-with-Babylon.js
[‡] https://github.com/WebGLInsights/WebGLInsights-1

[Feng 13] Dan Feng et al. "A Browser-Based Perceptual Experiment Platform for Visual Search Study in Augmented Reality System." *Ubiquitous Intelligence and Computing, 2013 IEEE 10th International Conference on and 10th International Conference on Autonomic and Trusted Computing (UIC/ATC).* IEEE, 2013.

[Wojciechowski 13] A. Wojciechowski. "Camera Navigation Support in a Virtual Environment." *Bulletin of the Polish Academy of Sciences: Technical Sciences* 61 (4): 871–884, 2013.

[Wood 02] Bradford J. Wood and Pouneh Razavi. "Virtual Endoscopy: A Promising New Technology." *American Family Physician* 66 (1): 107–112, 2002.

[Yannakakis 10] Georgios N. Yannakakis, Héctor P. Martínez, and Arnav Jhala. "Towards Affective Camera Control in Games." *User Modeling and User-Adapted Interaction* 20 (4): 313–340, 2010.

About the Contributors

Kamyar Abhari is a scientific developer at Synaptive Medical Inc., Toronto, Canada. He obtained his PhD and MSc in biomedical engineering from, respectively, Western University, London, Canada, and University of Manitoba, Winnipeg, Canada. His primary research interests include medical image visualization and human–computer interaction in image-guided surgeries.

Matthew Amato (@matt_amato) is a senior software architect, Analytical Graphics, Inc. He is a cofounder of Cesium, an open-source WebGL globe and map engine, as well as the inventor of CZML, a JSON schema for time-dynamic geospatial visualization.

Edward Angel (http://www.cs.unm.edu/~angel) is professor emeritus of computer science at the University of New Mexico (UNM). Ed Angel was the first UNM presidential teaching fellow. At UNM, he held joint appointments in computer science, electrical and computer engineering, and cinematic arts and was the founding director of the Art, Research, Technology, and Science Laboratory (ARTS Lab), a unique interdisciplinary center with educational, research, and economic development activities that span the range of digital media.

Ed's textbook, *Interactive Computer Graphics*, with Dave Shreiner, is now in its seventh edition. Ed and Dave thought that they were done with the textbook after the sixth edition but were both so intrigued with WebGL and its potential for teaching computer graphics that they did a seventh edition using WebGL exclusively. Ed has taught over 100 professional short courses, including WebGL courses at both SIGGRAPH and SIGGRAPH Asia.

Johannes Behr is head of the VCST Department, Fraunhofer IGD. He received his PhD degree in 2005. His areas of interest are virtual reality, real-time rendering, and 3D interaction techniques. Since 2005, Dr. Behr has worked at the Fraunhofer Institut für Graphische Datenverarbeitung (IGD, Germany) and leads the VCST group. Over the last 15 years, he has contributed to the OpenSG, InstantReality, and x3dom.org projects.

Nicolas Belmonte is a staff data visualization engineer at Uber. Previously, he led Twitter Interactive (interactive.twitter.com), a platform that delivers public-facing data visualizations around verticals like News, Government, Sports, and TV. Before Twitter, Nicolas created several open source frameworks for advanced data visualization, like the JavaScript Infovis Toolkit, and PhiloGL, that were used by organizations like Mozilla, Google, the White House, and Twitter. Today his interests range from advanced techniques in graphics to investigative journalism.

Florian Bösch (@pyalot • codeflow.org) is a freelance WebGL developer/consultant, Codeflow. After becoming a freelancer in 2008, Florian has specialized in WebGL services since 2011. He is servicing clients with WebGL-related needs from a variety of industries. He is the maintainer of webglstats.com and codeflow.org.

Nicholas Brancaccio (@pastasfuture • floored.com) is lead graphics engineer, Floored, Inc. Nicholas (Rhode Island School of Design, BFA, printmaking) spent his previous life getting poisoned by antique photographic chemistry and arguing that video games are art. One fateful day, while dehydrated in the perilous jungles of New Hampshire, he came across an outcropping. "Water!" he exclaimed. Making haste, he dived in. But this wasn't water at all; it was an ocean of hungry triangles. Before he knew it, he was fully consumed. Some time later, he was spotted on the alabaster dunes of New York City, mumbling a single phrase: "Draw teapots." And so he did. He is currently at Floored, Inc., responsible for the core rendering technologies in their in-house engine: Luma.

Diego Cantor-Rivera (@diegocantor • nucleojs.org • voxelent.com • imaging.robarts.ca/petergrp) is an engineer who is passionate about open source software and web technologies. Diego received his M.Eng. degree in systems and computer engineering from Universidad de Los Andes (Colombia), and his Ph.D. in biomedical engineering from Western University (Canada). Diego developed the first beating heart and the first brain cortical thickness maps in WebGL (2010). He is the author of the *WebGL Beginner's Guide* (Packt Publishing, 2012), which has sold internationally in English, Korean, and Chinese. He currently works with the Virtual Augmentation and Simulation for Surgery and Therapy (VASST) laboratory in Canada, where he develops virtual and augmented reality environments for neurosurgical teaching and training. He enjoys free time, cryptozoology, and other mythical entities.

Nicolas Capens (@c0d1f1ed) is a software engineer for Google and a member of Google's Chrome GPU team and a contributor to Android graphics tools. He is passionate about making 3D graphics more widely available and less restricted. His work on the ANGLE project helped create reliable WebGL support for Chrome on Windows. As the lead developer of SwiftShader, he enabled WebGL on systems with blacklisted GPUs or drivers. Through innovative multithreading, wide vectorization, and dynamic code specialization he continues to drive the convergence between CPU and GPU capabilities. Nicolas received his MSciEng degree in computer science from Ghent University in 2007.

David Catuhe (@deltakosh • http://blog.msdn.com/b/eternalcoding/) is principal program manager for IE and Web Standards, Microsoft. He is driving HTML5 and open web standards evangelization for Microsoft. He defines himself as a geek. He loves developing with JavaScript and HTML5 but also with DirectX, C#, C++, or even Kinect (he wrote a book about it that is available on Amazon). He is the father of Babylon.js and hand.js.

Paul Cheyrou-Lagrèze (@tuan_kuranes) is an R&D engineer at Sketchfab. Paul's experience ranges from C64 to Android's latest GPU. He made the first OpenGL-based html editor, Amaya, back in 2000 within the INRIA/W3C. Then he freelanced 15 years in the simulation industry in areas such as talking 3D avatars, multiprojector curve fitting, and surgery room simulation. He is interested in the broad subject of world computer simulation—not only 3D graphics, but also AI and physics. He is working now with the fabulous Sketchfab team, making 3D accessible to everyone, in the simplest way, yet awesome.

Patrick Cozzi (@pjcozzi • seas.upenn.edu/~pcozzi/) is principal graphics architect, Analytical Graphics, Inc., and a lecturer at the University of Pennsylvania. At Analytical Graphics, Inc. (AGI), Patrick leads the development of Cesium, an open-source WebGL virtual globe and map engine. He is the editor of *WebGL Insights*, coeditor of *OpenGL Insights*, coauthor of *3D Engine Design for Virtual Globes*, and a member of the editorial board for the *Journal of Computer Graphics Techniques*. Patrick frequently presents at SIGGRAPH and contributes to other book series. He is a member of Khronos and teaches GPU programming and architecture at the University of Pennsylvania, where he earned a master's degree in computer science.

Nick Desaulniers (@LostOracle • https://nickdesaulniers.github.io) is a software engineer fighting for the open web at Mozilla. When Nick is not helping third-party developers target Firefox OS, he's giving talks around the world about web technologies, learning new programming languages, or compiling C/C++ to JavaScript with Emscripten. Contributing to open-source software and an accessible Internet for all are some of the things that Nick is most passionate about. Nick received his BS in computer engineering from Rochester Institute of Technology and is an Eagle Scout.

Matt DesLauriers (@mattdesl), creative developer, Jam3, is a self-taught graphics programmer who studied film and media at Queen's University. Working at Jam3 in Toronto, he combines his love of film, programming, and design to produce interactive content for the web. When he is not tinkering with WebGL and open-source modules, he might be found at a local espresso shop or rock climbing gym.

Chris Dickinson (@isntitvacant • http://neversaw.us/), node core contributor, Walmart Labs, serves as a core committer to the Node.js and io.js projects on behalf of Walmart Labs. He is passionate about open-source, community-driven efforts to improve tooling, and writing parsers as a stress-relief mechanism. He is a maintainer of several projects within the stack.gl open-source organization, which is committed to building reusable, pluggable tools for WebGL.

Olli Etuaho (@oletus • oletus.fi) is a system software engineer, NVIDIA Corporation. Olli received his master's degree from Tampere University of Technology (TUT) in 2012, already equipped with some industry experience from game development and NVIDIA's mobile browser team. As a part of his work at NVIDIA, he has been an active contributor to the WebGL specification and the Chromium project, most recently helping with the WebGL 2 effort. As his passion project, he develops CooPaint, a WebGL-accelerated painting app.

Lin Feng is currently an associate professor, the director of the Bioinformatics Research Centre, and the program director of MSc (digital media technology) at the School of Computer Engineering, Nanyang Technological University. Dr. Lin Feng's research interest includes biomedical informatics, biomedical imaging and visualization, computer graphics, and high-performance computing. He has worked for more than 20 funded research projects since joining NTU and has published about 200 technical papers. He is a senior member of IEEE.

David Galeano (@davidgaleano • www.turbulenz.biz • ga.me) is a cofounder of Turbulenz Limited. As the cofounder of Turbulenz, David was directly involved in the creation of all the technology within the Turbulenz platform. Before creating Turbulenz in 2009, he spent 8 years at Criterion Software as senior software engineer working on RenderWare graphics, and at Electronic Arts as technical director of EA's internal rendering technology group. David originally worked for 6 years developing games at Dinamic Multimedia in Spain.

Jeff Gilbert (@jedagil) is a software engineer at Mozilla, where he leads the WebGL implementation for Gecko. He was first introduced to graphics and software engineering by patching and hacking together modifications for computer games. Years later, Jeff is on the other side of the fence, working to help developers author and port 3D content into the browser with WebGL.

Benoit Jacob is a software engineer at Google, formerly Mozilla. Until 2010, Benoit was a mathematician at the Universities of Paris and Toronto, and in his spare time developed open-source numerical software, such as the Eigen matrix library. He then joined Mozilla as a software engineer, to work on browser engine graphics and WebGL. In late 2014, he joined Google, going back to working on numerical software, specifically, Eigen.

Jacek Jankowski (@jacek_jankowski • http://grey-eminence.org) is a graphics software engineer at Intel and was previously a research fellow at INRIA and DERI. He has published more than 20 papers in the top-tier human–computer interaction, web technology, and computer graphics journals and conferences. Jacek is passionate about incorporating 3D graphics into the web user interface.

Brandon Jones (@tojiro • http://tojicode.com) is a software engineer on the Google Chrome GPU team, and has been chasing computer graphics as a hobby since he was a kid. Along the way he was distracted by the siren's song of web development and was overjoyed to find the two subjects collide in the form of WebGL.

Hugh S. Kennedy (@hughskennedy • http://hughsk.io) is an Australian software developer with a background in data visualization and installation art. He currently works for NodeSource, spending much of his time working on product development and design. He is one of the maintainers of stack.gl, an open and modular software ecosystem for WebGL built on top of browserify and npm.

Almar Klein (@almarklein • http://almarklein.org) is a freelance pythonista. After studying electrical engineering, he did a PhD in medical image analysis. Working with dynamic CT data sparked an interest in (3D) visualization, which resulted in the development of Visvis. Currently, Almar is a core developer of Vispy and collaborating with Continuum Analytics to lower the barriers of (scientific) computing by building the next-generation data analysis platform.

Alexander Kovelenov (www.blend4web.com) is a senior software engineer for Blend4Web. Alexander received his specialist degree in radio-frequency engineering at Bauman Moscow State Technical University (BMSTU). After that he worked for 3 years developing software for the Global Navigation Satellite System (GLONASS). In 2011, he joined the Blend4Web team as a senior software engineer. He is responsible for developing the engine architecture and the physics, sound, and visual programming systems.

Max Limper (@mlimper_cg • www.max-limper.de) is a researcher at Fraunhofer IGD/TU Darmstadt. As part of the Visual Computing System Technologies (VCST) group, he is developing rendering technology for academia and industry, such as InstantReality and X3DOM. His main focus is methods for polygonal mesh processing, aiming at optimized delivery and presentation of 3D meshes on the web. Max is also a PhD student at the Interactive Graphics Systems group (GRIS) at TU Darmstadt.

Mikola Lysenko (@mikolalysenko • 0fps.net) is a freelance computer programmer with a background in computational geometry. He specializes in computer aided design, data visualization, and GPU programming. Some of his projects include work for 3D Systems and plotly. Currently he is working on stack.gl, which is a modular ecosystem for developing WebGL applications. He also writes about geometry recreationally for the blog 0fps.

Ivan Lyubovnikov (www.blend4web.com) is a software engineer at Blend4Web. Ivan graduated from Tula State University, where he studied mathematics and computer science. His interests lie in web development and programming, especially in WebGL and HTML5 technologies. He joined the Blend4Web team in 2013 as a 3D engine programmer. He is responsible for developing major engine parts including, but not limited to shader processing pipeline, optimization, data processing, shadows, and visual effects.

Zhenyao Mo (https://www.linkedin.com/in/zhenyao) is a software engineer for Google Inc. He received his PhD degree in computer science from University of Southern California. For the past 5 years, he has been part of Google's Chrome GPU team, implementing and perfecting WebGL. Although constantly craving to write the coolest WebGL app ever, he never gets the chance. He accepts that his role is to enable other developers to have fun with WebGL.

Muhammad Mobeen Movania (http://cgv.dsu.edu.pk), assistant professor, DHA Suffa University, received his PhD degree in advanced computer graphics and visualization from Nanyang Technological University (NTU), Singapore, in 2012. He carried out research mainly in biomedical volume rendering and visualization in collaboration with the National Cancer Centre of Singapore (NCCS). After his graduation, he joined the Institute for Infocomm Research (I2R), a division of A-Star Singapore, as a research scientist. His responsibilities there were research and development in the areas of advanced computer graphics, augmented reality, and 3D animation.

Dr. Movania has published several international conference and journal papers in the area of computer graphics and visualization, including a poster at SIGGRAPH 2013. He has served as a reviewer for several recent OpenGL books including *OpenGL 4 Shading Language Cookbook* (second edition) and a video course, "Building Android Games with OpenGL ES Online Course," both published by Packt Publishing in 2013. He has written a book (*OpenGL Development Cookbook,* published by Packt Publishing in 2013) that details several applied recipes on using modern OpenGL. He has also authored a book chapter in *OpenGL Insights* published by AK Peters/ CRC Press in 2012). Dr. Movania is currently serving as an assistant professor in the Department of Computer Science at DHA Suffa University, Karachi, Pakistan. More information about his current research activities can be obtained from his research group web page.

Cedric Pinson (@trigrou) is the cofounder and CTO of Sketchfab—a platform for sharing 3D content online. A veteran of the game industry since 2001, he transitioned from a project leader in 3D client technology to launching his career as a freelance developer in 2008. Cedric's personal and freelance projects quickly led him to become a pioneer in WebGL technology. He developed OSG.JS—a framework implementing an OpenSceneGraph-like toolbox to interact with WebGL via JavaScript. Using this technology, he focused on the creation of a real-time, online 3D viewer. Cedric's innovation was quickly embraced by the 3D community and morphed into Sketchfab—a TechStars alum and venture-capital-backed company with over 15 employees in its New York and Paris offices.

Philip Rideout (@prideout • http://prideout.net) is a senior graphics developer in Berkeley, California. He helped build the industry's first GLSL compiler at 3Dlabs in the early 2000s. He is an alumnus of the developer tools team at NVIDIA, as well as Pixar's Research & Development division. He first started dreaming about computer graphics after a high school field trip to Intergraph Corporation.

Kevin Ring (www.kotachrome.com/kevin) is principal software engineer at National ICT Australia (NICTA). He recently moved halfway around the world to beautiful Sydney, Australia, to help NICTA build Australia's National Map and to spend just a little bit too much time on the beach. Previously, Kevin coauthored the book, *3D Engine Design for Virtual Globes,* with Patrick Cozzi, which helped launch the development of Cesium, the open-source virtual globe for the web. In between bouts of complaining about JavaScript, Kevin built Cesium's terrain and imagery rendering engine as well as a server-side tool to process hundreds of gigabytes of terrain data for efficient streaming to a web browser.

Evgeny Rodygin (www.blend4web.com) is a software engineer for Blend4Web. After receiving a specialist degree in mechanical engineering and a bachelor's degree in computer science at Bauman Moscow State Technical University (BMSTU), Evgeny worked as a design engineer of low-voltage systems in the Central Design Institute of Communication. Since 2012, Evgeny has been working with the Blend4Web team as a software engineer. He is a big fan of open-source software and Linux. His biggest interests are shader optimizations, artificial intelligence, and anything related to 3D graphics.

Cyrille Rossant (@cyrillerossant • http://cyrille.rossant.net) is a research associate at University College London. Dr. Rossant is a research data scientist and software engineer in neuroinformatics. A graduate of Ecole Normale Superieure, Paris, where he studied mathematics and computer science, he has also worked at Princeton University, University College London, and College de France. He is one of the lead developers of VisPy, an OpenGL-based high-performance visualization package in Python. He is the author of two books on Python for data science: *Learning IPython for Interactive Computing and Data Visualization* and *IPython Interactive Computing and Visualization Cookbook*, both produced at Packt Publishing.

Jeff Russell (@j3ffdr • marmoset.co) is an engineer and cofounder of Marmoset. Jeff has worked since 2012 on the company's technical foundation, focusing on graphics and engine development. His attention these days is focused on creating real-time graphics tools for artists. Prior to this he spent time in game development and virtual reality research.

Kenneth Russell (@gfxprogrammer • gfxprose.blogspot.com) is a software engineer on Chrome's GPU team at Google, Inc. and has been deeply interested in 3D graphics since taking his first college course on the topic. Among previous work, he interned at Silicon Graphics, working on the Cosmo Player VRML plug-in, and cofounded Sun Microsystems' Java Bindings to the OpenGL API (JOGL) project. He currently serves as the chair of the WebGL working group.

Tarek Sherif (@thsherif • http://tareksherif.net) is a 3D software engineer at BioDigital, where he's helping to build the BioDigital Human Platform, a WebGL-based publishing platform for visualizing anatomy, disease, treatments, and health information in interactive 3D. Previously, he was a Software Developer at the McGill Centre for Integrative Neuroscience, where he was lead designer and developer of BrainBrowser, a set of visualization tools for neuroimaging data built using WebGL and HTML5.

Dave Shreiner (@DaveShreiner), ACM/SIGGRAPH 2014 conference chair, has been working with OpenGL-related technologies for over 25 years. Dave authored the first commercial training course on OpenGL while at Silicon Graphics; has contributed to numerous books on computer graphics using OpenGL, OpenGL ES, and WebGL; and has collaborated with Ed Angel to help a multitude of programmers hone their OpenGL-related skills. He's been privileged to share his passion with thousands of developers.

Christian Stein (@_chrstein_) is a researcher with Fraunhofer IGD/TU Darmstadt. After collecting some industry experience as a software engineer in game development alongside his studies, Christian received his master's degree in computer science in 2014. In the VCST department of the Fraunhofer IGD, he is responsible for web-based rendering technology. His research focuses on spatial data structures and large-model rendering on the web.

Maik Thöner (@mthoener), a researcher with Fraunhofer IGD/TU Darmstadt, is part of the Visual Computing System Technologies Department of the Fraunhofer IGD, where he is researching and developing web-based rendering systems for large models. His main research interests include virtual reality, 3D in the web, distributed rendering, and hybrid rendering. He received his master's degree in computer science from the TU Darmstadt.

Vladimir Vukicevic is engineering director at Mozilla. Vladimir has been working on Firefox and open-web standards for over 10 years. During this time, he managed to successfully convince people that JavaScript and the web would be fast enough for 3D graphics to happen in the browser and thus WebGL was born. He chaired the initial WebGL working group within Khronos up until the finalization of the WebGL 1.0 standard. Since then, he has worked on graphics and performance problems with Firefox and Firefox OS and helped bring high-end games to the web via Emscripten and asm.js. Currently, Vladimir is focused on bringing VR capabilities to the web and, in general, making the web competitive with native desktop and mobile platforms.

Shannon Woods (@ShannonIn3D), is a software engineer at Google, and the project lead for ANGLE, an open-source graphics translation engine used by projects, from browsers to UI frameworks to mobile games to execute OpenGL ES across many platforms. Prior to her current work with ANGLE, she explored other corners of the 3D graphics world, developing software for game portability and real-time distributed simulation. She is a graduate of the University of Maryland, and enjoys close specification analysis, music, and teapots.